A Hands-On Course in Sensors Using the Arduino and Raspberry Pi

A Hands-On Course in Sensors using the Arduino and Raspberry Pi is the first book to give a practical and wide-ranging account of how to interface sensors and actuators with micro-controllers, Raspberry Pi and other control systems. The author describes the progression of raw signals through conditioning stages, digitization, data storage and presentation.

The collection, processing, and understanding of sensor data plays a central role in industrial and scientific activities. This book builds simplified models of large industrial or scientific installations that contain hardware and other building blocks, including services for databases, web servers, control systems, and messaging brokers. A range of case studies are included within the book, including a weather station, ground-vibration measurements, impedance measurements, interfacing medical sensors to web browsers, the profile of a laser beam, and a remote-controlled and fire-seeking robot.

This second edition has been updated throughout to reflect new hardware and software releases since the book was first published. Newly added features include the ESP32 microcontroller, several environmental and medical sensors, actuators for signal generation, as well as a chapter on websockets; all illustrated in new case studies.

This book is suitable for advanced undergraduate and graduate students taking hands-on laboratory courses in physics and engineering. Hobbyists in robotics clubs and other enthusiasts will also find this book of interest.

Features:

- Includes practical, hands-on exercises that can be conducted in student labs, or even at home
- Covers the latest software and hardware, and all code featured in examples is discussed in detail
- All steps are illustrated with practical examples and case studies to enhance learning

https://github.com/volkziem/HandsOnSensors2ed

Volker Ziemann obtained his PhD in accelerator physics from Dortmund University in 1990. After post-doctoral positions in Stanford at SLAC and in Geneva at CERN, where he worked on the design of the LHC, in 1995 he moved to Uppsala where he worked at the electron-cooler storage ring CELSIUS. In 2005 he moved to the physics department where he has since taught physics. He was responsible for several accelerator physics projects at CERN, DESY and XFEL. In 2014 he received the Thuréus prize from the Royal Society of Sciences in Uppsala.

Series in Sensors
Series Editors: Barry Jones and Haiying Huang

Other recent books in the series:

A Hands-On Course in Sensors Using the Arduino and Raspberry Pi

Second Edition

Volker Ziemann

CRC Press
Taylor & Francis Group
Boca Raton London New York

CRC Press is an imprint of the
Taylor & Francis Group, an **informa** business

First edition published 2023
by CRC Press
6000 Broken Sound Parkway NW, Suite 300, Boca Raton, FL 33487-2742

and by CRC Press
4 Park Square, Milton Park, Abingdon, Oxon, OX14 4RN

CRC Press is an imprint of Taylor & Francis Group, LLC

© 2024 Volker Ziemann

First edition published by CRC Press 2018

Library of Congress Cataloging-in-Publication Data

Names: Ziemann, Volker (Associate professor of physics), author.
Title: A hands-on course in sensors using the Arduino and Raspberry Pi /
Volker Ziemann.
Description: Second edition. | Boca Raton : CRC Press, 2023. | Series:
Series in sensors | Includes bibliographical references and index. |
Summary: "A Hands-On Course in Sensors using the Arduino and Raspberry
Pi is the first book to give a practical and wide-ranging account of how
to interface sensors and actuators with micro-controllers, Raspberry Pi
and other control systems. The author describes the progression of raw
signals through conditioning stages, digitization, data storage and
presentation. The collection, processing, and understanding of sensor
data plays a central role in industrial and scientific activities. This
book builds simplified models of large industrial or scientific
installations that contain hardware and other building blocks, including
services for databases, web servers, control systems, and messaging
brokers. A range of case studies are included within the book, including
a weather station, ground-vibration measurements, impedance
measurements, interfacing medical sensors to web browsers, the profile
of a laser beam, and a remote-controlled and fire-seeking robot. This
second edition has been updated throughout to reflect new hardware and
software releases since the book was first published. Newly added
features include the ESP32 microcontroller, several environmental and
medical sensors, actuators for signal generation, as well as a chapter
on websockets; all illustrated in new case studies. This book is
suitable for advanced undergraduate and graduate students taking
hands-on laboratory courses in physics and engineering. Hobbyists in
robotics clubs and other enthusiasts will also find this book of
interest"-- Provided by publisher.
Identifiers: LCCN 2023002031 | ISBN 9781032377483 (hardback) | ISBN
9781032376196 (paperback) | ISBN 9781003341703 (ebook)
Subjects: LCSH: Detectors. | Raspberry Pi (Computer) | Arduino
(Programmable controller) | Microcontrollers.
Classification: LCC TK7872.D48 Z54 2023 | DDC
006.2/54165--dc23/eng/20230216
LC record available at https://lccn.loc.gov/2023002031

ISBN: 978-1-032-37748-3 (hbk)
ISBN: 978-1-032-37619-6 (pbk)
ISBN: 978-1-003-34170-3 (ebk)

DOI: 10.1201/9781003341703

Typeset in CMR10 font
by KnowledgeWorks Global Ltd.

Publisher's note: This book has been prepared from camera-ready copy provided by the authors.*Publisher's note:* This book has been prepared from camera-ready copy provided by the authors.

Contents

List of Figures

Preface

TO THE SECOND EDITION

During the five years since the first edition was published both software and hardware have evolved, which makes updating the contents mandatory. At the same time, I seize the opportunity to expand the material and cover both new hardware and software features.

While the upgrade of the hardware from Raspberry Pi version 3 to version 4 requires no changes, the upgrades of the operating system are substantial and affect many chapters in the book. In particular, the initial setup and the network configuration changed, Python2 is phased out and replaced by Python3, and the MySQL database is replaced by MariaDB, to name few. Moreover, several software packages used in the first edition are no longer available and had to be replaced.

The Arduino and NodeMCU microcontrollers used in the first edition are still available, but I now also include examples with the more powerful ESP32 microcontroller, which offers on-board digital-to-analog converters and Bluetooth.

I replaced a few sensors covered in the first edition by newer and enhanced versions, for example, the BME680 environmental sensor replaces the BMP180 pressure sensor. Moreover, a number of additional circuits are covered; the TCS34725 color sensor, the SCD30 CO_2 sensor, and the AD9850 direct digital synthesizer. The latter warrants a new section in the book about the generation of periodic signals. Three sensors dedicated to medical applications are covered, the AD8232 electrocardiogram front end, the MAX30102 pulse oximeter, and the AD5933 bioimpedance network analyzer.

While all material from the first edition is still present in updated form, there are new chapters on websockets and a few extra projects. Websockets enable us to interface and control WLAN-capable microcontrollers from web browsers. We use them to build ESP32-powered data acquisition systems to interact with the three medical sensors mentioned in the previous paragraph.

Finally, software and color images from this book are available at `https://github.com/volkziem/HandsOnSensors2ed`.

TO THE FIRST EDITION

Some years ago three young students inquired about a moderately complex project to earn some credits. I happily agreed to supervise them and assigned moderately difficult tasks, namely, to build from scratch a data acquisition system for slow signals. I suggested to connect some sensors to an Arduino microcontroller and then write a program for the Arduino to interface the measurement values to the control system we use in our lab.

The students were very dedicated and a real joy to work with. They had the Arduino under control within a few hours and had the first sensors reporting their measurement values after the first day. Then they worked out a protocol that is compatible with our EPICS-based control system, and after discussions with our control systems experts and even more debugging, eventually the students had a prototype system working. After cleaning up their project, they had to give a presentation and write a report to earn their well-deserved credits.

I soon realized that there is a clear progression of the information generated by a sensor. The information bubbles upward through a sequence of microcontrollers and computers that provide data-handling, storage, and online presentation to a seminar presentation, and eventually ends up in a report. Understanding the path the measurement data take appeared like a useful concept to communicate to students. Moreover, I wanted the students to understand the details of the signal chain and how it *really* works. Therefore, I used the hands-on approach with programming the Arduino that serves as communication glue between the sensor and the control system. This proved beneficial for the students' understanding and was appreciated by them. The abstract concepts thus led to a very concrete realization. In the final stages of the project I coached the students on how to prepare a presentation for a seminar according to some simple guidelines, and eventually put the oral presentation into writing for a report to hand in and receive their credits.

This book is inspired by these students and their projects, but goes a step further and adds a number of additional topics such as signal conditioning, controlling actuators such as switches and motors, as well as control system setup, data storage, and networking. Please note that I cover only basic examples that are boiled down to the bare essentials in order to illustrate the main concepts and to get started quickly. Anyway, the concepts covered should come in handy when working with real-world data-acquisition tasks. I basically follow *Mrs. Robinson's guideline* of "help you learn to help yourself" (remember the Simon and Garfunkel song?) and try to fill the toolbox with practical know how. This know how should enable the reader to help herself and pick up datasheets and manuals to adapt the basics from this book to realize far more advanced projects.

User Guide

The main theme of the book is *From Sensor to Report,* and that should be the guiding principle of using it in the classroom, either in a student laboratory or as the basis for individual projects.

For a student laboratory I suggest installing the software with some of the more arcane instructions before starting the lab. This comprises turning the Raspi into a router (Section 5.4), installing the MySQL database clone MariaDB (Section 5.6.2), and installing EPICS (Section 6.1). The students should focus on the sensors and use the above systems as a background infrastructure. They should, on the other hand, understand the basic operation of the sensors, learn how to interface them to a microcontroller, and move the information to the next level on a different computer. This requires them to write network code, fill an SQL database, prepare the protocol files for EPICS, or present data on a web server. In the lab a knowledgeable supervisor, a "tutor," should be available to answer questions and guide the students. Using solderless breadboards in the lab enables the students to quickly arrive at a working system on which to base further experiments and try out new ideas.

A suitable scope for student projects, suitable for a single or a group of two students, is to connect a small number of sensors to an Arduino. Then they should be given a target system where they can publish the data. This can be a database, EPICS, MQTT, or a web page. After a prototype system is working, the students should present their system in a seminar and prepare a report.

All code and the corresponding images of the circuits on a breadboard for the first edition, prepared with Fritzing [1], are available on this book's github page https://github.com/volkziem/HandsOnSensors2ed.

Acknowledgments

This book only materialized because my students, Adam, Måns, and Frida, asked about "some project" and then completed it with such enthusiasm. Later, the students taking the course "Sensor to Report" (1FA349) at Uppsala University, which is based on this book, came with really cool project ideas; who could forget Annika's "Color to Sound" project[1] or Pedro's "Sourdough monitor" to name only a few. I gratefully acknowledge all students' contributions and input.

At the same time, I thank my colleagues Roger Ruber and Mattias Klintenberg, who read and commented on parts of the manuscript and provided essential feedback. Likewise I acknowledge help from Pawel Marciniewski with the electronics and Camilla Thulin with the photography.

This book is only possible thanks to the open source community that created the Arduino and Raspberry Pi ecosystems and the large number of people answering questions on Internet forums. I acknowledge the creators and maintainers of the Fritzing software. I relied on it to prepare many drawings to illustrate the wiring of circuits. A dagger[†] annotates the corresponding images.

I am indebted to Francesca McGowan, Rebecca Davies (for the first edition), and Danny Kielty (for the second edition), all of them at Taylor & Francis, for competently guiding me through the intricacies of writing and publishing a book.

Last but not least, I thank my family for putting up with me during the intense writing and editing periods of two editions. More often than not I was absent minded and showed a distinct lack of response to other matters.

---

[1] Annika Schlechter, *Color to Sound*, Elektor 1/2022, page 98.

## CHAPTER 1

Introduction

What is the path that the electrical signal from a sensor takes to end up in a report and how do we interface the different stages along the path? We address these questions because collecting sensor data, processing them, and deriving some understanding from the data plays an important role in many circumstances. One example is a utility company that gathers information about electricity, heating, and water in order to prepare statements for their customers and to estimate demand for their product in the future. Smart homes are another example; they measure temperatures or detect the presence of beverages in the refrigerator to adjust the thermostat in the first case, or to prepare a report for us to pick up some milk or beer on the way home in the second case. Quite generally, many Internet of Things (IoT) technologies share a common base with the topic of the book, but even large experimental collaborations such as the ATLAS [2] or CMS [3] that operate the huge detectors at the Large Hadron Collider (LHC) [4] at CERN [5] get their data from sensors that are buried deep inside the detectors. They sense currents from drift chambers where charged particles cause a discharge between wires at different potentials, or they cause electrical signals from semiconductor detectors, where they create electron–hole pairs that induce a current. Other examples are Hall sensors, to measure magnetic fields, and humidity sensors or barometric pressure sensors to detect variations in ambient conditions. All these sensors produce electrical signals that often need to be amplified or otherwise conditioned. This stage involves operational amplifiers and various filters to improve the signal-to-noise ratio. Once properly processed, the analog signals are passed on to analog-to-digital converters (ADC), where they are converted to a digital representation that is subsequently handled by computers. Often some of the computing power is located close to the sensor and is provided by microcontrollers that collect signals from nearby sensors and convert them to the underlying physical quantities, formatted to have a standardized output format. Thus they act as "communication glue" between the specific interface to the sensor and a more generic interface to a host computer that is usually located further away. The latter is the other end of the communication channel from the microcontroller and provides data storage and presentation, and sometimes also shows recent data for online monitoring. The host computer may run generic control-system software to provide a further abstraction layer towards higher-level software. Examples we discuss are MQTT [6], which is popular with IoT projects, the EPICS control system [7], commonly found in scientific laboratories, and *websockets* that allow us to use a web browser to communicate with data acquisition systems.

In this book we will build a system that contains all the ingredients also found in large scientific or industrial installations. In a sense it is a simplified model of a large installation, yet containing all the hardware and logical building blocks. In particular, we use the Arduino [8], ESP8266-based [9], and ESP32-based [10] microcontroller boards as

DOI: 10.1201/9781003341703-1

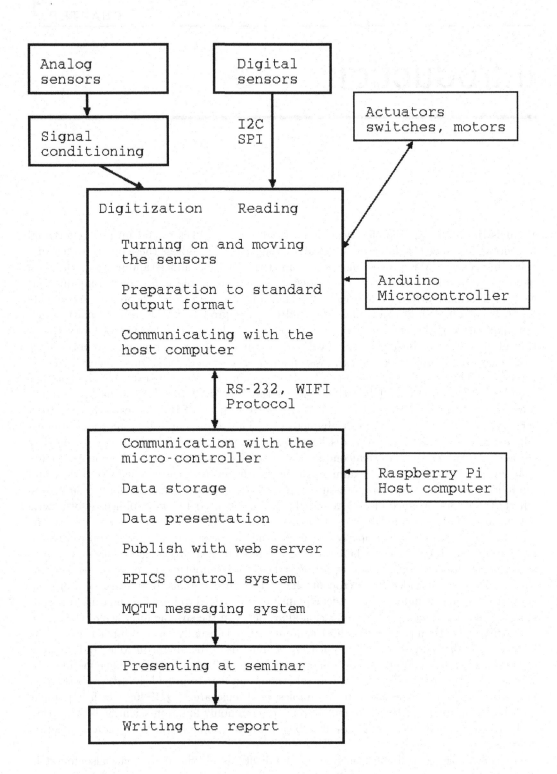

Figure 1.1 The outline of the book.

the local intelligence to control switches and motors to move the sensors around and enable reading them out and translating the signals to a format that allows communication with a host computer using a standardized protocol. This comprises the hardware channels, which can be USB, RS-232, Bluetooth, or WiFi as well as the logical protocol, such as a simple query-response protocol. As host computer we use a Raspberry Pi [11] which, in its most recent incarnation, features four processing cores and runs a standard Linux system with a huge base of available software, including web servers, the MATLAB^{TM}-clone Octave, and even Mathematica^{TM}, all without license charges.

The remainder of the book is organized as laid out in Figure 1.1. We first discuss a number of analog sensors and signal-conditioning methods, followed by a number of sensors that already provide their measurements in digital form, and the buses and protocols used. Next we discuss actuators. They switch things on and off, even those requiring large currents, and we learn how to control different types of motors that are sometimes needed as part of the measurement process. We then describe how to interface actuators and sensors with the Arduino, either by digitizing signals or by using the appropriate bus-interface. We go on to describe a program structure that permits the Arduino to support a simple query-response protocol in order to serve as a slave to a host computer. Next, we configure a Raspberry Pi as a standardized host computer. It will provide data storage in databases, and present the data in graphical form either using Octave and Python running locally on the Raspi or by publishing our measurement data with a web server, also running on the Raspi. We continue the discussion by installing the EPICS control system software and turn the Raspi into a full-blown control-system server that can join any other EPICS installation in a transparent way. We go on to discuss the MQTT message-passing system, which plays an important role in IoT applications. Websockets enable us to use web browsers, even on a smartphone, to directly communicate with the microcontrollers. Having assembled all the parts, we consider examples in which we build a weather station with distributed sensors, and systems to record ground vibrations, monitor the color of water, and measure the impedance of components. In more advanced examples, we build a data acquisition system that communicates directly with web browsers, then expand it to record electrocardiograms and other signals of medical relevance. Finally, we measure the width of the beam of a laser pointer, and a remote-controlled robot that also senses flames autonomously, moves to the fire, and sounds an alarm. We conclude by presenting guidelines about how to prepare a seminar presentation based on the examples and how to write a readable publication describing our data acquisition system using sensors, actuators, Arduino, and Raspberry Pi.

## CHAPTER 2

Sensors

A *sensor* is a device that converts a physical quantity to an electrical signal [12–15], and therefore either provides a voltage or a current, or causes a change of its resistance. More generally, the impedance of the sensor, which also comprises capacitive and inductive sensors, may change. We are thus faced with the task to measure either of these electrical quantities.

Below we discuss examples of the different types of sensors, but only show a selection of those available on the market. Searching the Internet for the physical quantity one wants to measure jointly with the keyword "sensor" will give the reader an idea of what is available. Once the sensor is identified, careful reading of the datasheet to learn about how to interface the sensor is mandatory.

2.1 ANALOG SENSORS

We start by considering resistance-based sensors.

2.1.1 Resistance-based sensors

Our first resistance-based sensor is a *light-dependent resistor* or LDR, shown on the left of Figure 2.1. It changes its resistance depending on the exposure to light, where the range depends on the device and typically extends from a few $100\,\Omega$ to $M\Omega$. Let us spend a few lines to reconcile the electronic schematics with the fundamental physics on a microscopic

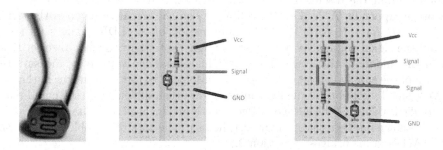

Figure 2.1 Image of a light-dependent resistor (LDR) on the left and how to connect it in a plain voltage-divider configuration (center[†]) and in a Wheatstone bridge (right[†]). A superscripted dagger[†] in the caption indicates that the image was created with Fritzing.

DOI: 10.1201/9781003341703-2

scale. The operating principle of this and many other sensors is based on the availability of electrons in the conduction band of a material. In good (wires) or bad (resistors) conductors, electrons partially fill the available states in the conduction band up to some energy, the Fermi level, whereas in insulators the Fermi level is located in between the completely filled valence band and the conduction band [16]. Therefore, no electrons are available in the conduction band. Furthermore, the energy-difference between the upper boundary of the valence band and the lower boundary of the conduction band, the *bandgap*, is large, while for semiconductors it is only on the order of electron-volt (eV).

In photoresistors the base material is often CdS, a semiconductor with a bandgap of about 2.4 eV. This energy equals that of photons of green light with a wavelength of about 500 nm. Therefore, green photons can elevate electrons from the valence to the conduction band and thus create electron–hole pairs. These now freely moving charge carriers conduct electric current and therefore increase the conductivity of the material. Figure 2.2 illustrates this in more detail. The upper figure shows a simplified band level scheme under dark conditions. The bold lines show the lower boundary of the conduction band and the upper boundary of the valence band. The dashed line shows the highest energy states that electrons occupy. For metals and resistors, this is close to the Fermi level. Conversely, in a semiconductor like CdS, the Fermi level lies between valence and conduction band. Yet, at room temperature, there are a few thermally excited electrons in the conduction band of CdS. We visualize this by the close proximity of the dashed line to the lower conduction-band boundary. In metals there are plenty of electrons in the conduction band, and the conductivity is high. In the resistor there are fewer electrons in the conduction band, or their mobility is impeded in other ways such that there is a shift in the Fermi level across the resistor. In dark conditions there are only very few thermally excited electrons in the conduction band of the LDR, and the conductivity is very low. Consequently, there is a large voltage-drop across the LDR and the measured voltage, which is the difference of Fermi levels between the measurement points. The measured voltage is therefore close to the full voltage delivered by the battery. If, on the other hand, the LDR is illuminated, the photons lift electrons from the valence band into the conduction band. This increases the conductivity and only a small voltage is dropped across the LDR. This consequently reduces the measured voltage, as shown in the lower graph of Figure 2.2. Note that we do not discuss the details of the interfaces between the different parts because that is beyond the scope of this book.

In the middle of Figure 2.1 we show how to connect a photoresistor in series with a resistor R_0 (here 10 kΩ) to create a voltage divider between the supply voltage V_{cc} and ground. From the discussion of voltage dividers and a short refresher of basic circuit theory in appendix A or [17], the voltage V_s on the signal terminal is then given by $V_s = V_{cc}R_{LDR}/(R_0 + R_{LDR})$. Thus, the illumination of the LDR changes its resistance R_{LDR} and the signal voltage varies correspondingly. Note that we have to select the resistor R_0 in the middle of the range of R_{LDR}. This causes the voltage we measure to vary around half of the supply voltage. Therefore, we also need to use a volt meter in that voltage range. Very small variations of the light intensity are then difficult to resolve and may need to be amplified. Using a Wheatstone bridge, where we compare the voltages in two resistor dividers, as shown on the right of Figure 2.1, helps to alleviate this problem. We expand on the use of Wheatstone bridges in Section 2.2.

Other resistance-based sensors are *resistance-based temperature detectors* (RTD) such as the PT100 temperature sensor, which is a calibrated platinum-based sensor with a resistance of exactly 100 Ω at 0 °C. It is based on the fact that the resistance of a very pure metal is determined only by scattering of electrons in the conduction band with phonons, which are vibrations of the ions that make up the crystal lattice of the metal. Moreover, higher temperatures cause stronger vibrations of the lattice, with correspondingly higher resistance.

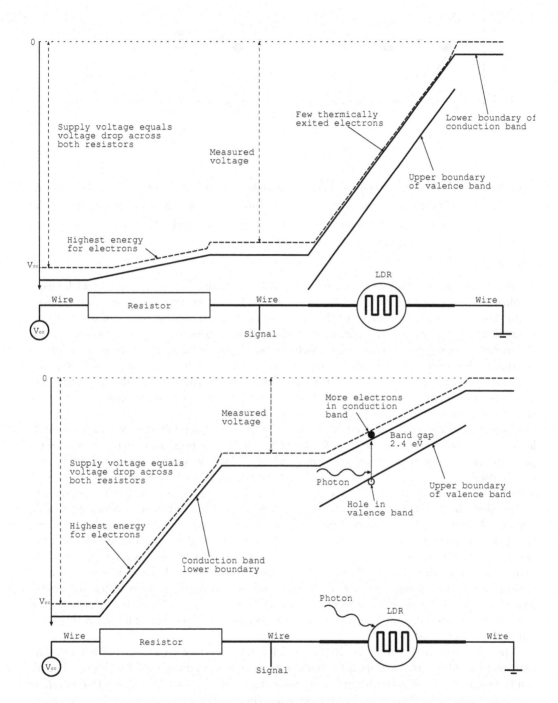

Figure 2.2 The band-level scheme and the schematics for the circuit with the LDR in a voltage divider. The upper graph illustrates dark conditions, and the lower graph shows conditions where the LDR is exposed to light. Note that the vertical axis by convention shows the energy of electrons. This causes the positive pole of the battery to have the most negative energy. See the text for a discussion.

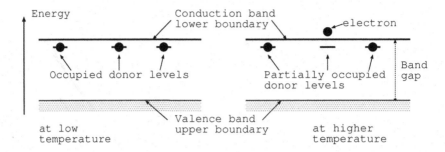

Figure 2.3 NTC resistors are doped semiconductors that have donor levels just below the conduction band. Increasing the temperature increases the kinetic energy of the electrons that allows them to occasionally jump into the conduction band, where they contribute to the conductivity of the material.

Intuitively one might think of the crystal ions at higher temperature to oscillate with larger amplitudes, creating a larger target for the electrons to scatter, thus impeding their motion. Since this is an intrinsic property of the material, calibration measurements of resistance as a function of temperature are universally valid for all sensors of the same metal, provided the metal is very pure and free of imperfections. Commercial sensors are often made of platinum wire wound on a ceramic support body. The PT100 sensors are connected to a calibrated current source, similar to the one shown on the right-hand side in Figure 2.17. The voltage drop across the sensor is then measured with a volt meter, just as any other resistance measurement.

Thermistors are resistors that have their temperature dependence deliberately made large. In *positive temperature calibration* (PTC) devices, the resistance increases with temperature, and in *negative temperature calibration* (NTC) devices, it decreases. PTCs are mostly used as protection devices that switch the resistance from a low- to a high-resistance state if a certain temperature is exceeded. They are based on polycrystalline materials that change their dielectric constant at a certain temperature, the Curie temperature, by a large amount. Above the Curie temperature, the state of the magnetic dipoles is disordered, and the dielectric constant is small. This causes the formation of large potential barriers between the crystal grains, which leads to a high resistance. Below the Curie temperature the molecular dipoles are aligned, the dielectric constant is large, and the resistance is low. A typical application of the PTC thermistor is a self-regulating heater, in which the heater also warms up the thermistor, which increases the resistance and limits the current to the heater until an equilibrium is found. PTCs can also be used to detect whether a threshold temperature is exceeded.

The converse thermistors are NTCs, which decrease their resistance with increasing temperature. They are often used for temperature sensing and are based on a doped semiconducting material that has occupied impurity donor levels below the conduction band, as shown in Figure 2.3. Increasing the temperature thermally excites these electrons to jump into the conduction band, thus increasing the conductivity. This effect is much larger than the reduction of the resistance due to the ions oscillating and impeding the motion of the electrons, which was responsible for the temperature dependence in the PT100 sensor. Both NTC and PTC thermistors are sensed by connecting them to a constant-current source and measuring the voltage drop across the thermistor.

A number of *position sensors*, either for rotary or linear position, are based on potentiometers. A potentiometer is a variable resistor where a slider moves up and down a

Figure 2.4 A rotary and a linear potentiometer (left) and a circuit illustrating the electric connections (right)[†].

resistance and shortens the distance of one end point to the slider, thereby reducing the resistance between two of the terminals. The distance between the slider and the other end point lengthens, causing the resistance between the slider and the other terminal to increase correspondingly. On the left-hand side of Figure 2.4 we show a rotary and a linear potentiometer with three connectors; the two end points are connected to dark wires, and the one controlled by the slider is connected to a lighter-colored wire. The schematic view on the right of Figure 2.4 explains the functionality; the slider controls a variable mid-point of a voltage divider and the output voltage interpolates from 0 to 5 V in this case. A variation of the potentiometer is a *joystick*, which is based on two orthogonally mounted potentiometers, controlled by a small stick. We simply need to measure the voltage on the output with respect to ground in order to determine the position of the slider or the stick. An image of a joystick is shown on the left of Figure 2.5.

Also, *fluid levels* of liquids with a small conductivity can be determined with a resistor whose resistance is varied by reducing the resistance between conducting stripes shown in Figure 2.5. The sensor is connected just as a potentiometer with the liquid level acting as the slider.

Applying an external force to a material changes its equilibrium shape, and therefore *strains* the material. An example is a stretched wire that gets longer and thinner if pulled. Consider the resistance $R = \rho L/A$ of the wire, given in terms of resistivity ρ, length L, and cross-section A, and consider its change. We see that increasing L and decreasing A increases the resistance R. Thus, the simple wire converts its deformation to a small change in the resistivity, and therefore serves as a *force-sensitive resistor* that is often used as one branch in a Wheatstone bridge. The small effect of only changing the geometry of the wire can be greatly enhanced by using heavily doped semiconductors instead. The latter also

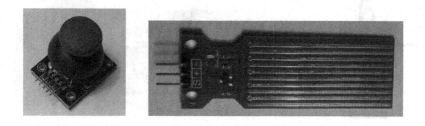

Figure 2.5 A joystick (left) and a fluid-level resistive sensor (right).

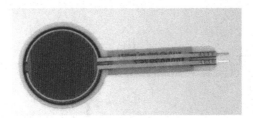

Figure 2.6 A strain gauge.

change their resistivity ρ by a large amount when strained. Alternatively, they are based on polymer thick-film technology. An example is the strain gauge shown in Figure 2.6.

The MQ-2 and other MQ-x are *gas sensors* sensitive to different types of gas, depending on their type specification. They are based on a semiconductor substrate with a thin surface layer of polycrystalline SnO_2 tin oxide that is either sputtered or deposited by evaporation. The specificity to various gases depends on the temperature of the active area, which is adjusted by a heater, by the method of deposition, or by small additions of other materials, such as palladium, gold, or platinum. The variation of the resistance depends on the grain boundaries of the polycrystalline SnO_2 and on the insulating oxide layer between the grains that is affected by the adsorbed gases. On the left of Figure 2.7 we illustrate the working principle. The heater is located under the SnO_2 active layer and powered by passing a current through it. The resistance, which depends on the concentration of the specific gas, can be measured between the terminals labeled 1 and 2. A device mounted on a small breakout board is shown on the right of Figure 2.7. The heated sensing area is located under a protective metallic cover.

After this short selection of resistance-based sensors we progress to discuss sensors that report a voltage directly.

2.1.2 Voltage-based sensors

An example of a sensor that directly produces a voltage at its output pin is the LM35 *temperature sensor*, which is a silicon-bandgap temperature sensor. The operating principle is based on passing known currents I_n with $n = 1, 2$ with current densities j_n across the base-emitter junctions of two bipolar transistors, and comparing their respective volt-

Figure 2.7 Schematic of an MQ-x gas detector (left) and a sensor mounted on a small breakout board (right).

age drops $V_{BE,n}$. The voltage difference is proportional to the temperature. This is easily understandable by inverting the current–voltage curve for the diode of the base-emitter junction

$$j_n = A(T) \left[e^{(eV_{BE,n} - E_g)/kT} - 1 \right]$$

where $E_g = 1.2\,\mathrm{V}$ is the bandgap energy of silicon, k is the Boltzmann constant, and T the absolute temperature in Kelvin. $A(T)$ is a device-specific constant with moderate temperature dependence. Assuming that both transistors are located on the same substrate and have the same temperature, we solve for two current densities j_1 and j_2 and obtain for the voltage difference $\Delta V_{BE} = V_{BE,1} - V_{BE,2} = kT/e \ln(j_1/j_2)$. In the LM35, the base-emitter diodes of the two transistors have different areas such that the ratio of the areas determines the current densities, provided that the same macroscopic current passes through the two transistors. There are operational amplifiers on the same substrate to provide signal conditioning such that the LM35 produces an output voltage V_s that is related to the temperature T by $V_s = T/100$. Here V_s is measured in volts and the temperature in degrees Celsius, such that a temperature of $23\,^{\circ}\mathrm{C}$ results in a voltage of $0.23\,\mathrm{V}$. The LM35 has three pins; one is connected to ground, one to the supply voltage, and the third one carries the voltage V_s that is proportional to the temperature. Note the polarity for connecting the LM35 in Figure 2.8. The flat surface is pointing to the wires on the left-hand side.

Thermocouples are temperature sensors that are based on the effects of temperature and temperature gradient on conductors made of different materials. Directly at the junction of the conductors, the Peltier effect causes a current that depends on the temperature. This happens at the points labeled by their respective temperatures T_1 and T_2 on the top left in Figure 2.9. On the wire segments a temperature gradient causes an additional current to flow, the Thomson effect. And finally, joining the two junctions and the wires causes a current to circulate, provided the loop is closed. This is called the Seebeck effect. If the loop is open, as shown at the top left of Figure 2.9, a voltage U develops at the end terminals as a consequence of the Peltier, Thomson, and Seebeck effects. In practice, one junction, say at T_1, is held at known and constant temperature, for example, by immersing the junction in ice water. Then the voltage U is related to the temperature difference $T_2 - T_1$ of the sensing end at T_2 and the reference temperature T_1. The magnitude of the voltage generated depends on the combination of metals and is typically on the order of $50\,\mu\mathrm{V}/^{\circ}\mathrm{C}$.

In a *thermopile* a number of wire segments of materials A and B are connected in series, as shown on the bottom left in Figure 2.9. This increases the sensitivity of the device to temperature differences. Thermopiles are often found in devices sensing heat and infrared radiation, such as thermal imaging devices or contact-free thermometers. The sensor used in the latter is shown on the right of Figure 2.9.

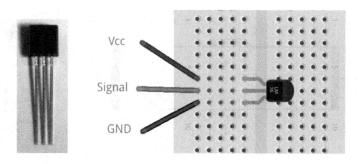

Figure 2.8 Image of an LM35 temperature sensor (left) and how to connect it (right[†]).

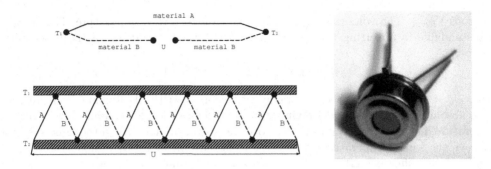

Figure 2.9 The schematic of a thermocouple (top left), of a thermopile (bottom left), and an image of an MLX90614 contact-free thermometer (right).

Some crystals and ceramics react to external stresses by producing a *piezoelectric* voltage between opposite sides of the material, as a consequence of rearranging charges within their crystal structure. The resulting voltages reach several kV and can be used to produce sparks in ignition circuits or in old-fashioned vinyl record players. There, a "crystal"-stylus is squeezed in the grooves of the record, and the generated voltages are amplified and made audible as sound. In scientific applications, piezoelectric sensors are used to measure pressures or forces.

The speed of angular motion is easily sensed by a DC electrical motor that is operated backward as a generator. Instead of applying a voltage to turn the axis of the motor, turning the axis induces an induction voltage in the motor coils that is proportional to the angular velocity. Attached to a propeller that is turned by either a flowing liquid or a gas, such a device can measure *flow rates*.

Hall sensors produce a voltage that is proportional to the *magnetic induction B*. Their mode of operation is explained in Figure 2.10 and is based on passing a known current I through a semiconductor. In the presence of a magnetic field, the Lorentz force deflects the charge carriers – electrons and holes – towards perpendicularly mounted electrodes (shaded). This creates a potential difference (a voltage) between the electrodes, which causes a transverse electric field that counteracts the deflection from the Lorentz force such that the following charge carriers can move towards the exit electrode undeflected. In equilibrium, the voltage difference between the upper and lower electrodes is proportional to the magnetic induction B and can be measured with a voltmeter. The A1324 Hall sensor, shown on the right in Figure 2.10, has signal-conditioning circuitry on board and only needs three pins

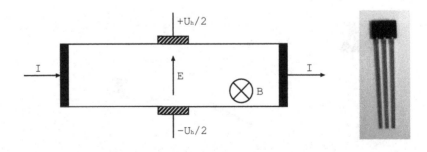

Figure 2.10 Schematic of a Hall sensor (left) and the A1324 sensor (right).

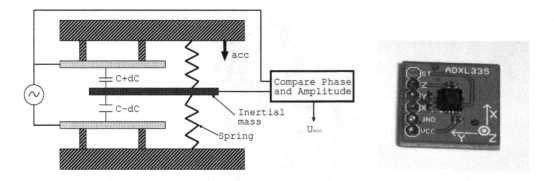

Figure 2.11 The operational principle of an ADXL accelerometer (left) and an ADXL335 mounted on a breadboard (right).

for ground, supply voltage, and output voltage. The latter is proportional to the magnetic field, with a sensitivity of 50 mV/mT centered at 2.5 V when no field is present.

The ADXL335 is a three-axis integrated *acceleration* sensor based on micromachined structures on a silicon substrate where an inertial mass is suspended by springs [14]. The inertial mass is part of an assembly of capacitors driven by an AC voltage that is used to measure the imbalance of a capacitive voltage divider. As opposed to our simplified model with only one capacitor doublet, the real device uses a large number of interleaved capacitor doublets in order to increase the sensitivity. Figure 2.11 illustrates the principle of operation for a single direction. On the left in Figure 2.11 there is an AC-voltage generator that drives the light-grey capacitor plates. The dark-grey inertial mass is placed halfway between the driven plates, and in the absence of acceleration, the capacitances between the two light grey plates and the inertial mass are equal. Any acceleration introduces an imbalance in the capacitances that affects the voltage level on the inertial mass. Comparing the phase and amplitude of that signal with that of the driving AC signal yields direction and magnitude of the acceleration. After some signal processing, it is then made available as U_{acc} on one of the output pins of the ADXL335 in the range from 0 to 3 V. This voltage is proportional to the acceleration in the range from -3 g to +3 g and is updated at a rate of about 100 times per second.

The SM-24, shown in Figure 2.12, is another type of accelerometer called a *geophone*. It is based on a coil connected to the housing by springs embedded in a magnetic field generated by permanent magnets that are attached to the housing. If the housing moves, the coil remains stationary due to its inertia, and a voltage is induced in the coil, which is proportional to the velocity and can be measured. The sensitivity is 28.8 V/(m/s) and the device operates in the range of 10–240 Hz.

Microphones convert sound to electrical signals and can be classified as sensors. Two major classes are on the market. *Dynamic microphones* operate similarly to the geophones. A coil, attached to a membrane, is excited by sound, moves in a magnetic field, and induces an induction voltage in the coil that is amplified and measured. In *electret microphones*, the membrane constitutes one electrode of a capacitor. If it moves, the capacitance changes, and the amount of charge stored on the capacitor is pushed on and off the capacitor and creates a current that is amplified and measured.

The electret microphone serves as a nice example with which to turn to current-based sensors.

Figure 2.12 An SM-24 geophone with a diameter of about 30 mm.

2.1.3 Current-based sensors

The BPW34, shown on the left-hand side in Figure 2.14, is a *pin diode* that generates a current of up to $100\,\text{nA}$ depending on the irradiance of up to mW/cm^2. Pin diodes are similar to conventional diodes and consist of a semiconductor with a *pn* junction. It is created by doping semiconductor material, often silicon based, with a material that has either five valence electrons, in which case it becomes *n*-doped, or three valence electrons, in which case it becomes *p*-doped. Pin diodes, however, have an additional layer of un-doped, intrinsically ('*i*') conducting silicon in order to increase the target area for the photons to produce additional charge carriers. One operational mode of the pin diode, called photoconductive, is illustrated on the left-hand side of Figure 2.13, which shows a simplified energy-band diagram of a reverse-biased diode. Note that by convention, the upwards energy axis corresponds to the potential energy of *electrons* that is lowest at the most positive voltage and that is found on the left-hand side. In the figure the cathode (*n* side) is therefore at higher voltage than the anode (*p* side) and results in all charge carriers being pulled out of the intermediate zone; electrons to the left and holes to the right. This results in the diode blocking any current flow. The extra layer of un-doped silicon provides extra potential charge carriers that act as targets for photons, having energy higher than the band-gap. These photons create electron–hole pairs by lifting electrons from the valence

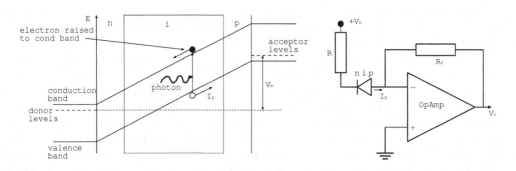

Figure 2.13 Energy-band diagram (left) and circuit (right) of a reverse-biased pin diode in photoconductive mode.

Figure 2.14 Image of a BPW34 pin diode on the left and two phototransistors on the right; an IR-sensitive BPX38 and an SFH3310, sensitive to visible light.

band into the conduction band, as indicated in Figure 2.13. The applied voltage, which is more positive on the left-hand side, causes the electrons to move to the left and the holes move to the right. Combined, this constitutes a current I_p. Adding a colored film of plastic – a filter – in front of the diode makes the response color sensitive. Using red, green, and blue filters thus produces the color sensor used in the TCS34725 that we discuss below. We mention in passing that ionizing radiation, such as high-energy photons and gamma rays as well as charged particles with high energies, create electron–hole pairs. This makes pin-diodes suitable as radiation detectors. The circuit on the right-hand side of Figure 2.13 shows an operational amplifier that converts the current I_p flowing towards its negative input port into a voltage $V_o = -R_f I_p$ on its output port. This use of the op-amp is referred to as a *transimpedance amplifier*. We will cover this and other uses in more detail in the coming sections.

Phototransistors such as the BPX38 or SFH3310, both shown on the right of Figure 2.14, are similar to normal transistors, but their base-collector diode is a reverse-biased photodiode, similar to the one described in the previous paragraph. It causes a current to flow as a consequence of impinging photons. The base-emitter diode is already forward biased and will ensure that the collector–emitter connection becomes conducting. Moreover, often there is a lens to increase the number of photons impinging onto the base terminal with the photosensitive area. Phototransistors sensitive to special spectral ranges such as infrared radiation, by suitably choosing their band gap, can be used as flame detectors.

The sensors in imaging applications such as cameras are *charge-coupled devices*, or CCDs, which are similar to a pin diode attached to a small capacitor, one for each pixel of the camera. Exposure to light transfers a small charge to the capacitor. The often large number of pixels are read out sequentially by transferring the charge from on capacitor to the one closer to the external readout port. A bucket chain to transfer water that is emptied at the end point comes to mind. In a CCD, once the charge reaches the output port, it is passed through a resistor, where it creates a voltage drop that is measured.

Solar cells operate in a similar fashion to photodiodes in photovoltaic mode where they provide a voltage to a load. They are, however, optimized to absorb as large a part of the spectrum as possible, and also to have a large absorbing area, in order to maximize the electric power available to the load.

After this brief overview of different analog sensors, we need to address how to prepare the signals such that they can be easily interfaced. This preparatory stage is referred to as signal conditioning.

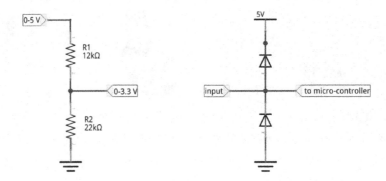

Figure 2.15 On the left we have a voltage divider to reduce the input voltage of 0–5 V to 0–3.3 V. The right circuit shows the use of clamping diodes to protect the input of the microcontroller to lie between ground and 5 V[†].

2.2 SIGNAL CONDITIONING

The analog signals from the sensors can be too high or too low. They can be too noisy or otherwise inadequate to directly feed to an analog-to-digital converter (ADC) or a microcontroller. In many cases some signal conditioning is needed, and we will discuss some common methods in the following sections.

2.2.1 Voltage divider

In case the input voltage of the sensor exceeds the input range of what the microcontroller can handle, we have to reduce the voltage by a *voltage divider,* which consists of two fixed-value resistors. A typical example is to reduce the range from 0–5 V to 0–3.3 V, which is accomplished by a combining two resistors with a ratio of $R_1/(R_1 + R_2) = 3.3/5 = 0.66$. A close example is shown in Figure 2.15, with a 12 and a $22\,k\Omega$ resistor. The ratio is $22/34 = 0.65$, which is close to the desired ratio. Other combinations with larger and smaller values will work as well, as long as the ratio is correct. The sum of both resistors should not be too small because that will draw a larger current from the 0–5 V voltage source, and, depending on the internal resistance of the source, may affect the measured values.

In order to improve the measurement sensitivity, resistance-based sensors are often wired in a Wheatstone-bridge configuration. An example is shown on the right of Figure 2.1. Here the voltage divider on the left side of the breadboard provides half the supply voltage at its center tap (the lower-signal wire) because the two resistors are equal. The voltage divider on the right-hand side is the same one, we encountered previously, with the upper-signal wire connected to the point between the upper resistor and the LDR. Normally one would choose the resistance of the right resistor to be in the middle of the range of interest of the LDR, such that voltage-difference between the signal wires is close to zero, indicating mid-range. In this way, depending on the light exposure, the voltage difference between the wires varies around zero and the sign tells us whether the exposure is lower or higher than the expected mid-range value. Since we now deal with voltages that vary around zero, it is easier to amplify that voltage in order to increase the sensitivity. For example, when using a plain voltmeter, we can use a smaller voltage range.

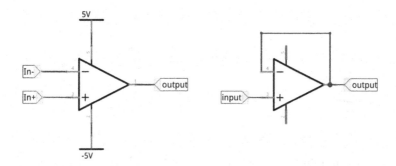

Figure 2.16 A bare operational amplifier (left) and wired as a line buffer (right). We omitted the supply wires on the latter schematics†.

In case we use a piezo-based sensor, the generated voltages can be much higher than is acceptable in the following circuit such as a microcontroller. In such a situation, *clamping diodes*, as shown on the right of Figure 2.15, are used. If the input voltage is between 0 and 5 V, the diodes are blocking, and the signal is passed on to the microcontroller. If, on the other hand, the input voltage exceeds 5 V plus the forward voltage drop, the diode starts conducting and shorts the input to the upper power rail of 5 V. If the input voltage is below 0 V, the lower diode starts conducting and shorts the input to ground. In either case, the voltage delivered to the microcontroller is limited to 0-5 V plus or minus the forward diode voltage drop. Many integrated circuits including microcontrollers have built-in clamping diodes. The Raspberry Pi, however, is a notable exception.

2.2.2 Amplifiers

Very small electrical signals usually need to be amplified to reach levels adequate for further processing. The standard device to achieve this is an *operational amplifier* or op-amp, shown on the left of Figure 2.16. There are two input ports on the left, one labeled "plus," one labeled "minus," and one output port. The latter delivers a voltage that depends on the difference between the two input ports. In an ideal op-amp, the amplification factor is infinite and we usually use some feedback mechanism to obtain a deterministic behavior, as discussed below. Normally, op-amps require both positive and negative supply voltage, even though sometimes it is possible to tie the negative supply rail to ground, in which case only unipolar signals can be amplified.

Before discussing different circuits, we need to describe three basic principles that characterize op-amps.

- The input impedance of the input ports is "infinite," which means that no current flows into the op-amp and it will not load the upstream circuitry.

- The op-amp tries to reduce the difference between the input ports $V_+ - V_-$ to zero.

- The amplification without feedback is quasi-infinite, and the output has a very low impedance and can provide high output currents.

These simple rules will help us to design and understand the following circuits, but consult for example [17–19] for a more extensive discussion.

We start by considering a *line buffer*, which is typically used as an impedance converter to transform the output of a sensor with a high impedance to a low-impedance signal that

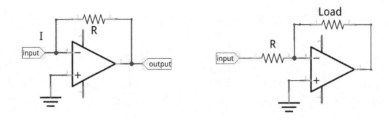

Figure 2.17 A transimpedance amplifier (left) and a voltage-controlled current source (right)[†].

is less susceptible to noise. We show the circuit on the right-hand side in Figure 2.16, where the output of the op-amp is fed back onto the negative input port. By using the second op-amp rule, we see that the op-amp tries to make the positive and negative input ports equal, but the negative is tied to the output, which forces the output to follow the positive input. Moreover, by the first rule, the input impedance is high and no current (or at least very small current) is drawn from the sensor, while the output impedance is low and can provide a high current. Note that we omitted the connections to the supply rails on the right of Figure 2.16 and will do so henceforth in order to avoid cluttering the schematics.

Adding a resistor R to the feedback branch, as shown on the left-hand side in Figure 2.17, forces a current I, entering the circuit from the left, to pass R where it causes a voltage drop $V_- - V_{out} = IR$. Since the positive input terminal of the op-amp is tied to ground, we have $V_+ = 0$ and the second op-amp rule also forces V_- to zero, such that we find that the output voltage V_{out} is given by $V_{out} = -IR$. The feedback resistor R thus converts the input current I to the output voltage V_{out}. This is the reason that the circuit is referred to as a *transimpedance amplifier*. Note that we already met one in Figure 2.13.

Adding a resistor to the negative input terminal turns the previous circuit into a *voltage-controlled current source*, shown on the right-hand side in Figure 2.17. The grounded positive input of the op-amp forces the voltage on its negative input terminal to be zero, which causes a current through the resistor to be $I = V_{in}/R$. And this current must pass through the load such as a PT100 temperature sensor discussed in Section 2.1.1.

Let us consider this circuit a little further by calculating the voltage at the output of the op-amp, as shown on the right-hand side in Figure 2.18. The input current I_{in} only flows through the input resistor R_4 and the feedback resistor R_5 because the input impedance of the op-amp is essentially infinite and no current flows into the input ports. But the

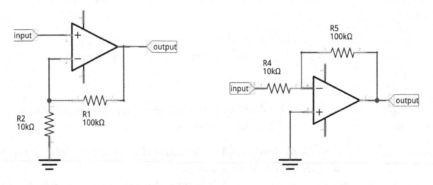

Figure 2.18 A non-inverting (left) and inverting (right) amplifier[†].

equality of the current in the input and the feedback resistor implies $I = (V_{in} - V_-)/R_4 = (V_- - V_{out})/R_5$. Moreover, we observe that the positive input port is grounded, which forces the negative input port to be on ground potential as well. This implies $V_- = 0$, and the relation between input and output voltage becomes $V_{out} = -V_{in}R_5/R_4$, where the negative sign indicates that the amplifier is inverting. Note that we can add several input resistors R_4 in parallel with one end connected to the negative input port. This allows us to add the currents passing through the parallel copies of R_4, and we obtain a summing amplifier. The resistor values were simply chosen to be in a reasonable range. They need to be determined for each application.

If we want to amplify a voltage without inverting it, we use the circuit shown on the left-hand side in Figure 2.18. The two resistors R_1 and R_2 constitute a voltage divider that forces the negative input voltage to be $V_- = V_{out}R_2/(R_1 + R_2)$. But, at the same time, the op-amp forces $V_- = V_+$, which, after solving for V_{out}, leads to $V_{out} = V_+(R_1 + R_2)/R_2$, where $(R_1 + R_2)/R_2 = 1 + R_1/R_2$ is the amplification factor. Since the output voltage V_{out} has the same sign as the input voltage V_+ on the positive input port, this configuration is called a non-inverting amplifier.

Sometimes the signal one wants to measure changes around a non-zero baseline. Examples are the Hall sensor A1324 from the previous section, where zero magnetic field produces $2.5\,\text{V}$ and the magnetic field added or subtracted from that value depending on its polarity. In order to increase the resolution, we want to amplify not the signal, but the difference of the signal to the baseline. In other words, we need a circuit to subtract the baseline and amplify the difference. A differential amplifier as shown in Figure 2.19 accomplishes this feat, provided that $R_2 = R_4$ and $R_1 = R_2$. The output voltage V_{out} in that case is given by $V_{out} = (V_2 - V_1)R_2/R_1$. Adding a potentiometer that adjusts V_1 between the positive and negative supply rail subtracts it as baseline voltage.

A circuit similar to the one in Figure 2.19 but with line buffers on the inputs is called an *instrumentation amplifier*, and an example is shown on the left of Figure 2.20. Normally it is not necessary to build instrumentation amplifiers from discrete components. There are ready-made circuits available such as the AD620 or INA131. Moreover, the AD8232 is a special-purpose instrumentation amplifier for *electrocardiography* applications. It amplifies and processes the weak signals picked up by electrodes attached to a patient's chest and outputs the characteristic traces seen on heart-rate monitor screens. It provides signal-conditioning circuitry to filter out and to compensate external disturbances, such as the $50\,\text{Hz}$ line frequency ($60\,\text{Hz}$ in the US).

Often, sensors produce small voltages that vary around zero, but the analog-to-digital converter (ADC) requires an input range of 0 to $5\,\text{V}$. We thus face the problem of amplifying a bipolar signal and changing the baseline level to mid range of the ADC. We show a circuit that amplifies by approximately a factor of 10 in Figure 2.21. The amplification is mainly determined by the ratio of the feedback resistor R_3 to the input resistors R_1 and R_2. The voltage divider of R_6 and R_8 provides the mid-range offset voltage. The average level of the output voltage crucially depends on the tolerances of the resistors, and in order to place the level safely in mid range, we use the capacitor C_1 to first remove the DC level of the output signal before adjusting it properly to mid-range, with the voltage divider consisting of R_4 and R_5.

In case we need to amplify input signals that vary over a huge range of values, a *logarithmic amplifier* such as the one on the right of Figure 2.20 is a useful circuit. It can be shown that the relation between input and output voltage is $V_{out} = -V_t \ln(V_1/I_sR)$, where V_t is the thermal voltage and I_s the saturation current of the diode. Swapping the diode and the resistor results in an exponential amplifier.

A close relative to the operational amplifier is the *comparator*. It can be visualized as an op-amp with very large, even infinite, amplification, whose output port saturates at the

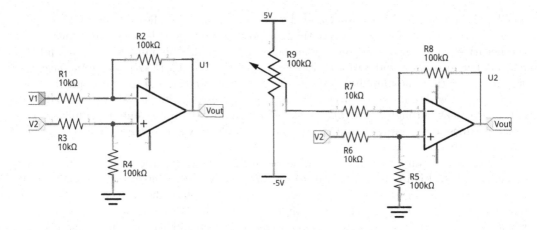

Figure 2.19 A difference amplifier (left) and the same circuit with adjustable V_1 (right) that is used to subtract the baseline[†].

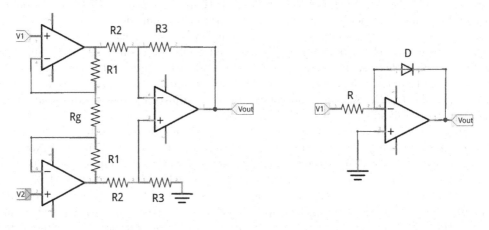

Figure 2.20 Instrumentation amplifier (left) and logarithmic amplifier (right)[†].

power rails. If the voltage at the positive input terminal of a comparator is larger than that on the negative input terminal, the output voltage is very close to the positive supply voltage. On the other hand, if the voltage on the positive input terminal is lower than that on the negative, the output is close to ground potential. In this way it translates the input voltages to a binary digital state. A comparator may therefore be considered as a 1-bit analog-to-digital converter, and we will see in a later section how it is used to extend the number of bits of the conversion. Some comparators have the threshold when switching from low to high output configured to be slightly higher than the threshold switching from the high output state back to the low one. This small hysteresis prevents the output from switching back and forth uncontrollably, should the voltages on the two input terminal be very close.

After the basics of signal amplification, we will now address the question of how to reduce noise in the circuits and decrease the sensitivity of a circuit in an undesired frequency range. This is the realm of filters.

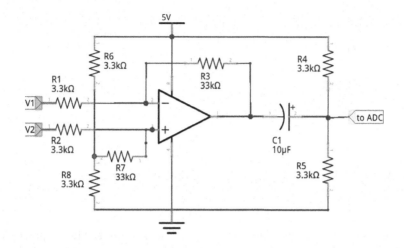

Figure 2.21 Amplifying a weak signal and shifting it to mid-range between the power rails[†].

2.2.3 Filters

The task of filters is to remove certain frequencies from an electrical signal such as all high frequencies, in which case the filter is called a low-pass filter. An example is a low-pass filter that removes "hissing" in audio-signals. The converse filter is a high-pass filter that removes low frequencies. An example is a anti-rumble filter found in old vinyl record players. If we know that the desired signal contains only a certain range of frequencies, and we wish to remove all others, we use a band-pass filter. An example is the IF filter found in radios that are based on the super-heterodyne principle. And finally there are filters that remove only frequencies in a narrow band. They are called band-stop or notch filters. An example is a filter that removes the omnipresent 50 Hz or 60 Hz hum coming from the power grid.

We first consider a *low-pass filter*, which in the simplest incarnation is a frequency-dependent voltage divider made of a resistor with resistance R and a capacitor with capacitance C, as shown on the left of Figure 2.22. For a refresher of basic concepts circuit theory, such as impedance, please consult Appendix A. In our circuit, the capacitor has an impedance $1/i\omega C$, which gets smaller with increasing frequency $\omega = 2\pi f$. Intuitively, the higher frequencies are shorted to ground. If we build a voltage divider as shown on the left of Figure 2.22, the output voltage V_{out} is given by $V_{out} = (V_{in}/i\omega C)/(R + 1/i\omega C) = V_{in}/(1+i\omega RC)$, and we see that it is attenuated with increasing frequency ω and, conversely,

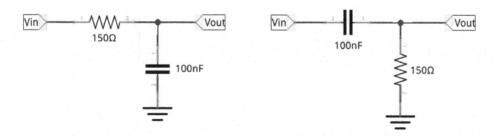

Figure 2.22 A simple low-pass (left) and high-pass (right) filter. As illustration we chose the component values to give $f_c = 10\,\mathrm{kHz}$[†].

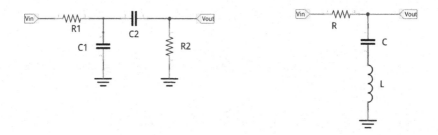

Figure 2.23 A simple band-pass (left) and band-stop (right) filter[†].

the low frequencies are unaffected, hence the name low-pass filter. The frequency where the signal amplitude is attenuated by a factor $\sqrt{2}$ is given by $\omega_c = 2\pi f_c = 1/RC$, and the imaginary unit i indicates that there is a phase shift between input and output voltage that depends on the frequency. Equivalently, a low-pass filter can be constructed with an inductor with impedance $i\omega L$ and resistor R, but in many operational situations the inductances have values that are difficult to find; therefore filters are usually constructed from resistors and capacitors. The frequency dependence of the filter asymptotically behaves as the first inverse power of ω, and the filter is called a single-pole filter. Cascading two such filters results in two-pole low-pass filters that exhibit a steeper frequency-dependence of $1/\omega^2$.

Swapping the resistor and capacitor in the low-pass filter results in the *high-pass* filter shown on the right of Figure 2.22, which has a frequency dependence of $i\omega RC/(1 + i\omega RC)$ that attenuates low frequencies due to the factor ω in the numerator and approaches unity as $\omega \gg \omega_c = 1/RC$.

If we want to filter out everything except a small range of frequencies, we need a *band-pass* filter that we can most easily construct by a combination of a low- and a high-pass filter, as shown on the left of Figure 2.23. Here we have to ensure that the cutoff-frequency $1/R_1C_1$ of the initial low-pass filter is higher than that of the high-pass filter, given by $1/R_2C_2$.

In case we need to reject a certain perturbing frequency, we implement a *band-stop* or notch-filter, which is shown on the right of Figure 2.23. The combination of inductance L and capacitance C has a resonance at the frequency $\omega_c^2 = 1/LC$ where their series resistance

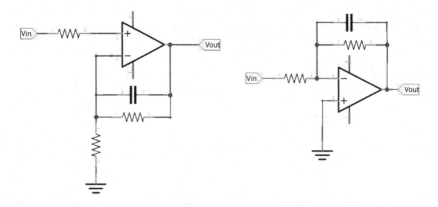

Figure 2.24 An active non-inverting (left) and inverting (right) low-pass filter[†].

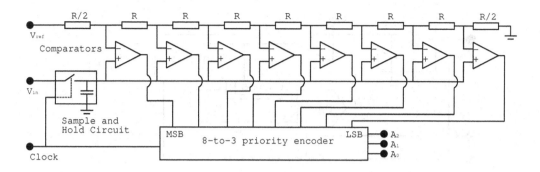

Figure 2.25 The operating principle of a 3-bit flash ADC.

vanishes, and a signal close to this frequency is shorted to ground and not passed on to the output.

The filters discussed so far are passive filters that only depend on resistors, capacitors, and inductors, and will only be able to attenuate unwanted frequencies. Sometimes, however, we need to combine amplification of a signal with filtering. The simplest solution is placing an operational amplifier immediately after the filter. Alternatively, one may include a capacitor in the feedback branch of the amplifier, as shown in Figure 2.24.

There is a huge amount of expertise in designing filters, both passive and active, documented in the literature. For a practical overview, see [20].

2.2.4 Analog-to-digital conversion

Normally we want to convert analog signals to a digital format in order to process them on a computer. The device that achieves this feat is an analog-to-digital converter (ADC), which can be conceptually understood as a very rapidly measuring voltmeter. It consists of a sample-and-hold circuit that keeps the voltage constant for a short time while it is measured and digitized using a number of different operational principles.

In a *flash ADC* the voltmeter is based on large number of comparators that compare the voltage against a sequence of voltages derived from a resistor ladder. An additional circuit encodes the output from the comparators into a binary representation. Since all comparators operate in parallel, flash ADCs operate at the highest conversion rate, in excess of 10^9 samples per second. This high conversion rate comes at a price, though, because for a resolution of n bits, 2^n comparators are required. We illustrate the operating principle of a 3-bit flash ADC in Figure 2.25. On the left there are inputs for the reference voltage V_{ref}, the voltage to be converted V_{in}, and a clock signal. First, the input voltage is held constant for the duration of the conversion in a sample-and-hold circuit, which consists of a switch and a small holding capacitor. The voltages to compare to are produced by a resistive voltage divider, shown on the top. The voltages from the divider are routed to the negative input terminal of the comparators. All of the latter then compare the sampled-and-held input voltage, simultaneously. The outputs of the comparators are connected to the inputs of a priority encoder that translates this into a 3-bit binary representation at the output pins A_0, A_1, and A_2. The conversion process and synchronous operation of sampling and encoding is coordinated with an externally supplied clock signal. Flash ADCs are normally available with resolution of 8 bits or less, and are relatively expensive. To sample at lower rates, we can use less expensive ADCs such as those discussed in the following paragraph.

Successive approximation ADCs provide higher resolutions with larger number of bits at lower cost, albeit at a lower conversion rate. They replace the resistor ladder and the large

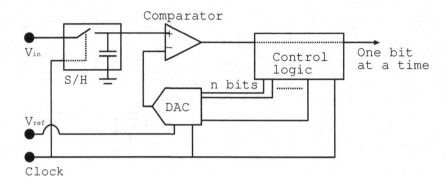

Figure 2.26 The operating principle of a successive approximation ADC.

number of comparators with a single comparator and a digital-to-analog converter (DAC) to dynamically adjust the reference voltage to compare to. We discuss the operational principle of DACs in Section 3.13, but here we just mention that they are circuits that translate a digital word with n bits to an analog voltage, and they can do this very quickly. They play a central role in the operation of a successive approximation ADC whose operating mechanism is illustrated in Figure 2.26, where the input voltage is held constant during the conversion in a sample-and-hold circuit. Then the voltage is passed to a comparator whose negative input terminal is determined by the output voltage of the DAC. In the first comparison, the n–bit input word is set to the B'1000... which results in half the reference voltage V_{ref}, where we prepend the letter B to identify the binary representation of a number. If the input voltage V_{in} is smaller, the output of the comparator is at ground level, or logically a low state. The first, most significant bit (MSB), is therefore "0." The control logic then sets the DAC to B'0100... and compares again on the next clock cycle. If V_{in} is larger than $(1/4)V_{ref}$, the comparator output will go to high-voltage level, and the next bit to pass to the output is therefore "1," and the DAC receives the input B'0110... to compare the next step in the bisection sequence. If this is repeated n times the ADC clocks out all bits of the conversion from MSB to least significant bit (LSB). We see that in this case we trade conversion speed for simpler hardware. Most ADCs we use later in this book are of the successive approximation type. We note in passing that often a multiplexer switch is placed before the sample-and-hold circuitry, which allows us to select different input voltages with a single ADC shared among the different channels. Only one channel can be converted at the highest conversion rate.

It is possible to find a compromise between conversion speed and hardware complexity by combining, for example, two 4-bit flash ADCs, one additional DAC, and advanced control logic to a so-called pipelined flash ADC. They can sustain a continuous conversion-rate determined by the 4-bit ADC. Even though each conversion requires two clock cycles, the first ADC can already start converting the next sample while the second ADC still works on the four least significant bits of the previous sample. Most high-speed ADCs for radio-frequency applications with more than 8 bits use this method.

Instead of a very high conversion rate, it is often desirable to achieve high resolution; in other words, more bits, albeit at a lower conversion rate. Devices that fulfill this requirement are *delta-sigma ADCs* whose simplified operating principle is illustrated in Figure 2.27. It is based on adding quantized current pulses to the negative input terminal of the op-amp such that its output is forced to zero voltage. Note that this terminal is a virtual ground, because the positive input terminal is grounded, and operational amplifiers always strive to

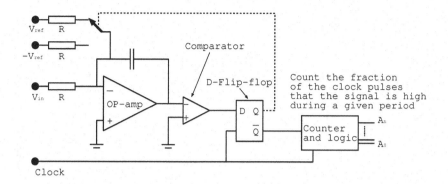

Figure 2.27 The operating principle of a delta-sigma ADC.

make their input voltages equal. The op-amp is configured as an integrator and sums the currents over time. Forcing the accumulated charge on the capacitor to zero is achieved by the feedback, shown as a dashed line from the non-inverting output of the flip-flop to the switch that either adds a positive or negative current of magnitude V_{ref}/R into the inverting input terminal of the op-amp. The purpose of the flip-flop is to produce well-defined time-steps, equal to the clock-frequency, for the injected current pulses. All we have to do in the end is to count the number of clock cycles when the output is high and divide by the total number of elapsed cycles. This results in a digitized word with the number of bits given by the time we chose to average. And that time can be quite long. Assume that we use a clock of 10 MHz and sample for 0.1 s such that 10^6 clock pulses happen, which results in a resolution of 20 bits because $10^6 \approx 2^{20}$. Delta-sigma ADCs derive their name because the small quantized difference (delta) current are summed (sigma), together with the current from the input voltage V_{in}, in the integrator. The high resolution makes delta-sigma ADC a good choice to measure the often small voltages in Wheatstone bridges. We need to keep in mind, however, that this increased resolution comes at the price of a rather low conversion rate; often only a few tens of conversions per second are possible.

Despite having a large number of bits, ADCs add noise to the measurements, because they cannot measure voltage differences smaller than that corresponding to the least significant bit. Thus, they introduce a *quantization error*, which could be reduced by choosing an ADC with a larger number of bits. But even then the quantization causes a slightly inaccurate representation of the signal amplitudes.

Sampling the signal at discrete instances in time causes a second potential misrepresentation, because signals of frequencies f higher than half the sampling frequency $f_s/2$, also called the Nyquist frequency, cannot be distinguished from signals with frequency $nf_s - f$ or $nf_s + f$. The origin of this ambiguity is illustrated in Figure 2.28. On the left we display a sinusoidal signal with frequency $f = 0.15f_s$ (dashed) and $f = (1 - 0.15)f_s$ (solid), and on the right the solid line is a sine with frequency $f = (1 + 0.15)f_s$. We see that at the sampling instances, indicated by the squares, the curves have the same values, which makes them indistinguishable if only sampled at rate f_s. In the frequency domain, shown in Figure 2.29, we find that a signal S_1 that lies between the Nyquist frequency $f_s/2$ and f_s is recorded by the ADC at frequency s_1, which is S_1 mirrored at the Nyquist boundary. The signal S_2 that lies above f_s is observed as signal s_2. The zone between zero and $f_s/2$ is commonly called the base band or first Nyquist zone, and the appearance of higher-frequency signals in the base band is called *aliasing*. Since the original frequency of the aliased signals is unknown, they are often regarded as noise. A simple way to prevent aliasing is to use analog low-pass filters with cutoff frequencies below the Nyquist frequency

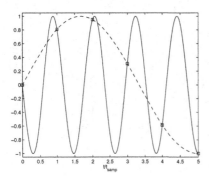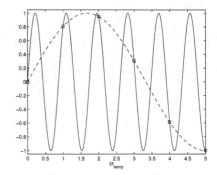

Figure 2.28 Sampling a signal in different Nyquist zones. The dashed line displays a frequency of $0.15f_s$ and the solid lines show signals with frequency $(1-0.15)f_s$ on the left and $(1+0.15)f_s$ on the right. Note that at the times when the signal is sampled (indicated by boxes), the signals are indistinguishable.

before passing the signals to the ADC. Filters such as those discussed in Section 2.2.3 are often sufficient.

Note that ADCs are often built into microcontrollers and sensors with digital interfaces, such as those discussed in later sections. After this treatment of the ADC, the workhorse of digital data-acquisition systems, we need to look at the task of providing power to our circuits.

2.2.5 Supply voltage

Of course, our sensors and also the microcontroller require electric power to operate, and that is normally supplied by a power source. A common type of power supply uses a transformer to step down the 220 V or 120 V AC voltage from the wall to a commonly used voltage range of around 5–20 V. This depends on the rating and winding ratio of primary to secondary winding of the transformer. Since most electronics circuits require DC voltages, we need to rectify the AC voltage from the secondary winding of the transformer. The simplest way to do this is to use a single rectifying diode with adequate power rating, as

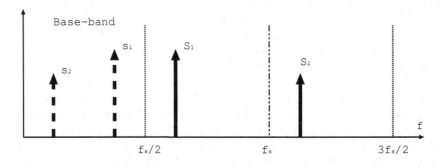

Figure 2.29 The dashed signals s_1 and s_2 are the images aliased into the base band of the original signals S_1 and S_2 in higher Nyquist zones.

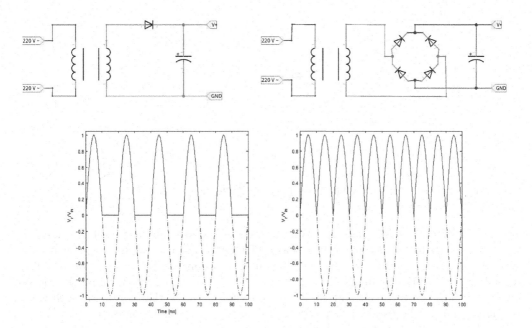

Figure 2.30 Schematics of very simple power supply circuits[†].

shown on the left of Figure 2.30. The diode only lets voltages pass in one direction, and effectively cuts off the negative half-cycle of the AC voltage. The ratio of the voltages before and after the diode are shown on the plot below. The output in that case is very bumpy, and can be smoothed somewhat by adding a large electrolytic capacitor with a few 100s of μF. It is charged during the positive half-cycle and is discharged during the negative, where it supplies charge to the powered circuit. The more current the load draws, the bigger the capacitor needs to be. Note also that the power in the negative half-cycle is dissipated in the diodes, which heats them up and may make a heat sink for the diodes necessary.

A better solution is to use a bridge-rectifier circuit made of four diodes with adequate power rating, as shown on the right side of Figure 2.30. The diodes are placed in such a way that during the positive half-cycle, one set of diodes conducts, and during the other half-cycle, the other set of diodes conducts. Essentially the circuit works by transforming the sine of the AC to the absolute value of the sine, as can be seen on the plot below the circuit where we show the ratio of the voltages before and after the bridge rectifier. We still need the smoothing capacitor, but for the same capacitance, the ripple on the output voltage is much smaller.

The transformers and capacitors, operating at the line frequency of 50 Hz (60 Hz in the US) are rather bulky. It is advantageous to use higher frequencies, which makes the use of much smaller transformers and capacitors possible. These high frequencies are created by a transistor that rapidly switches the rectified primary voltage on and off before passing it through the transformer and rectifying the voltage once again. These *switch-mode power supplies* are more efficient and smaller than old-fashioned power supplies and are used in practically all phone chargers and computer power supplies.

Any remaining ripple of the output voltage can be reduced further by using a linear voltage regulator based on a temperature-stabilized voltage source with a bandgap voltage reference. In such a circuit, the known temperature gradient of the output voltage from a configuration with two transistors of different size is balanced by the opposite temperature

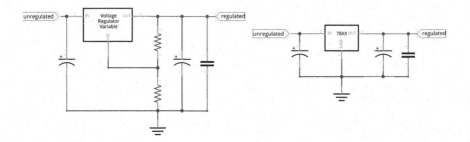

Figure 2.31   Variable-voltage (left) and fixed-voltage (right) regulator circuits[†].

gradient of suitably chosen resistors. We discussed a similar configuration in Section 2.1.2 when discussing the LM35 temperature sensor. The result is a very stable output voltage on the order of 1.25 V, near the bandgap of silicon. The commercially available linear voltage regulators, such as the commonly used LM317, usually have three pins for input IN, output OUT, and adjustment ADJ, and they have internal circuitry that forces the voltage on the output pin to the bandgap reference voltage of 1.25 V. It is straightforward to use an external voltage divider with suitably chosen ratio to select any output voltage up to some maximum current specified in the datasheet. On the left of Figure 2.31, we show how a voltage divider with two resistors is used to set the desired output voltage. The larger capacitors have capacitances of a few μF and guarantee stable operation of the circuit. The smaller capacitors of typically 100 nF are used to absorb any high-frequency glitches from the load.

For many standard operating voltages, such as 3.3, 5, 9, and 12 V, ready-made linear voltage regulators are available. They do not require the voltage divider, and most have the numbers 7803, 7805, 7809, or 7812 in their name. The last two digits specify the output voltage. The same series with 79xx exist for negative output voltages. These small circuits are very handy for obtaining well-regulated output voltages. Also, if a power supply already gives 5 V, but 3.3 V are required for a sub-circuit, we use a 7803 or similar. Normally the voltage regulators require about 2 V higher input voltage than their specified output voltage, unless we choose a special *low-dropout voltage* (LDO) regulator, which only requires about 0.5 V overhead. The 3.3 V regulator MCP1700 is a member of this category.

Of course, we can also power an electronic circuit from batteries or rechargeable batteries. Of special interest are lithium-ion or lithium-polymer rechargeable, because they can provide very high currents, which is particularly important for devices, such as WLAN circuits or motors. The lithium-ion cells provide voltages in multiples of nominally 3.7 V, and require special dis- and recharging circuitry, because their high intrinsic energy density requires special care to prevent under- and over-charging, which can physically damage the battery.

For electronic circuits that are used near a computer, we can use power drawn from the USB port. Many RS-232-to-serial converter circuits provide up to about 500 mA, which is often sufficient for small circuits. To make circuits very portable in the field, we can also use solar cells and so-called supercapacitors as temporary power source. In the laboratory, we have, of course, bench power supplies with adjustable output voltage, and current-limiting circuitry which prevents excessive currents that might damage our electronics.

After discussing analog sensors, analog-to-digital converters, and power supplies, we now turn to sensors that directly produce digital signals.

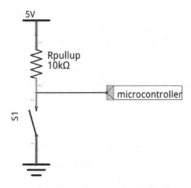

Figure 2.32 Connecting a switch or button with a pull-up resistor[†].

2.3 DIGITAL SENSORS

We now address sensors that do not require an external ADC, but report their measurement values directly to the microcontroller in digital from. Some sensors already have ADCs built in, while others do not need one. We start with the latter, of which the most prominent examples are buttons and switches.

2.3.1 Buttons and switches

The simplest digital sensor is certainly a *switch* that is either closed or open, or a *button* where the open or close state is only activated temporarily. We use these terms interchangeably. Our task is to sense their state in a reliable way, and this is normally done with a pull-up resistor that is connected to the supply voltage in the way shown in Figure 2.32. In this way the sensing pin on the microcontroller can reliably detect the supply voltage. Only if the switch S1 is pressed does the voltage on the pin drop to zero or ground level. Only a small current, determined by the magnitude of the resistor, flows when the switch is closed. The actual value of the resistor is uncritical, but values around 10–30 kΩ are usually reasonable. Without the pull-up resistor, the voltage potential on the pin is undefined when the switch is open, and determined by stray capacitances in the system. So, it is recommended to always use a pull-up resistor when sensing the state of a switch. Note that swapping the position of the resistor and the switch, the resistor functions as a pull-down resistor, and the sensing level is zero unless the switch is closed.

We need to point out that mechanical switches have the undesirable characteristic of bouncing. Closing the switch is often accompanied by a fast on-off sequence. Fast microcontrollers or other computers act so rapidly that they are easily fooled by sensing multiple switch-closures instead of a single one. Sometimes a small time-delay is introduced after the first switch closure is detected. In this way, only a single closing event is accounted for, and others during the short timeout period are ignored. Optionally, an analog debouncer, which is a simple low-pass filter, as discussed in Section 2.2.3, may be used.

There are a number of other sensors that act just like a switch and can be sensed in the same way. An example is a *reed switch*, which is sensitive to magnetic fields and closes a switch if some field level is exceeded. A further example is a *tilt switch*, which senses the position of a small conducting sphere that horizontally rolls back and forth, and closes a contact when it bumps into one of the extreme positions.

A variation on the theme of switches are *rotary encoders*, which use two switches, A and B, that open and close periodically as an axis is turned. The direction of the rotation can

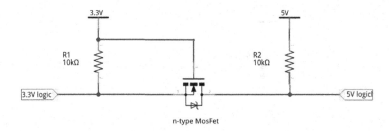

Figure 2.33 Level shifter circuitry using a n-type MOSFET. The source of the MOSFET is connected to the 3.3 V logic and the drain to the 5 V logic[†].

be detected, because each switch produces a regular pattern of on and off, but the timing of the two switches is shifted by 1/4 of the period length. Essentially, one switch produces a cosine-like pattern and the other a sine-like pattern. From knowing both, we can determine the direction of rotation by observing whether switch A leads switch B or vice versa. But the fundamental sensing process is based on the same mechanism as shown in Figure 2.32.

2.3.2 On/off devices

A number of sensors provide a *voltage level* to inform the microcontroller about their state or change of state. They can be thought of as a switch with a built-in pull-up resistor and can be sensed in the same way.

A problem can occur if the operating voltage level of the sensor and the microcontroller do not agree. Nowadays many sensors operate on levels of 2.5–3.3 V, and microcontrollers on levels from 2.5 to 5 V. Sensing higher external voltages such as those used in cars (12 V) or industrial control applications (24 or 48 V) requires some adjustment of the voltage level, to prevent damaging either the sensor or the microcontroller. There are level-changing chips available, such as the 74LVC245, but in many cases a simple voltage divider with two resistors is adequate. Yet, it only works if there is signal flowing from the high to the low-voltage side. In case a bidirectional signal flow is necessary, such as on the data line of the I2C bus, the solution shown in Figure 2.33, based on an n-type MOSFET transistor, is easy to implement. In the first case, when both logic signals are high, the MOSFET is not conducting, because the voltage difference between gate and source is close to zero. In the second case, if the 3.3 V logic is controlling and the signal is pulled low, the difference between gate and source is positive and the MOSFET conducts, such that even the 5 V logic level is pulled low. In the third case, when the 5 V logic is controlling and the 5 V logic signal is pulled low, the built-in diode (visible in the schematics) conducts, and causes the voltage of the source to drop to about 0.7 V. At this point the gate source voltage drop is sufficiently large to fully cause the MOSFET to conduct, which also pulls the 3.3 V logic level low.

But let us return to the sensors. A prominent device that reports its state through a changing voltage level is a *PIR proximity sensor*, shown on the right of Figure 2.34. It senses the change in the infrared radiation level, which announces the presence of living beings. The sensors are based on collecting the incident infrared radiation with a Fresnel lens, which is the dome visible in Figure 2.34, on a pyroelectric sensor. The lens is often made of polyethylene, a material chosen for its low absorption of IR radiation. One side of the sensor is consequently warmed up and expands, which causes buckling of the piezo- or pyroelectric material, often a polymer film. Two effects contribute to the voltage between

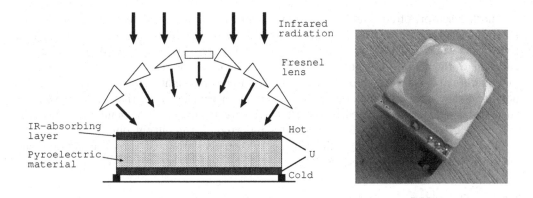

Figure 2.34 Schematic view of a PIR sensor (left) and the hardware (right).

the upper and lower plates. First, the buckling strains the material and causes a piezoelectric voltage. Second, heat flows from the hot to the cold side and adds a pyroelectric voltage. The schematic setup of the device is illustrated on the left of Figure 2.34. The tiny voltage that is generated is subsequently amplified and exposed to the surrounding electronics on an output pin that goes from the low to the high voltage level once it is triggered by the presence of a person, and stays there for a programmable (usually by a small potentiometer) amount of time.

Several sensors report their measurement value encoded as a voltage pulse, with the duration of the pulse proportional to the value. We therefore need to measure the duration of that pulse with adequate accuracy. One device that employs this mode is an HC-SR04 *distance sensor*, shown in Figure 2.35, which operates like a sonar. It emits a short ultrasonic (40 kHz) sound burst and records the duration until the echo arrives, which is the round-trip time Δt of the sound burst. The distance L is given by this duration, and the speed of sound (approximately $v = 340\,\text{m/s}$) by $L = \Delta t/2v$, or, in convenient units, $L[\text{cm}] = 0.017\Delta t[\mu s]$. A pin on the device goes high when the pulse is emitted, and returns to low once the echo arrives or some specified timeout expires. Somewhat more advanced models using the same method, but having a larger range, are the LV-EZx sensors. They support other modes of reporting the distance as well, such as an analog voltage proportional to the duration, or direct reporting as RS-232 signals, but more on that below.

Figure 2.35 HR-SR04 distance sensor.

Another sensor, often found in modern consumer electronics, detects the *infrared signals* from the remote control. It is not strictly a scientific sensor, but still an interesting device that works remarkably reliably, because it rejects disturbing environmental effects and changes the channel on the TV only when a button on the remote is pressed. Typically they operate at a wavelength of 940 nm, which matches the light-emitting diode on the remote, and they have a built-in optical band-pass filter that lets only that wavelength pass. Then the signals are modulated by 38 kHz carrier frequency, which is demodulated on the sensor, such that only the base-band signal is reported on the output pin.

Now we turn to sensors that directly report their measurement values in digital form, and start by discussing I2C devices.

2.3.3 I2C devices

A large number of sensors have some logic built in and support a high-level communication protocol. An example is the *I2C* protocol operating on the I2C bus. The physical connection to devices supporting I2C only needs four wires: Ground, supply voltage Vcc, clock SCL, and data SDA. The latter two require a pull-up resistor, which is often already included in the microcontroller that also serves as the I2C busmaster to orchestrate the communication. It configures the sensor, initiates a measurement, and then reads data from the sensor. Physically, the communication is based on a synchronous serial protocol, where the data line is sampled every time the clock line changes from a high level to a low level. The protocol is standardized, and we will not go into details, but mention that the I2C devices and also I2C sensors have a number of registers internal that can be written to in order to configure the sensor, or read from in order to retrieve sensor data. The communication is entirely based on exchanging digital signals, and, as mentioned before, coordinated by the busmaster which normally is a microcontroller. Several devices can share the same SDA and SCL lines, because each device has its own address and responds only to those messages intended for it by specifying the device address.

The BMP280, or the enhanced BME680, shown on the left-hand side in Figure 2.36, have *barometric pressure* sensors on board. These sensors are based on measuring the strain caused by the deformation of a membrane that separates an evacuated test-volume and the outside air pressure [14] with a piezoresistive strain gauge, as illustrated in Figure 2.37. The entire assembly is directly built into a silicon substrate, and the strain gauge is created by suitable doping a small section. A temperature sensor with a resolution of 0.01 °C or

Figure 2.36 A BME680 environmental sensor and the HYT-221 humidity sensor.

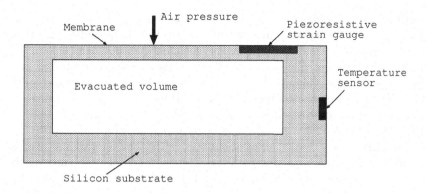

Figure 2.37 Illustration of the operational principle of a barometric pressure sensor.

better is included because the semiconductor-based strain gauge is temperature dependent. Further signal conditioning and processing circuitry is included in the assembly as well, such that the device communicates with a microcontroller via an I2C-bus on address 0x76 or 0x77. The measuring range is from 300 to 1100 hPa (mbar) with a relative accuracy of ±0.01 hPa and an absolute accuracy of about 1 hPa. The enhanced version BME680 is an *environmental sensor*. Apart from barometric pressure and temperature, it also measures the humidity and the presence of *volatile organic compounds* (VOC) with a miniaturized version of the device shown in Figure 2.7.

The dependence of the barometric pressure on the signal from the primary sensor, the strain gauge, is rather intricate and may vary from one device to another due to manufacturing tolerances. Each device, like many other sensors as well, therefore needs to be *calibrated* by exposing it to known conditions, here the pressure, and recording device-specific constants that allow us to accurately determine the pressure from the primary measurements. In the case of the BMP280 or BME680, a built-in ADC reports a value related to the resistance of the strain gauge measured in a bridge circuit. The calibration constants, determined during the manufacturing and calibration process, are stored in memory on the chip. The datasheet describes a detailed procedure to obtain the pressure based on the value reported from the ADC and the calibration constants.

The SCD30 is a *nondispersive infrared* (NDIR) sensor that determines the relative concentration of carbon dioxide (CO_2) molecules by measuring the absorption of infrared light with a wavelength of $4.3\,\mu m$ in a sample volume. The principle is illustrated on the left-hand side in Figure 2.38. Light from an infrared diode passes through the sample volume and

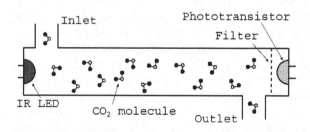

Figure 2.38 Nondispersive infrared sensing and the SCD30 CO_2 sensor.

a wavelength-selective optical filter, placed just before a phototransitor that detects the residual intensity; more CO_2 molecules simply absorb more light and reduce the intensity. The sensor, mounted on a small breadboard, is shown on the right-hand side in Figure 2.38. A small on-board microcontroller converts the intensity to the concentration in parts-per-million (ppm) that is accessible via I2C bus at address 0x61. Apart from the concentration, the SCD30 even measures the temperature and the humidity, both of which affect CO_2 sensing and are used to improve the resolution accuracy.

The HYT221 and HYT939 are sensors that measure the *relative humidity* in the range from 0 to 100 %RH with a resolution of about 0.02 %RH. The operational principle of the measurement is based on a capacitor with a dielectric, made of a polymer as the sensing medium. The polymer is highly hygroscopic and easily absorbs water, which changes the relative dielectric constant ε_r by a large amount, because ε_r of the dry material is much smaller than that of water, which is about 80. Correspondingly, the capacitance C, being proportional to ε_r, changes by a large amount as well. This change of the capacitance is determined in a Wheatstone bridge with capacitors in two branches. For the calculation of the relative humidity, knowledge of the temperature is required, and provided by a built-in temperature sensor. On board, the raw data are compensated for a number of non-linearities, post-processed, and made available in registers that are accessible via an I2C interface. The I2C address of the sensor is hard-wired to 0x28 and a device is shown on the right-hand side in Figure 2.36.

The HMC5883 is a three-axis magnetic field sensor that is based on the change of the resistance due to the *magnetoresistive* effect. The active material is a long meandering strip of a nickel-iron alloy on a nonconducting substrate. The change of resistance is due to the change of the spin–orbit coupling of valence electrons in the material. This affects the ease with which the conduction electrons propagate through the material and consequently the resistance. The active material is part of a Wheatstone bridge whose output voltage is conditioned and amplified before being digitized with an on-board ADC and made available on the I2C bus. The HMC5883L measures the field along three axes with a 12-bit (4192 steps) resolution in the range of ±8 gauss or ±800 μT. It is typically used in compass applications for mobile phones. The noise floor of the measurements is on the order of 0.2 μT and up to 160 measurements/s can be taken.

The MPU-6050 is a three-axis motion-detection chip consisting of an *accelerometer and gyroscope* to measure acceleration and angular velocities in three spatial dimensions. The accelerometer is based on the same sensing principle as the ADXL335 discussed previously and illustrated in Figure 2.11, where a spring-suspended inertial mass changes its position when accelerated and varies capacitances that are measured in a bridge circuit. In the MPU-6050, however, digitizer and post-processing digital circuitry are added to provide the measurement data in digital form. The operational principle of the *gyroscope* [14] on the MPU-6050 is illustrated in Figure 2.39, where a small mass, suspended by springs on an inner frame, is forced to oscillate in the x direction by a micromachined electrostatic motor. The mass therefore has a velocity component v_x in the same direction. If the entire assembly rotates around the z axis, which points into the paper, a Coriolis force, given by the crossproduct of the velocity vector and the rotation vector, will point along the y axis. And this force moves the inner frame in the y direction, against the force of the springs that connect the inner to the outer frame. This motion subsequently produces an imbalance in the sensing capacitors, very much like what happens in the accelerometer, and is sensed in much the same way as well. Also, here the analog voltages are digitized on the chip, further processed, and made available on an I2C bus. The performance of the device is rather impressive. It measures the acceleration in various ranges between ±2 g and ±16 g with 16-bit resolution and a rate of up to 1 kHz. The rotation speed can be measured between ±250°/s and ±2000°/s with a resolution of about 10 to 100°/s and a rate of up

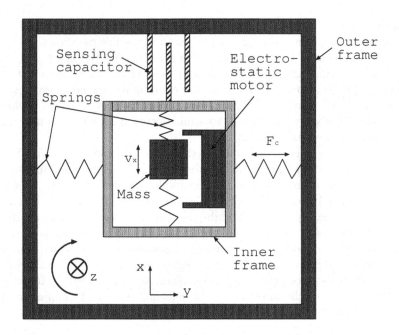

Figure 2.39 The operational principle of one gyroscope in the MPU-6050.

to 8 kHz. The I2C address of the device is 0xA0 or 0xA1, depending on the state of IO-pin AD0. An enhanced version of the MPU-6050 is the new ICM-20948 sensor with I2C address 0x68 or 0x69. It combines the functionality of accelerometer and gyroscope with that of a three-axis AK09916 Hall sensor, all internally aligned. The magnetic sensor has its own I2C address 0x0C, but shares the I2C pins with the accelerometer and the gyroscope.

The MAX30102, mounted on a small breakout board called MAXREFDES117 and sold by the manufacturer of the circuit, is shown in the middle of Figure 2.40. It is a *pulse-oximeter* that determines the heart-rate and the saturation level of oxygen in blood by using a phototransistor to measure the relative absorption of the light emitted by a red LED and an infrared LED. Infrared light is predominantly absorbed by oxygenated hemoglobin, which is the molecule that transports oxygen in our blood. Red light, on the other hand, is

Figure 2.40 An MPU-6050 accelerometer, a MAX30102 pulse oximeter, and a TCS34725 color sensor, mounted on their respective breadboard.

predominantly absorbed by hemoglobin without attached oxygen molecules. By measuring the two absorptions many times during a heart beat it is possible to extract the time-varying pulsation of the oxygenated blood coming in the arteries from the heart. Determining the ratio of two time-varying signals has the benefit of reducing many systematic variations among different individuals that would otherwise compromise the accuracy of the measurements. The MAX30102 provides the raw signal from the phototransistor while much of the numerical postprocessing is relegated to a microcontroller that communicates with the device listening at address 0x57 via I2C.

The AD5933 is an example of a bioimpedance measurement system. It contains a frequency-adjustable current source and a subsystem to synchronously measure the voltage drop across an unknown impedance, for example the part of a human body between electrodes. Frequencies below 1 kHz only probe the uppermost layer of the skin, because the skin behaves like the capacitor in a high-pass filter and blocks the lower frequencies to enter deeper into the body. With higher excitation frequencies we probe deeper into the body and determine the amount of body fat by measuring the rest, which is mostly water. Frequencies around 10 kHz measure the water outside of cells because cell membranes also act like a capacitors that block the lower frequencies, but increasing the frequencies above 50 kHz also determines the the water inside the cells and thus the total amount of water in the body. Like the other devices in this section is the AD5933 accessible via I2C at address 0x0D.

The TCS34725, shown on the right in Figure 2.40, is a *color sensor* that uses multiple photodiodes to determine the color of light by sensing the current from photo diodes behind optical filters that predominantly let one color pass, three diodes for red, three for green, and three for blue. Moreover, three unfiltered diodes provide information about light intensity. Internally, the photocurrents from the diodes are first integrated, then digitized with a 16-bit ADC, before they are made available on an I2C bus at address 0x29.

The sensor chips are usually very small and difficult to work with, but luckily there are so-called breakout boards available that route the pins of the sensors to normally spaced (2.5 mm spacing) pins that can be attached, for example, to solderless breadboards. Figure 2.40 shows an MPU-6050, MAX30102, and a TCS34725 mounted on breakout boards.

2.3.4   SPI devices

The SPI interface is a synchronous serial communication bus, similar to the I2C bus, but it can operate at much higher speed and is therefore often used for devices that require the continuous transfer of large amounts of data, such as displays or audio equipment. SPI communication requires one master on the bus, a role normally taken by a microcontroller. The sensors are typically slave devices. They need at least six wires to connect: ground and supply voltage, the clock CLK, one line to send information from the master to the slave (MOSI, for master-out slave-in), one line for the reverse direction (MISO for master-in slave-out), and a chip-select line CS to identify the currently active slave. CLK, MISO, and MOSI lines can be shared among many slaves, but each slave requires its individual CS line.

Some of the devices in the I2C section support this interface as well, and the interface can normally be selected by setting a pin on the device high or low. Details can be found in the datasheets.

In a later project we connect the analog-to-digital converter MCP3304 to a microcontroller via SPI communication. It has eight single-ended 12-bit input channels but can also be configured to use two input channels as differential input. This circuit detects whether one or the other input is larger and thereby provides an additional sign bit. Thus, it provides one extra bit to obtain a 13-bit resolution. At the same time, the input range is extended to plus or minus times the supply voltage, despite operating from a unipolar power supply.

Figure 2.41   The TFmini optical distance sensor.

This chip or its sibling, the MCP3208, are strong candidates to expand the number and resolution of analog input channels for many microcontrollers.

2.3.5   RS-232 devices

Several devices report their measurement values by sending them via the asynchronous RS-232 protocol. Originally, the physical medium for the communication channel used a current loop, but nowadays most sensors operate on 3.3 or 5 V levels. The communication happens point-to-point between two partners who have agreed on a communication speed, which is often 9600 baud or 115200 baud. Three wires are required at the minimum, one for ground potential, one, labeled TX for transmitting from device A to device B, and another one, labeled RX, for transmitting in the reverse direction. To establish communication, cables connect the TX pin on one device to RX on the other and vice-versa. Naturally, ground pins need to be connected as well.

One device that supports RS-232 communication is the LV-EZx distance sensor we discussed earlier. It can be configured to send the distance measured as an ASCII string that can be read in a terminal program. Even the TFmini distance sensor, shown in Figure 2.41 reports the measured distance via RS-232. It emits a modulated infra-red light pulse through one lens, records it through a second lens, and determines the phase shift of a reflected signal with respect to the emitted signal. From this it determines the time-of-flight to the nearest obstacle in the line of sight.

Sensors that query the *global positioning system* (GPS) use a small patch antenna on the sensor to pick up signals from a number of satellites placed in geostationary orbits, which broadcast their position and timing information with high precision. On-board electronics that normally comprise a microcontroller use triangulation in order to determine the position of the sensor with high accuracy and convert that information to an ASCII string containing the position in a standardized format, called *NMEA*. The string is written, typically once per second, to an RS-232 serial line, where it is straightforward to read and decode.

2.3.6   Other sensors

Apart from the standardized protocols, there exist a number of communication standards that device vendors come up with. Examples of such devices are the *relative humidity sensor*, DHT22, and the DHT11; the latter is shown on the right in Figure 2.42. In order to determine the relative humidity, the temperature must be known and it is determined with a thermistor. To measure the humidity they use a capacitive humidity sensor. The operational principle is based on using a small capacitor with an exposed dielectric that has a large affinity to attract water, which changes the relative dielectric constant ε_r and the

Figure 2.42 A GPS receiver on the left and a DHT11 humidity sensor on a breadboard on the right.

capacitance by a large amount. This is determined in a Wheatstone bridge with capacitors in two branches. Internally, the DHT sensors have this, and further circuitry such as an analog-to-digital converter on board to calculate the relative humidity. They provide the data using a non-standard, though documented, digital interface that we discuss further in Section 4.4.5, where we connect a DHT11 to a microcontroller.

The DS18b20 temperature sensor, as was the LM35 discussed earlier, is based on a bandgap temperature sensor. Here, however, ancillary digital electronics and signal processing circuitry is added on the chip such that the measurement value is postprocessed and made available using the so-called Dallas 1-wire bus protocol. The 1-wire protocol uses only ground and a single additional wire to transmit power and information to and from the device.

The air quality can be characterized by the density of microscopic particles suspended in air. In Figure 2.43 we show two such sensors. On the left we see a PPD42NS particle sensor. Inside this detector a resistor heats the air, which causes the air with the suspended dust particles to rise and pass through the light emitted by an infrared diode. There the dust particles scatter the light onto a phototransistor, which pulls an output pin to low potential. After signal conditioning and amplification, a cleaned-up signal is available. It is low when

Figure 2.43 A Shinyei PPD42NS particle sensor (left) and a GP2Y1010AU0F dust sensor (right).

particles scatter light, and high otherwise. The device is calibrated such that the ratio of time at low signal to total time can be translated to particles per liter. The GP2Y1010AU0F, shown on the right in Figure 2.43, works in a similar way. It also detects light scattered off of dust particles, but it periodically turns the infrared diode on and integrates the signal from the phototransistor and one has to sample the output value 0.28 ms after the LED was turned on. The difference between the signal with LED on versus off provides reasonable rejection of ambient light. The performance of both dust sensors can be improved if we place them in the airstream created by a fan, which we need to turn on and off.

And that brings us to actuators, devices such as switches or motors that cause some change in external conditions. Sometimes they are part of the measurement process, such as the fan mentioned in the previous paragraph, or we need to move the sensor to where we want to measure, which typically requires motors.

QUESTIONS AND PROJECT IDEAS

1. Where is the Fermi level in Figure 2.3?

2. Discuss different ways of measuring the capacitance of a capacitor.

3. Discuss different ways of measuring the inductance.

4. Research the possibility of how to use a pin diode as a detector for *ionizing radiation*.

5. Like pin diodes, LEDs generate a small current when illuminated. Use this effect to build a *photometer* from a number of LEDs with different colors to sense different parts of the optical spectrum.

6. Discuss a system to *measure the rotation speed* of a wheel with a small magnet and a reed switch. Can you design it such a way that you can determine both speed and direction of the rotation?

7. Discuss methods to determine the *wind direction*.

8. Discuss how to measure the amount *rain during a day*.

9. Research how to measure the *concentration of sugar* (dextrose) in water by observing the polarization change as a function of amount of sugar dissolved. Which of the described sensors can be used?

10. Build a discrete 3-bit flash ADC from a resistor ladder and external components, such as the LF198 sample-and-hold circuit, two LM339 quad-comparators, and a 74HC147 8-to-3 priority encoder.

11. Find out how an *electronic nose* works.

## CHAPTER 3

Actuators

Even though sensors are the main topic of this book, sometimes we need to turn devices on and off, or we need to move a sensor very accurately, with much higher precision than we can achieve by hand. In other cases, the sensor is not at the location where we need to measure some quantity. In such situations we need an actuator to move it in a controlled way. Here *actuator* is the generic term for a device that controls external parameters. Examples are motors, valves, or switches, and we start the discussion with the latter.

3.1 SWITCHES

Turning an electric signal on and off is very easily done by toggling an output pin of a microcontroller, as we shall discuss in quite some detail in the next chapter. A typical microcontroller can provide rather limited currents on the order of a few mA. This is normally sufficient to control a single light-emitting diode, an LED, which typically draws less than 20 mA.

3.1.1 Light-emitting diodes and optocouplers

An LED, shown on the left in Figure 3.1, is similar to a conventional diode and consists of a semiconductor with a pn junction. If it is forward biased, with the n-terminal connected to the lower potential, the charge carriers, electrons, and holes are pulled into the space-charge

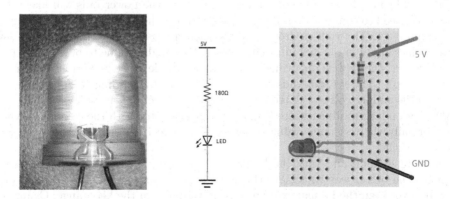

Figure 3.1 A close-up of a light-emitting diode is shown on the left. In the center is the schematic of connecting an LED and the same circuit on a breadboard[†].

DOI: 10.1201/9781003341703-3

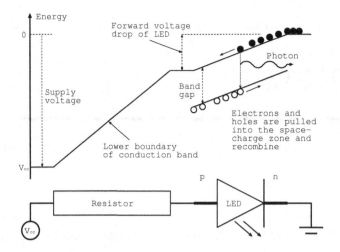

Figure 3.2 The physics of an LED.

zone, provided the voltage is higher than the voltage drop of the diode. This situation is opposite to the situation encountered in Figure 2.13, where the diode was reverse biased. The simplified band-level scheme and the corresponding circuit for the forward-biased LED is shown in Figure 3.2. We only show the lower boundary of the conduction band, and the upper boundary of the valence band inside the diode. Here both electrons and holes are pulled into the space-charge zone and have a chance to recombine. The energy that is released by the electrons dropping from the conduction band to the valence band is emitted as a photon. Note that the forward voltage drop of the LED is about the size of the bandgap and the resistor is needed to limit the current. LEDs emitting different colors have different bandgaps, and correspondingly, different forward voltage drops, where shorter wavelengths corresponds to larger bandgaps.

As already seen in Figures 3.1 and 3.2, LEDs connect to a circuit with two wires, the short one is the cathode and needs to be connected to ground, or the more negative potential. The cathode is normally indicated by a flattened face of the hemispherical housing. The other wire is the anode and is connected to the more positive potential, provided we want the LED to light up. But simply connecting the LED to the power rails will likely destroy the diode. We need to connect a resistor in series with the diode, as shown in the center and right on Figure 3.1. The voltage drop V_d across most LEDs is between 1.5 and 3 V, where the lower values apply to red and infrared LED, and the higher value to blue and ultraviolet LEDs. Moreover, the typical operating current of the diode must be limited to some value below $I = 20\,\text{mA}$. The resistor thus needs to be specified according to $R > (V_c - V_d)/I$, where V_c is the supply voltage. A red LED thus works nicely on $V_c = 5\,\text{V}$ with $R = 180\,\Omega$ or $220\,\Omega$, or even larger values where I picked a larger resistor with the commonly available values. Normally the resistance value is not very critical; if chosen too large, the LED is simply less bright.

In order to dynamically vary the brightness, changing the resistor value is rather inconvenient. The better way to achieve this is by rapidly turning the diode on and off at a rate much faster than the human eye can resolve, typically in the kHz range. Changing the duty cycle of the on- versus off-time proportionally changes the brightness of the LED. The added benefit of this method, called *pulse-width modulation,* is that the dissipated energy is reduced by the same ratio.

If an LED or any other pulse-width modulated device needs to be galvanically separated from the control electronics, we use an optocoupler. This is a small integrated circuit that consists of an LED and a phototransistor. As discussed in Section 2.1.3, a phototransistor behaves like a normal transistor with a high current amplification, but instead of passing a current into the base terminal of the transistor, the LED illuminates the region with depleted charges in the transistor and creates a large number of electron–hole pairs. Thus, the transistor becomes conducting and switches on a device on the "other" side of the circuit, without being electrically connected. This implies that only the on- or off-state or digital signals are communicated across the optocoupler. Typically they are used if devices located at different electrical potentials need to be switched on or off, or to prevent electric perturbations from the "other" side. We note in passing that the MIDI communication between musical instruments uses optocouplers at their respective inputs. This removes electrical disturbances and protects the musicians from electrocution.

So far we only switched LEDs on and off, but we will look at how to control large currents and loads.

3.1.2 Large currents

Switching large currents and voltages requires us to "amplify" the small current that the microcontroller provides, and transistors of various flavors do just that. Namely, they amplify the current flowing into their base terminal by a factor β, which is specified in the datasheet of the transistor to a value that flows across the collector-emitter terminals. Refer to the left side of Figure 3.3 for the naming and location of the respective terminals of an NPN-transistor. If the amplification β is large enough, moderate base-currents cause the output current to saturate the transistor such that it behaves like a switch that turns on the connection from collector to emitter.

In the example shown in Figure 3.3, the $1\,\mathrm{k}\Omega$ resistor on the base limits the input current across the base-emitter link to $5\,\mathrm{mA}$, provided the control voltage is $5\,\mathrm{V}$. If the transistor has an amplification of 100, which is typical for many small signal transistors, it can switch up to $500\,\mathrm{mA}$, which is more than is actually needed in this case. The actual current flowing through the collector-emitter link is determined by the $12\,\mathrm{V}$ supply voltage to the LED and the $680\,\Omega$ resistor limiting the current. Note that using the transistor also decouples the voltage level of the controlling circuitry and the supply voltage, here $12\,\mathrm{V}$, of the consumer, the LED. This way of operating a transistor is called *open collector,* because we can think

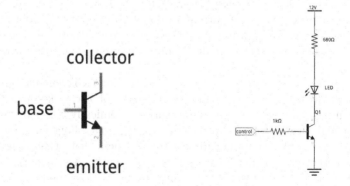

Figure 3.3 The terminals of an NPN transistor (left), and using an NPN transistor as switch (right)[†].

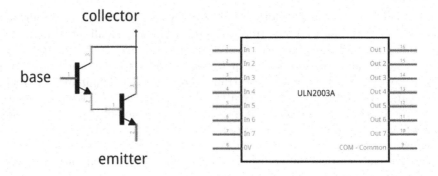

Figure 3.4 Two NPN transistors connected to form a Darlington pair (left) and a ULN2003 Darlington array (right)[†].

of the collector terminal as a generic connection point for an (almost) arbitrary load that is connected to its own supply voltage. If the emitter-collector link is in the nonconducting state, no current flows, and the load is turned off. If a positive current flows into the base terminal, the emitter-collector link becomes conducting, the lower terminal of the load is shorted to ground, and a current flows from the load's power supply through the load and turns it on.

Since switching is very common, small switching currents are often desirable. This requires a transistor with a large amplification. One way to achieve this is to connect two transistors as a *Darlington* pair, as shown on the left in Figure 3.4. Such a pair has the current amplification of approximately the product of the two individual transistors, and is often used in switching applications, which accounts for the availability of integrated circuits that pack seven or eight Darlington pairs into a single package. An example, the ULN2003, is shown on the right in Figure 3.4. The seven input terminals on the left-hand side of the package are the base terminals of the Darlington pairs, and the output terminals on the right-hand side are the corresponding open collector terminals. The ground connection is located at the lower left, and the external positive power supply voltage is connected to the terminal labeled COM. This specific chip can switch up to seven times 500 mA, and the supply voltage for the load can be up to 40 V. The datasheets provide a lot more detailed information.

In case we need to switch very high voltages of up to 1 kV or very large currents, we use MOSFETs. They require almost no current to flow into the gate in order to switch. Only the capacitance between the terminals needs to be charged, and this creates an electric field that pulls charge carriers into the depletion zone of the MOSFET, which causes it to conduct.

Using transistors makes switching unipolar voltages very convenient, but if we need to switch AC household voltages on or off, such as lamps or the wake-up radio, we need a relay. Relays consist of a small electromagnet that mechanically closes or opens a contact. This achieves a high degree of separation between the controlling circuitry and the load circuit. The two sides only communicate via the magnetic field that toggles the switch, depending on whether current flows through the coil or not. A schematic image is shown on the left in Figure 3.5. There we see the coil with its two terminals on the left, and the switch on the right-hand side that toggles between the two terminals. Normally we need to prepend a transistor to switch the coil on and off because this requires a larger current than the microcontroller provides. Moreover, we need to pay attention, because the coil is an inductive load, and turning it off causes a large induction voltage that may damage the

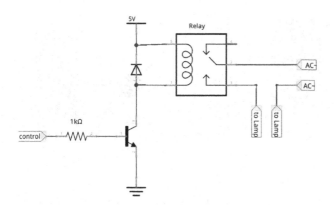

Figure 3.5 Schematic illustrating the functionality of a relay (left[†]) and an image of a relay (right).

transistor or other parts of the circuit. This can easily be prevented with a *flyback diode*, also shown in Figure 3.5, bypassing the coil in the relay in the normally non-conducting polarity with cathode pointing towards the positive supply voltage. Any voltage in the reverse direction passes through the diode, rather than the transistor. Care needs to be taken with respect to power rating of the diode, and the speed. Normally Schottky diodes, which are particularly fast-switching, are used.

The terminals on the high-power side can handle both AC and DC voltages. The relay is very similar to a normal switch in the way that it establishes an electrical contact between two terminals. In Figure 3.5, it closes or opens the contact between the AC supply and a lamp, but the lamp can be replaced by any other device that needs to be turned on and off.

Please note that the relay in the image can actually handle up to 240 V AC, but we need to stress that this is a potentially lethal voltage. This should only be handled by suitably trained and qualified personnel. If in doubt, do not try it yourself!

Now we have several means at our disposal to switch devices and actuators on and off. Some of the most prominent and useful actuators are motors.

3.2 MOTORS

Motors in general translate chemical or electrical energy into mechanical energy, normally by rotating an axle in order to provide torque. Here, and in subsequent sections, we address electrical motors only. They come in a wide variety of types and can be powered by both AC and DC voltages. One of the most common types, and the one we will use, is a conventional DC motor with a commutator.

3.2.1 DC motors

The operating principle of a DC motor is most easily explained with the help of Figure 3.6. It is based on static magnetic field that is stationary in space, the *stator*, and one or more coils, electrically excited by an external voltage source. They are forced to rotate by reversing the polarity of the exciting voltage synchronously with the rotation. The rotating part is called the *rotor*, and the synchronous switch is the *commutator*. The torque on the rotor is provided by the Lorentz force that the electrons of the electric current experience in the external magnetic field. After half a turn, the external magnetic field has the "wrong"

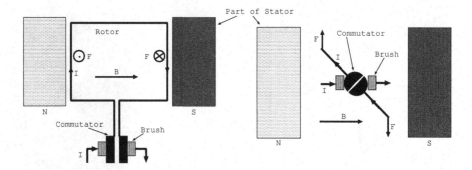

Figure 3.6 Basic operation principle of a DC motor. The magnetic north pole of the stator is to the left and the south pole to the right of the coil. Current is supplied to the brushes where it enters the commutator and passes through the rotor-coil, where the Lorentz force causes a force on the wire, moving it upwards. After half a turn, the commutator has rotated and reverses the polarity such that the current on the left side of the coil points in the same direction as before.

polarity to continue forcing the coil to rotate, and would brake the rotation. This is prevented by reversing the polarity of the current flowing at just that point in time which allows the rotation to continue. The polarity reversal is effected by the commutator, which is shown as the two D-shaped objects that rotate between externally fixed electrodes, called *brushes*. Moreover, reversing the supply voltage causes the motor to rotate in the opposite direction.

The sliding contact between the brushes and the commutator sometimes causes sparking at the moment of polarity reversal, because the brushes briefly connect the two D-shaped objects and thereby short circuit the supply voltage. The sparks then cause electrical disturbances that may be partially alleviated by connecting the brushes with a small, say 100 nF, capacitor. This also reduces the creation of ozone with its distinct smell, which often accompanies the operation of DC motors. In commercially available motors these effects are minimized, and other motor designs are used that circumvent these deficiencies. For example, in some *brushless DC motors*, the magnetic field of the rotor is provided by permanent magnets whose position is continuously monitored with a Hall sensor. This information is used in an electronic control unit to periodically excite the current through the coils of the stator in order to maintain the rotation. The complexity of the control unit places these motors outside the scope of this book, and we use normal DC motors for our projects despite their shortcomings.

We can adjust the torque and the speed of the motor by varying voltage applied to the coils. A higher voltage makes the motor turn faster, and reasonable voltages are determined by the resistance of the coils. Often, however, it is advantageous to limit the voltages and control the currents that flow through the coils instead. Moreover, since the rotor has a finite inertia due to its mass, we can use pulse-width modulation to control the total power delivered to the coils in order to adjust torque and speed. In the fleeting moments when no current flows, the inertia maintains the rotation.

In case you have a model railway, you may know that in the early days the speed of the trains was regulated by a variable transformer, and it was very difficult to run the trains at very slow speeds with voltages too low to reliably excite the coils. Modern model-railway controllers use pulse-width modulation to regulate the speed. In that way, the full supply voltage and current are always flowing, which reliably excites the coil, just not all the time. Operating the trains at slow speed now works much more reliably.

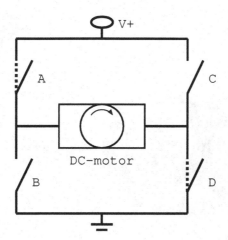

Figure 3.7 Schematic illustrating the functionality of an H bridge.

In this context we have to keep in mind that turning the coils rapidly on and off also rapidly changes the magnetic field, and this causes a voltage, the backward electromagnetic force, or *back-emf* that opposes the driving voltage. Thus, the faster the motor turns, the more the back-emf reduces the voltage, and thereby also the current flowing through the coils. This results in a reduced torque and for a given coil resistance we can either run at a slower speed or increase the driving voltage in order to maintain the required torque.

As with the model trains, we also want to run the motor both forward and backward. Therefore we need to control the polarity of the supply voltage. Instead of reversing the supply cables by hand, which is not a very convenient option, we use an H bridge whose functionality is easily explained with the help of Figure 3.7. In the center of the figure a DC motor is located, with its supply leads extending to either side. The unipolar supply voltage is connected to the upper terminal labeled V+ and to ground. Depending on the position of the switches A, B, C, and D, the current flows through the motor. In case A and D are engaged, it flows from left to right, This is indicated by the dotted lines and the sense of rotation indicated by the small arrow in the motor. In case B and C are engaged, while A and D are open, the current flows in the opposite direction, causing the motor to turn in the opposite direction as well. Thus, by suitably toggling the four switches, we can adjust the sense of rotation of the motor at will. We only have to ensure that only one of the switches A and B is engaged at a given instance, because otherwise the supply voltage is short circuited.

Of course, we can replace the switches with transistors and arrive at a system that is easily controllable by a microcontroller. And the situation is made even simpler, because ready-made integrated circuits that implement H bridges are available. The L293D is one of them, and we will use it in later chapters to control motors. Finally, we add speed control by pulse-width modulation. We either add a switching transistor between the upper terminal and the supply voltage, or we directly modulate the input terminals that control the switches A, B, C, and D.

At this point we can adjust the speed and direction of a motor, which is essential for moving from one place to another, and the velocity at which this happens. On the other hand, if we only want to change the position or angle by a small amount, such as the rudder of a boat, we need means to directly adjust the position, rather than the velocity. Next we discuss two types of systems that do this, servomotors and stepper motors.

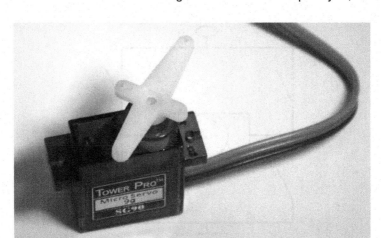

Figure 3.8   A small model-servo.

3.2.2   Servomotors and model-servos

The term *servo* refers to using the motor in conjunction with a position encoder in a *closed-loop* feedback, or servo, loop [21]. Here the motor speed is continuously adjusted to reduce the difference between the desired position and the actual position, as reported by the encoder. This type of servomotor is often used in industrial applications, and requires an elaborate control system with a PID controller [21] and powerdrive-electronics. Servomotors are used in large industrial robots and applications that require high accuracy of positioning. Matching the parameters of the controller to the desired performance of the control loop in terms of accuracy, speed, and acceptable overshoot requires expert attention, and is beyond the scope of this book.

Instead, we use the modest cousin of the servomotor, the *model-servo,* often simply referred to as a *servo,* shown in Figure 3.8. It is often encountered in radio-controlled cars and planes but can also appear in robotic applications, such as controlling the position or angle of smaller robotic arms. Servos have a potentiometer mounted on the rotating shaft that is used as an encoder. It provides information about the position of the shaft, which typically has a rotation range of 0–180 degrees. The resistance is compared to a voltage derived from an electric signal corresponding to the desired position. The difference signal is amplified and used to turn a DC motor in the direction that minimizes the difference. This constitutes a simple proportional controller. Normally a small gearbox between the motor axis and the shaft reduces the speed but increases the precision of control and the torque. Note that the servo connects with three wires to the outside. Two of the wires are ground and supply voltage, nominally 4.8 V. They are colored black and red, respectively. The third wire, yellow or white in many cases, carries the information about the desired position.

The information about the desired position is communicated to the servo as a pulse-width modulated signal, as shown in Figure 3.9. The servo expects a train of pulses with a spacing of 20 ms. Each individual pulse has a duration between 1 and 2 ms, with a 1.5 ms duration specifying the mid-position. Producing such pulse patterns is easily done with a microcontroller, and we will do that in later chapters.

Servomotors and servos use a closed-loop system to achieve a high accuracy of positioning, while stepper motors achieve this without closed-loop feedback.

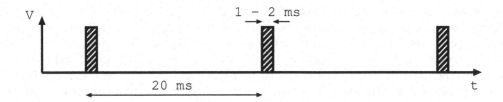

Figure 3.9 The timing of the control signal for the servo.

3.2.3 Stepper motors

Stepper motors change the angular position of a shaft in small discrete steps, such that counting the steps gives the position. In this way, no feedback or servomechanism is needed to achieve high repeatability, and the motor can be operated open loop.

In Figure 3.10, we illustrate how this is accomplished in a permanent-magnet stepper motor. The static part of the motor consists of an iron yoke and four coils, of which two each are operated in series. Here the upper and lower coils, labeled A and B, and the right and left coils, labeled C and D, are two such pairs. On the rotor a moderately large number of permanent magnets are assembled. In the figure we only use six magnets, labeled 1 through 6, with the black end indicating the north pole. They are oriented radially and their polarity alternates.

We assume that initially only coils A and B are excited in such a way that the upper coil behaves as a magnetic north pole and the lower as a south pole. In that case, the magnetic field lines pass through both upper and lower coils, pass through the magnets in the rotor, and return through the surrounding yoke. With this coil excitation the rotor is indeed oriented, as shown in Figure 3.10, because the upper coil behaves as a north pole and attracts the south pole of the upper permanent magnet, labeled 1. Consequently, they are as close as possible. The converse is true for the bottom coil and the magnet labeled 4. In order to turn the rotor by 30 degrees in the counterclockwise direction, we have to turn

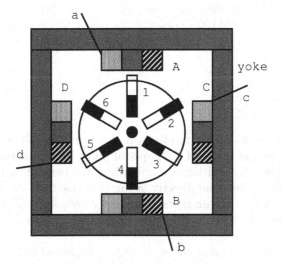

Figure 3.10 Schematics of a stepper motor. Note that the coils are denoted by uppercase letters and the corresponding terminals with lowercase letters.

coils A and B off and coils C and D on. Thus, coil C on the right-hand side is a north pole that attracts the permanent magnet labeled 3, which has a south pole pointing outwards. Again, the converse is true for coil D on the left and magnet 6. The rotor will then move until magnet 6 faces coil D. We continue turning the rotor by turning coils C and D off and reversing the polarity of coils A and B, such that the upper coil acts as a south pole that attracts magnet 2. Continuing in this way, we find that we can turn the motor in a number of steps of 30 degrees each by periodically exciting the coils in the following pattern:

```
terminal a: 1000 1000 1...
         b: 0010 0010 0...
         c: 0100 0100 0...
         d: 0001 0001 0...
```

where time runs from left to right. Here, 1 indicates that the terminal is connected to the positive power supply voltage, and 0 indicates that the coil is connected to ground potential. Stepping through the sequence backward causes the motor to turn in the other direction. Observing that time slots 1 and 3 show reversed polarity for the coils A and B implies that we need an H bridge to implement the polarity reversal. We will return to this implementation in Section 4.5.

It is instructive to analyze the excitation pattern of the four terminals a through d a little more closely. The first observation is that a sequence of four time slots repeats itself over and over again. Second, we observe that a fundamental pattern is the excitation of terminal a with the sequence 1000. The excitation of terminal b is shifted to the right by two time slots which we may interpret as 180 degrees out of phase if we assume that 360 degrees corresponds to the four time slots. Likewise, terminal c and d are 90 and 270 degrees out of phase with the excitation of terminal a. This interpretation will prove useful later when we write our own stepper motor driver and discuss other modes of operating the motor, such as half- and microstepping modes.

Note that in the excitation pattern only one coil is excited at a time. But we can also operate the motor with two coils excited simultaneously, one coil pulling one permanent magnet and the other coil pushing another magnet. In this way the torque of the motor can be increased by about 40 %, albeit at the expense of doubling the required power, because there are two coils excited at the same time. The pattern with increased torque is based on the fundamental sequence 1100. Applying the appropriate phasing between the four terminals from above, we arrive at the following pattern to excite the terminals:

```
terminal a: 1100 1100 1...
         b: 0011 0011 0...
         c: 0110 0110 0...
         d: 1001 1001 1...
```

where we see that the excitation of terminals b through d is shifted by 2, 1, and 3 time slots with respect to that of terminal a. A special point to note is that exciting two coils simultaneously will cause the positions towards which the permanent magnet will point to lie in between the coils, rather than directly facing the coils. Since we now have the single-coil excitation pattern where the magnets point at the coils and the double-coil excitation pattern where it points in between the coils, we might consider a combination of both patterns...

...and find that by interleaving the steps of the single-coil and double-coil excitation patterns, we obtain a sequence where the step size is halved. This mode of operation is called half-step mode. The pattern in which we have to excite the coils is the following:

```
terminal a: 11100000 11100000 1...
```

```
b: 00001110 00001110 0...
c: 00111000 00111000 0...
d: 10000011 10000011 1...
```

where the period length is 8 time slots with a fundamental excitation pattern of 11100000. The excitation sequence of terminal b is the fundamental shifted by half the period length, or four time slots, or 180 degrees. Note that here, 360 degrees, correspond to the eight time slots. And the sequence for terminals c and d are shifted by two and six time slots, respectively.

Note the general pattern: terminals a and b are excited by a cosine-like sequence, which is evident by calculating the voltage applied to coils A and B, which is a-b

```
a - b =  1  1  1  0 -1 -1 -1  0
```

which admittedly is a poor approximation of a cosine. Likewise the voltage applied to coils C and D is c-d

```
c - d = -1  0  1  1  1  0 -1 -1
```

and that is an equally poor approximation of a sine. Apparently, the signals applied to the coil pairs roughly follow a cosine and a sine-like excitation. It is easy to see how we can generalize this and use better approximations of the cosine- and sine-like excitations of the coils, which is called *microstepping* the motors.

In microstepping modes, the permanent magnets on the rotor are moved to several intermediate positions in between coils. This finer control comes at the expense of requiring a more advanced driver to control the output current in finer steps than turning it on and off. An example of such a driver is the DRV8825 circuit, which exposes very few control pins for direction and stepping as well as selecting which microstepping mode to use. Internally it generates the appropriate excitation pattern of the currents that are applied to the coils.

In the above example, the stepper motor uses only six permanent magnets, resulting in an angular increment of 30 degrees per step, or 12 steps per revolution, but it is easy to see that we can increase the number of magnets to reduce the step size significantly. Typical step sizes for commercially available motors are 200 or 400 steps per revolution.

Here we need to point out that assigning one phase to coil terminals a and b and the other to terminals c and d is only a convention, and some vendors prefer to enumerate the coils and associated terminals sequentially when looking at the motor in Figure 3.10. This amounts to swapping the labels of terminals b and c and implies that we may have to swap two cables from the driver circuit to the motor in order to make it turn. It is best to check the datasheet for the motor to find out what convention is used for that particular motor.

Furthermore, stepper motors come with different numbers of connecting cables that expose the terminals of the coils. On the left in Figure 3.11, the situation is depicted where all eight terminals are exposed. Sometimes the center taps are internally connected, in which case only four terminals are exposed, and this corresponds to the situation in the discussion earlier in this section. On the right, the center terminals are internally connected and are exposed, in which case the total number of exposed wires is six. If the two central connections are internally joined, five wires are exposed. Normally the exposed wires are color coded and identified in the datasheet, but measuring the resistance between the exposed wires allows us to determine the internal wiring of the motor experimentally.

In the previous stepper motor examples, we always reverse the polarity of the coil excitation, which requires a drive circuit with an H bridge. Stepper motors, however, can also be operated without an H bridge in unipolar mode. In Figure 3.12 we show the connection for a unipolar stepper motor that has center taps of the coils wired together, as discussed in the previous paragraph, and exposes only one terminal that is connected to the positive

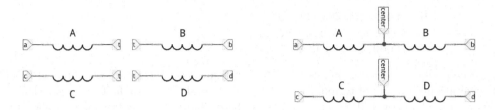

Figure 3.11 Wiring of the coils in a bipolar stepper motor (left) where the respective center terminals are often connected. In a unipolar motor (right) the center terminals are exposed[†].

voltage of the motor power supply. The other terminals of the coils are the collector of an NPN or Darlington transistor and are bypassed with flyback diodes. In a ULN2003 circuit, the diodes are already built in. Placing a positive voltage to one of the base terminals a, b, c, or d causes the corresponding transistor to conduct and the coil to be excited.

For unipolar stepper motors, we can use the same excitation pattern we use for bipolar motors. For example, the excitation pattern for the full-step mode with larger torque will also turn a unipolar stepper motor in one way

```
terminal a: 1100 1100 1...
         b: 0011 0011 0...
         c: 0110 0110 0...
         d: 1001 1001 1...
```

Reversing the order causes the motor to turn the other way around. Half- and microstepping modes can be implemented in the same way.

All devices above were based on switching voltages or currents fully on and fully off, but occasionally, careful adjustment of a control voltage is required; for example, to set the control input of a power supply that determines its output current.

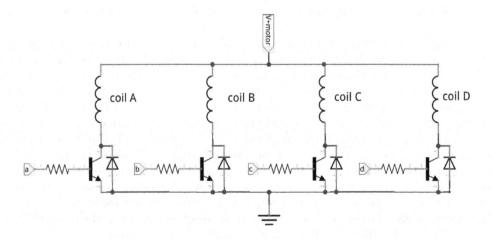

Figure 3.12 Connection for a unipolar stepper motor[†].

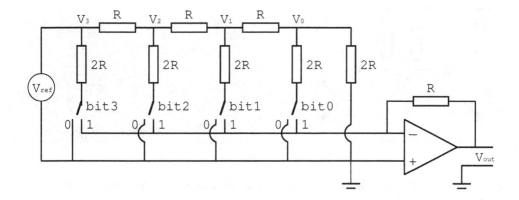

Figure 3.13 Operational principle of a digital-to-analog converter.

3.3 ANALOG VOLTAGES

A simple way to produce a constant voltage is to low-pass filter a pulse-width modulated output voltage. The filter reduces the on–off variation and results in an average voltage level corresponding to the peak voltage times the duty factor of the pulse-width modulation. Varying the duty factor will produce a time-varying output voltage. There are, however, limitations to this method, mainly due to the quality of the output voltage, which is likely to have some ripple left over from the original modulation.

A more reliable way is to use a *digital-to-analog converter* (DAC), which is a device that converts a digital word with a given length, such as 8 bits or 12 bits, to a discrete voltage with 256 or 4096 intermediate levels, respectively. The operational principle is illustrated in Figure 3.13 for a 4-bit DAC using an $R - 2R$ resistor network and an operational amplifier. The switches labeled bit3 to bit0 represent the digital input word, and the inverting input of the operational amplifier sums the currents that are added by toggling the respective switches. The inverting input is a virtual ground because the op-amp forces it to be the same potential as the noninverting input. This implies that the current flowing through all resistors is constant, and independent of the position of the switches; if the switch is in position 0, the current flows to the real ground, and if it is in position 1, it flows into the virtual ground. This implies that the voltages $V_3, \ldots, V_0$ are constant and are given by $V_{k-1} = V_k/2$, which is easy to see for V_1 and V_0. The two right-most resistors with values $2R$ in parallel combine to a single resistor of value R. But this means that the node with V_0 is sandwiched between two resistors of value R, of which one is connected to a point with voltage V_1, thus $V_0 = V_1/2$. Since all the voltages V_n are fixed the current flowing into the inverting input of the operational amplifier, provided that bit number k is set, is given by $V_k/2R$, with $V_k = V_{ref}/2^{3-k}$. And all currents coming from the four branches add up, and the operational amplifier converts the current to an output voltage V_{out}.

Most DACs use the $R - 2R$ ladder resistor network, with a current-adding operational amplifier but add a digital front end with semiconductor switches that are controlled from a parallel bus, I2C, or SPI. An example of the latter is the MCP4921, a 12-bit DAC that is controlled via an SPI-compatible interface.

3.4 PERIODIC SIGNALS

Occasionally circuits need to be periodically switched on and off or otherwise stimulated by periodic signals. Before the advent of integrated circuits, astable multivibrators were used

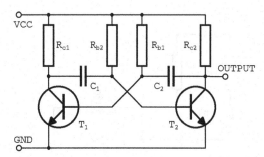

Figure 3.14 A multivibrator circuit (left) and a breakout board with AD9850 direct digital synthesizer (right).

for this purpose. The left-hand side in Figure 3.14 shows a basic example. Only one of the two transistors is conducting at any time. At the time just before T_1 switches on the left plate of capacitor C_1 had been charged through R_{c1} to be more positive than its right plate. But as T_1 starts conducting, C_1 discharges through R_{b2} until the base voltage of T_2 exceeds the voltage at which start conducting and discharges C_2. At this point the roles of the two transistors and capacitors are reversed. Neither of the two states is stable, which causes the voltage at the output to oscillate. If the two resistors R_{b1} and R_{b2} connected to the base of the transistor and the two capacitors C_1 and C_2 are equal, the oscillation frequency of the output signal is approximately given by $f \approx 0.7/R_{b1}C_1$. Today, circuits with discrete components are largely replaced by special-purpose chips, such as the widely-used NE555, which recently celebrated its fiftieth birthday. It requires very few external components to generate periodic output signals over a wide range of frequencies.

Most modern microcontrollers operate at such high clock rates that makes toggling output pins under software control a convenient option. We will return to this possibility in later chapters. If, on the other hand, other than rectangular signals are needed we can resort to *direct digital synthesis*. It is based on approximating the desired waveform – let's assume it is sinusoidal – by a table of integer values that we store in memory of the controller and successively pass the values to a DAC. To reduce the frequency, we wait a little while before outputting the next value from the table. To increase the frequency, we jump over intermediate samples. For low frequencies, this is straightforward to implement in software. Luckily, for frequencies in the multi-MHz range, ready-made circuits, such as the AD9850, which produces sine and rectangular output up to 40 MHz, are available. Mounted on a breakout board, it is shown on the right-hand side in Figure 3.14. Programming the AD9850 is based on a synchronous serial protocol with a clock and a data line.

The analog signals, switches, and motors we discussed so far are likely the most-used actuators, but there are others, and we consider a selection of them in the following section.

3.5 OTHER ACTUATORS

In order to open doors or to move devices between well-defined end stops, *solenoid magnets* are used. They use soft-iron cores with surrounding magnetic coils whose current is turned on or off. They are conceptually not much different from an LED, except that the current required is much higher, and that requires a larger power supply. Moreover, a solenoid is a large inductive load, and turning it off causes a large back-emf that requires a large flyback protection diode, just as we discussed in the case of relays and DC motors in Section 3.1.2.

Figure 3.15 Sketch of a butterfly valve.

The flow of liquids and gases can be controlled with pumps and valves. Pumps are usually based on motors driving paddle wheels, propellers, or turbines that move the liquid or gas from a region of low pressure to that of high pressure. The motors are controlled in the same way as discussed before. Flow rates are controlled by valves, which constrict the aperture through which the medium flows. An example is a butterfly valve, shown in Figure 3.15. The aperture can be restricted by the shaded aperture, which has the same diameter as the inner diameter of the pipe. Rotating the shaded aperture then adjusts the flow rate from fully closed, in which case the shaded aperture covers the entire inner diameter of the pipe, to fully open, in which case it stands perpendicular to the flow of the liquid or gas. To change its state we can either use a servo for smaller systems, a stepper motor, or a large servomotor, in case we need to constrict the flow in an oil pipeline.

A common means for controlling large machinery uses hydraulics, based on controlling the flow of hydraulic oil. It uses pumps to move the oil into cylinders that cause a linear travel, or it is pumped through a turbine or propeller-based device to cause rotary motions. In either case the original pressure is provided by pumps, based on electric motors. The position of valves is also controlled by the same devices we discussed earlier in this chapter.

And finally, we may consider humans as part of the larger system, and they need to be actuated. For this purpose we may use various attention-grabbing devices, such as piezoelectric buzzers, sirens, and loudspeakers, as well as optical devices comprising LEDs, NeoPixel color LEDs, liquid crystal displays, or small displays such as those found on printers or washing machines that report the current state of the apparatus.

By now we have covered quite a selection of actuators and sensors and can turn things on, move them around, and measure a number of physical quantities. The next task is to interface this variety of sensors and actuators with their multitude of interfaces, and provide a standardized interface to communicate with the external world. This is the task of the microcontroller, and the flavors we will discuss in the following chapter are Arduinos and ESPs.

QUESTIONS AND PROJECT IDEAS

1. What is the purpose of a fly-back diode? When do you need it?

2. The TV images are badly disturbed as soon as you turn on a DC motor. What is the cause and how can you alleviate the problem?

3. How do you control the speed of a DC motor if you need to run it at low speed?

4. When do you need to use an H bridge driver to control a DC motor?

5. Can you use a Darlington driver, such as the ULN2003, to control a DC motor with pulse-width modulation?

6. When would you use a DC motor, and when is a model-servo the better option? Discuss!

7. Use a multimeter to determine the resistance between the wires of a stepper motor and identify the internal wiring.

8. Discuss the pros and cons of using unipolar versus bipolar stepper motors.

9. Discuss microstepping and draw the currents in the coils as a function of time.

10. We plan to control a device by low-pass filtering a pulse-width modulated signal. How does the ripple of the control voltage depend on the input impedance of the device? How does the capacitor enter the discussion? What happens if we plan to apply fast changes? Discuss!

11. Investigate what a *thyristor* is and where it is used.

12. Investigate what a *triac* is and where it is used.

13. Investigate what an *IGBT* is and where it is used.

14. Design a small crane to lift cargo from a model boat onto a quay. Which actuators and which sensors do you need? Motivate their use.

15. Design a mechanism that wakes you up with almost certainty. Which actuators do you use and why?

## CHAPTER 4

Microcontroller: Arduino

The microcontrollers we use in this book come are all supported by the Arduino *integrated development environment* (IDE), which provides an easy-to-use programming environment that allows us to quickly start developing software. There are limitations, but it is a wonderful system to start learning about microcontrollers.

4.1 HARDWARE

The original Arduinos are based on Atmel microcontrollers and we mostly discuss the Arduino UNO. Support for additional families of controllers, based on ESP8266 and ESP32 microcontrollers with built-in WLAN support was later integrated into the Arduino IDE. These controllers can be therefore programmed in much the same way as UNOs. But let us start with the UNOs.

4.1.1 Arduino UNO

An Arduino UNO is shown in Figure 4.1, where the main component is the ATmega328p microcontroller from Atmel (now Microchip^{TM}), which is the large chip with 28 legs in the image. It is a controller with 8-bit-wide registers, and operates at a clock frequency of 16 MHz. It has 32 kB RAM memory and 1 kB non-volatile EEPROM memory, which can be used to store persistent data that need to survive tuning off and on the supply voltage. There are three timers on board, which are basically counters that count clock cycles and are programmable to perform some action, once a counter reaches some value. The UNO interacts with its environment through 13 digital input–output (IO) pins, of which most can be configured to be either input or output and have software-configurable pull-up resistors. Several of the pins are configurable to support I2C, SPI, and RS-232 communication. Moreover, there are six analog input pins. They measure voltages of up to the supply voltage of 5 V. Optionally, an internal reference voltage source provides a 1.1 V reference. All digital and analog pins are routed to pin headers that are visible on both sides of the Arduino printed circuit board (PCB) in Figure 4.1. Furthermore, the built-in hardware RS-232 port is connected to an RS-232-to-USB converter that allows communication and programming from a host computer. There is no WiFi, Bluetooth, or Ethernet support on the UNO board, but extension boards to provide this functionality, so-called *shields*, are available for mounting directly onto the pin headers.

One can describe the Arduino UNO as having the intelligence of a washing machine. It keeps time with the timers, it can sense voltages from, for example, temperature sensors, and it can turn motors or pumps on or off, depending on whether some condition is met. In this way it can also provide the glue logic to interface sensors (analog, I2C, SPI, and

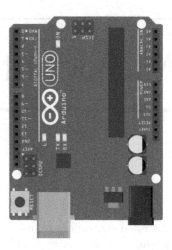

Figure 4.1 An Arduino UNO[†].

other) to the host computer, and that is the mode in which we will use the UNO later on. In passing, we mention that there is a large selection of Arduino boards, from those with a smaller footprint, the *Micro* and the *Nano* over the *Mega* with its large number of IO pins to the *MKR*, *RP2040*, and *Portenta* boards with more powerful CPUs and built-in wireless interfaces.

In the next section, however, we address a separate group of microcontrollers, the ESP8266 and the ESP32.

4.1.2 ESP8266, NodeMCU, and ESP32

The ESP8266 was released by the Chinese company Espressif in 2014 followed by the ESP32 in 2016. These controllers quickly became one of the most exciting platforms for Internet of Things (IoT) projects because they have built-in and ready-to-use WiFi support, including a small patch antenna. The software development kit (SDK) from Espressif was soon integrated into the Arduino IDE, and today the ESP8266 and ESP32 can be programmed using the Arduino ecosystem as easily as the original Arduinos.

But let us now look at the ESP8266 hardware, which comes in different incarnations. We will not only focus on the more elaborate NodeMCU board but also mention the basic ESP-01 in this section. Images of the ESP-01 and NodeMCU hardware are shown at the left and center of Figure 4.2, respectively. The microcontroller on both chips is essentially the same, only the number of internal pins that are routed to pins on the circuit boards is different. Internally, the ESP8266 chips feature 32-bit RISC CPUs that normally operate at 80 MHz or 160 MHz, and have 64 kB of RAM for instructions and 96 kB for data. There are 16 general-purpose IO (GPIO) pins that can be configured as input or output pin, as well as a single analog input with a 10-bit ADC. The controller supports I2C and SPI communication as well as RS-232. All this is similar to the Arduino, except it runs faster, and the internal registers are wider (32 instead of 8 bit in the ATmegas), but the really cool feature is the built-in WiFi support. It fulfills the specifications for IEEE 802.11 b/g/n with WPA authentication, which is what most wireless networks use today. So basically, we are able to connect an ESP8266 to any wireless network.

Figure 4.2 The ESP-01 (left), NodeMCU (middle), and ESP32 (right).

Two years after the ESP8266, the more powerful dual-core ESP32 microcontroller was released. It has practically all the features, including WiFi, of the earlier model, but runs at higher clock speeds of up to 240 MHz, has 320 kB of RAM, and Bluetooth (BLE) support on board. Particularly attractive for data acquisition applications are up to eighteen 12-bit ADC channels and two eight-bit digital-to-analog converters (DAC). In recent years a number of variants of the ESP32 with different numbers of cores and other capabilities appeared.

After having introduced the hardware, we need to move on and start programming. We therefore provide a quick-start guide for the Arduino IDE in the following section.

4.2 GETTING STARTED

Despite the availability of the new Arduino IDE version 2, we suggest to use the most recent version 1, which is 1.8.19 at the time of writing, because some features needed in later chapters are not yet available in version 2. Both versions are available from

`https://www.arduino.cc/en/software`

for the most common operating systems, including Linux, Mac, and Windows. Basically, we go to the webpage, select the version appropriate for our computer type, and download the software package. Then we install it by following the instructions given on the website. Once the installation completes, we start the IDE by clicking the icon, or start it from the command line by typing `arduino` followed by `Enter`, which should open the IDE and display a window similar to the one shown in Figure 4.3. If not, create a "New" *sketch,* which is what Arduino programs are called, by selecting "New" from the "File" menu.

Here we already see the general structure of Arduino programs (or synonymously "sketches"). There is a `setup()` function, which is *executed once,* immediately after power is turned on. In this routine all initialization housekeeping is done, such as defining a pin to be output or input, and configuring the serial line. Once the `setup()` function completes, the `loop()` function is called repetitively, such that once it completes, it is called again and so forth, until power is turned off. Note that the programming language supported by the Arduino IDE is very similar to the C language. There are, however, a number of extensions to provide access to the specific hardware, such as the ADC.

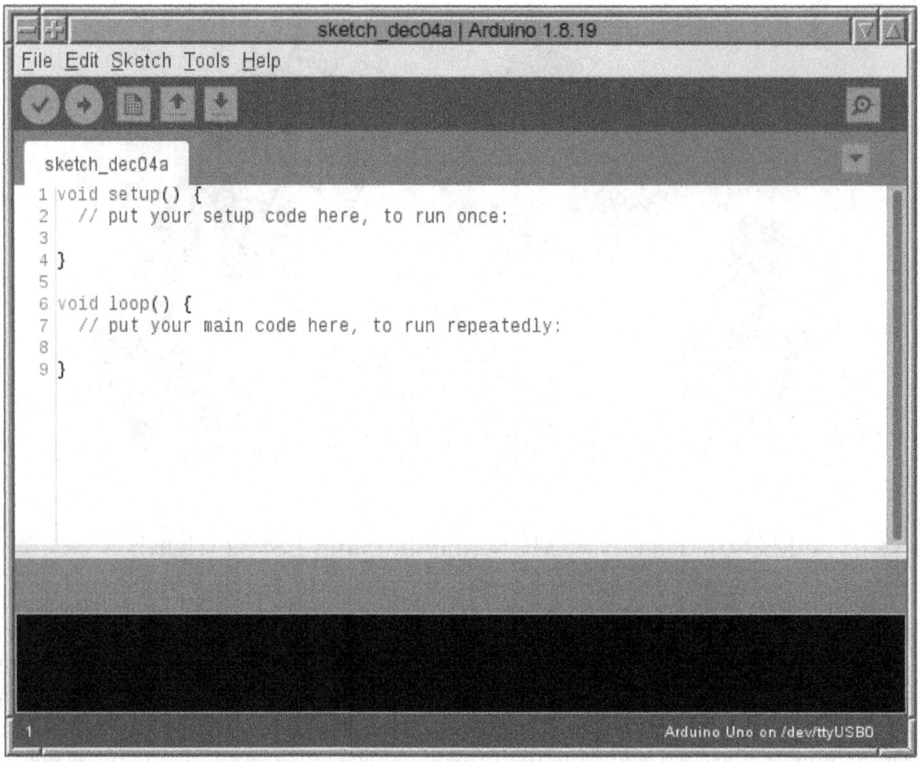

Figure 4.3 The Arduino IDE.

Now we connect the Arduino UNO to any USB port on the host computer where the Arduino IDE is running, and select Arduino UNO from the Tools→Board menu. This step tells the IDE for which processor the compiler will generate code, as well as some hardware-specific definitions such as the names of IO pins that we can use when programming. At this point the IDE "knows" what the hardware is, but we still need to tell the IDE to which USB port the UNO is connected. This we do in the Tools→Port menu, where normally the serial port to which the UNO is connected automatically appears and can be selected. On Linux this often is /dev/ttyUSB0 or /dev/ttyACM0. On a Windows computer it is COMx where x is some number.

At this point we could start programming the UNO, even though the example program in Figure 4.3 only contains empty functions. But before writing programs for the Arduino, we want to install support files for ESP8266-based microcontrollers. They are easily installed by opening the File→ Preferences menu and adding http://arduino.esp8266.com/stable/package_esp8266com_index.json to the "Additional Boards Manager URL" text box and clicking the "OK" button. Then open the Tools→Board menu, open the "Boards Manager," and find the "esp8266" platform. Then select the newest version and click on the "Install" button. Once installation completes, select the "ESP826 Boards" platform from the Tools→Board menu. For the ESP-01 the entry "Generic ESP8266 Module" is a good choice and for the NodeMCU platform it is "NodeMCU 1.0."

Installing support for ESP32 processors follows the same procedure. Separated by a comma, we add the following address https://raw.githubusercontent.com/espressif/arduino-esp32/gh-pages/package_esp32_index.json to the "Additional Boards Manager URL" text box in the File→ Preferences menu. We then open the Tools→Board

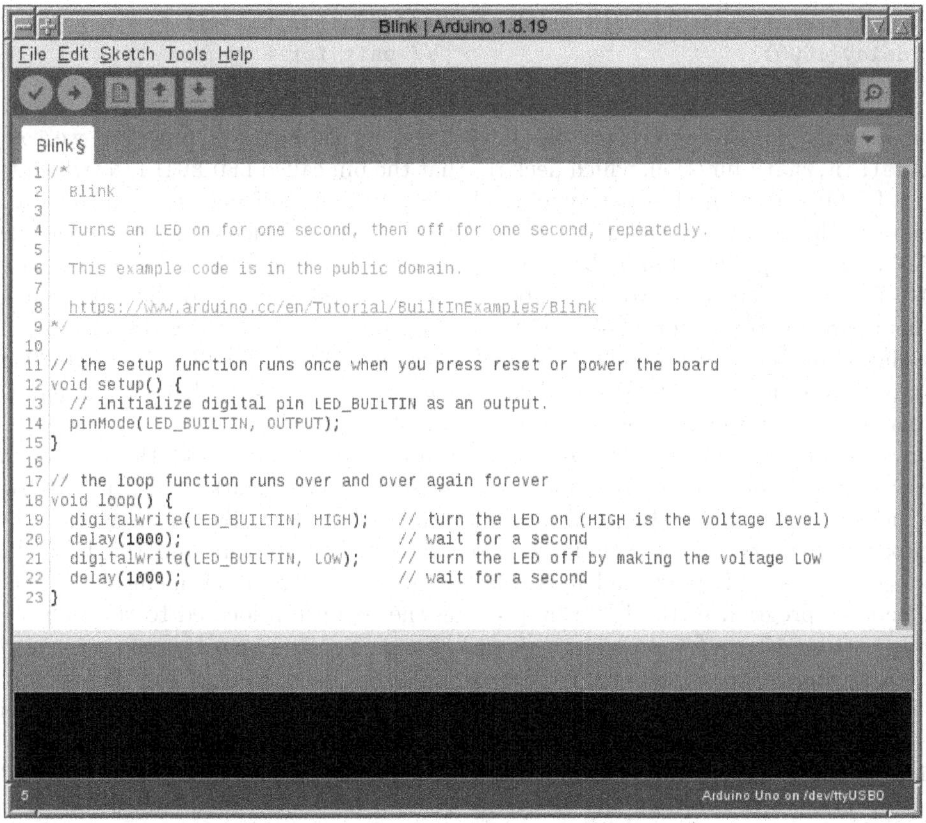

Figure 4.4 The "Blink" sketch.

menu, search for the "esp32" platform, and install the latest version, which might take a little while. Once finished, we select the model – the one shown on the right-hand side in Figure 4.2 is a "ESP32 Dev Module" – from the list in the "ESP32 Boards" submenu.

Once the correct platform (UNO, plain ESP8266, NodeMCU, or ESP32) is selected – we will tacitly assume an UNO is connected – we are ready to write programs (sketches) and download them to the hardware.

4.3   HELLO WORLD, BLINK

Our first sketch is the standard for almost any new controller, namely to cause an LED to blink. Examples to interface hardware such as sensors can be found in the File→Examples menu, where the example to blink an LED is located under the 01.Basics→ Blink menu item. Once selected, a new window opens with the example code. Figure 4.4 shows the bare code, without many comment lines that are preceded by //. It is reproduced here.

```
void setup() {
  pinMode(LED_BUILTIN, OUTPUT);
}
void loop() {
  digitalWrite(LED_BUILTIN, HIGH);   // turn the LED on
  delay(1000);                       // wait for a second
```

```
    digitalWrite(LED_BUILTIN, LOW);    // turn the LED off
    delay(1000);                       // wait for a second
}
```

Let us step though this sketch one line at a time. In the `setup()` function, we call the `pinMode(Pin,What)` function, which declares that the pin called LED_BUILTIN (which is pin 13 on a UNO) will be used as output pin. That is all the initializing we do in the `setup()` function. In the `loop()` we call `digitalWrite(Pin,State)`, which causes the controller to put 5 V onto the specified pin if the requested state is HIGH and 0 V if the requested state is LOW. The latter happens two lines later. In between the changes to the output pin, we tell the controller to wait for a specified number of milliseconds in the `delay(time_in_ms)` function. So, all the loop function does is turn on the LED on pin 13, wait for 1000 ms or 1 s, turn it off, wait again for 1 s, and then start all over again. Note that all commands are terminated by a semicolon, as is customary in the C-language.

Once we are satisfied with our program, we can compile it and check that the syntax is correct and we did not forget any semicolons. Pressing the checkmark symbol just below the `File` menu entry compiles the program and reports progress in the status window below the program window. If the compilation completes without errors and the UNO board is connected to a USB port, and it is selected in the `Tools→Port` menu entry, we can download the program to the UNO by pressing the → button located to the right of the compile button. After a few seconds, a small LED on the UNO board should start to blink once per second. Now we can change, for example, the delay time in the sketch, compile and download again, and observe whether the LED blinks according to the newly specified times.

Here we only used very few commands, such as `pinMode` or `digitalWrite`, but there are many more, and checking the Reference section on the Arduino website at `https://www.arduino.cc/reference/en/` is very inspirational. Moreover, clicking on the different commands opens a new page, with explanations and examples on how to use the functions. A further source of wisdom is the "Documentation" tab from the menu list on the top of `https://www.arduino.cc/`, which offers a large number of tutorials and examples. Keeping these resources in mind will help us to quickly locate information to make things work. Moreover, there are a large number of books on various Arduino projects, and compendia of tips and tricks such as [22].

After the first step to make the microcontroller follow our bidding, we move on and read sensors.

4.4   INTERFACING SENSORS

Here, the main task of the Arduino microcontroller is to read sensors, with their different, often idiosyncratic interfaces, and convert the measured values to a standard representation that is communicated to a host computer. We start with the simplest sensor, a button or a switch.

4.4.1   Button

In Figure 4.5, we show how to connect a button between ground and pin 2 of the Arduino, such that pressing the button will cause pin 2 to read 0 V or LOW. The following listing shows a sketch that causes the built-in LED on pin 13 to light up, if the button is pressed.

```
// button_press, V. Ziemann, 161013
void setup() {
  pinMode(2,INPUT_PULLUP);   // button, default=HIGH
```

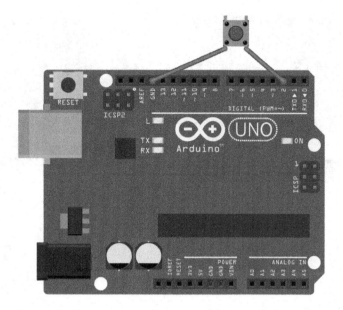

Figure 4.5 Interfacing a button to the Arduino[†].

```
  pinMode(13,OUTPUT);        // LED
}
void loop() {
  if (digitalRead(2)==LOW) {
    digitalWrite(13,HIGH);
  } else {
    digitalWrite(13,LOW);
  }
  delay(10);
}
```

The program follows the normal scheme of initializing the hardware in the setup() function. We first declare that pin 2 is an input pin, and we enable the internal pull-up resistor that causes a well-defined "high" state on the pin if no button is pressed. Pin 13 with the LED is declared as output, so we can turn it on and off from within the sketch. In the loop() function, we check the state of the button by calling the digitalRead(Pin) function, and test whether pin 2 is LOW. If this is the case (note the double equal sign == in the comparison), the LED on pin 13 is turned on by calling digitalWrite(13,HIGH). If pin 2 is found not to be LOW, the LED on pin 13 is tuned off with digitalWrite(13,LOW);. Note the braces { and } and their use to define blocks of code in the if (..) { } else { } construction. After the if statement, a short delay() ensures that mechanical button bounces are ignored and that the processor has a little time for its internal affairs. This is not absolutely necessary, but good style.

Sensing the state of switches that indicate open doors or windows follows exactly the same scheme, and we can sense up to 13 different switches (if we do not use the LED on pin 13) and react in some manner.

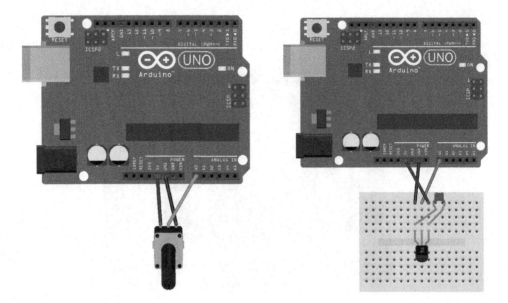

Figure 4.6 Interfacing a potentiometer (left) or an LM35 temperature sensor (right) to the Arduino[†].

4.4.2 Analog input

In an earlier chapter, we saw that some sensors require us to measure an analog voltage because they change their resistance. We place them in a voltage divider and then read the voltage change on the central tap, as shown on the left in Figure 2.1. Others, such as the LM35 temperature sensor, directly produce a voltage and in this section we show how to interface these sensors to the Arduino. First, we want to measure a voltage from one of the analog input pins; here we use A0. The left of Figure 4.6 shows how to connect a potentiometer as a variable voltage divider to the Arduino. One end of the potentiometer is connected to ground, the other end to the supply voltage, and the wiper with the variable tap of the potentiometer is connected to pin A0. Turning the axis of the potentiometer causes the voltage on the center tap to vary between ground and the 5 V supply voltage.

The sketch to read the voltage and report it via the serial cable (that is connected to the USB port) to the host computer is shown next.

```
// Analog and serial communication, V. Ziemann, 161130
int inp,val;
void setup() {
  Serial.begin(9600);   // baud rate
}
void loop() {
  if (Serial.available()) {
    inp=Serial.read();              // read character from serial
    val=analogRead(0);              // read analog
    Serial.print("Value is ");      // and report back
    Serial.println(val);
  }
  delay(50);                        // wait 50 ms
}
```

Since we want to temporarily store integer values, we have to allocate memory for the variables, which is done in the line containing int inp,val; as it declares two integer variables named inp and val. In the setup() function, we declare that we want to use the built-in serial (RS-232) hardware port with a speed of 9600 baud, which is about 1000 characters per second. The receive (RX) and transmit (TX) lines of the serial port are connected to pin 0 and pin 1, but also routed to the RS232-USB converter on the UNO board that is connected to the host computer. In the loop() function, we first check whether some communication from the host computer has arrived, and if that is the case, we read the character with the Serial.read(); command, but do nothing further with it. In the next line we read the analog pin A0 using the built-in ADC, and store the value in the integer variable val. The value returned from the 10-bit ADC is a number between 0 and 1023 ($= 2^{10} - 1$) and *not* the voltage. Later we will show how to convert this value to a voltage. The result of the measurement is then sent via the serial line by Serial.print() and Serial.println() commands. The difference between the two is that the former does not send a carriage return character at the end of the message, and the latter does. After the if() { } statement at the end of the loop, again a small delay is added. Now we can compile and download the sketch to the Arduino, unless some syntax error crept in. If that is the case, we need to check for missing semicolons or spelling errors, especially the "camelHump-spelling" and capital letters in the middle of the commands. Care is advisable, but the IDE uses syntax highlighting, which makes such errors reasonably easy to spot. After all errors are corrected, we download the sketch to the hardware.

Once the sketch is running on the Arduino, we want to test it, and for that we need to have access to the other end of the serial communication channel on the host computer. The easiest way is to open the "Serial Monitor" by pressing the icon that looks like a looking glass at the top right in the Arduino IDE. Clicking on it opens a simple terminal program that has an input text box at the top. After ensuring that the baud rate in the text box agrees with the one specified in the sketch, we enter any character and press "Enter." If all is well, the Arduino should report with a line Value is nnnn in the text box below the entry box. Now turn the potentiometer and enter a character followed by "Enter," and see how the returned value changes.

So far, so good, but would it not be nice if we could read out more analog ports and even ask for which ones to read. This task is addressed in the following sketch. It is rather similar to the previous one, so we only discuss the new features.

```
// Analog and Serial communication, v2, V. Ziemann, 161130
int inp,val;
void setup() {
   Serial.begin(9600);  // baud rate
}
void loop() {
   if (Serial.available()) {
      inp=Serial.read();       // read serial line
      if (inp=='s') {
        val=analogRead(0);     // read analog
        Serial.print("A0= ");  // and report back
        Serial.println(val);
      } else if (inp=='t') {
        val=analogRead(1);
        Serial.print("A1= "); Serial.println(val);
      } else {
        Serial.println("unknown command");
```

```
      }
    }
    delay(50);                    // wait 50 ms
  }
```

In this sketch we actually test which character was sent on the serial line, and depending on whether it is an "s" or a "t" the value from analog pin A0 or A1 is returned. The characters sent back to the host via the serial line contain both the pin read and the value. If a character different from "s" or "t" is received by the Arduino, the string "unknown command" is returned via the serial line. In this way, we build a simple query-response protocol to make the Arduino do different things, depending on the character we send.

In the next example we assume that an LM35 temperature sensor is connected to analog pin A0 in the way shown on the right in Figure 4.6. Note that we add a 100 nF capacitor across the positive supply voltage and ground pins of the LM35, to increase the stability of the circuit. Note that it is good practice to add such a decoupling or bypass capacitor close to the power pins of each chip in a circuit. In the following, we often do not show these capacitors on the circuit diagrams in order to make them less cluttered, but the capacitors should be included in the hardware. The software to interface the LM35 is slightly more elaborate than in previous examples. It allows us to send somewhat descriptive multicharacter commands to the Arduino and receive a response that echos the request followed by the value. The basic idea is that we request a parameter, say A0, by sending the parameter name with a question-mark appended, such that the request looks like this: A0?. The Arduino subsequently replies with A0 <value>. In the following sketch we implement this, but also that the Arduino first reads a full line, in the sense that it waits for a carriage-return and a line-feed character before responding.

```
// query-response, V. Ziemann, 161013
int val;
float temp;
char line[30];
void setup() {
  analogReference(INTERNAL);   // 1.1V internal ADC reference
  Serial.begin (9600);
  while (!Serial) {;}
}
void loop() {
  if (Serial.available()) {
    Serial.readStringUntil('\n').toCharArray(line,30);
    if (strstr(line,"A0?")==line) {
      val=analogRead(0);
      Serial.print("A0 "); Serial.println(val);
    } else if (strstr(line,"A1?")==line) {
      val=analogRead(1);
      Serial.print("A1 "); Serial.println(val);
    } else if (strstr(line,"T?")==line) {
      temp=1.1*100*analogRead(0)/1023;
      Serial.print("T "); Serial.println(temp);
    } else {
      Serial.println("unknown");
    }
  }
}
```

In the first few lines, we declare the variables, and in particular, a 30-character buffer `line` to hold the multicharacter query. In the `setup()` function, we use the internal 1.1 V reference voltage for the ADC and initialize the Serial communication channel to operate at 9600 baud. The `while(!Serial)` statement waits for the initialization of the Serial library to complete, and is only needed on some newer Arduinos. We include it just for safety. In the `loop()` function, we wait for characters to be available before reading an entire line until the end-of-line character `\n`. The `readStringUntil` returns a `String` but we want to use a character array instead, and need to convert the type in the appended `toCharArray` method. At this point the character array `line` contains the request that was sent from the host computer to the Arduino; for example, `A0?`. The following construction of `if (strstr(line,"A0?")==line)` and similar commands checks whether `line` starts with the characters `A0?` by comparing the pointer to the first occurrence of `A0?` in the array `line[]` to the pointer to the start of `line[]`. The commands that follow are executed provided the result evaluates to true, or else the subsequent comparison in the `else if` sequence is tested. In the example, we test for `"A0?"`, `"A1?"`, and `"T?"` to reply with the raw values of the analog pins 0 and 1 or with the value from analog pin 0 converted to degrees Celsius. Consult the discussion in Section 2.1.2 regarding the conversion coefficient. If a command arrives, and it is not in the list, the Arduino replies with `unknown command`. The way to specify the query with an appended question mark and the reply to echo the query with value appended is borrowed from the SCPI [23] command-language that is commonly used to remotely control measurement equipment via GPIB, also known as IEEE-488, or VXI.

As a final example, we discuss a simple *temperature logger* that again uses the hardware setup as shown on the right side of Figure 4.6, but sends the temperature value continuously to the host computer via the serial line. The sketch is shown here:

```
// ultra-simple LM35 temperature logger, V. Ziemann, 161201
float temp;
void setup() {
  analogReference(INTERNAL);  // 1.1V internal ADC reference
  Serial.begin(9600);
  while (!Serial) {;}  // wait for serial to initialize
}
void loop() {
  temp=1.1*100*analogRead(0)/1023.0;
  Serial.println(temp);
  delay(1000);
}
```

where the `setup()` function equals that from the previous example, but the `loop()` function measures the analog pin and converts the digital word to the temperature, and sends that to the host computer. Then it waits 1000 ms before repeating the same. In the Serial Monitor, that can be started by clicking on the button at the top right in the IDE; we see that the temperature value in Celsius is reported once per second in the lower text box provided the baud rate selected in the selection box at the lower right matches the baud rate specified in the sketch.

In the Arduino IDE, however, there is also a graphical plotting tool available. We select it by choosing the `Tools→Serial Plotter` menu item. This opens a window and shows the values received on the serial line in graphic format, provided the baud rate is selected to match the one specified in the sketch. Note that the `Serial Plotter` can show several measurement values simultaneously as traces in different colors. This functionality, built in to the Arduino IDE, may serve as a simple logger for temperatures or other measured values, if a quick solution to record some values as a function of time is needed.

So far, we managed to read analog voltages, and even devised a simple protocol to report the measurement values to the host computer. In the next section, we interface I2C-based sensors to the microcontroller, and use the same query-response protocol to communicate with the host.

4.4.3 I2C

And we continue with another temperature sensor, but a special one that is used by nurses in hospitals. It is based on a contact-free thermometer that takes your temperature by pointing a gadget into your inner ear. An example is the MLX90614 infrared thermometer shown on the right in Figure 2.9. It uses a thermopile to deduce the temperature from the picked-up infrared radiation and it reports the temperature to our Arduino via an I2C interface. Here we point out that the MLX90614 comes in versions for 5 V and 3.3 V supply voltage. In this example we use the 5 V version, labeled `AAA`. The device has four pins – check the datasheet to find out which pin does what – for ground, supply voltage, and the I2C lines SDA as SCL. The latter two are connected to pin A4 and A5 on the UNO because the I2C data and clock lines are routed to the same output pins as the analog pins A4 and A5. These pins cannot be used for analog measurements, in case they are used for I2C connectivity. We show how to connect the sensor in Figure 4.7, and illustrate how to talk to it in the following sketch.

```
// Read MLX90614 IRthermometer, V. Ziemann, 170717
#include <Wire.h>
const int MLX90614=0x5A;   // I2C address
float getTemperature(uint8_t addr) { //.....getTemperature
  uint16_t val;
  uint8_t crc;
  Wire.beginTransmission(MLX90614);
  Wire.write(addr);        // address
  Wire.endTransmission(false);
  Wire.requestFrom(MLX90614,3);
  val = Wire.read();
  val |= (Wire.read()<<8);
  crc=Wire.read();         // not used
  return -273.15+0.02*(float)val;
}
void setup() { //...................................setup
  Serial.begin(9600);
  Wire.begin();
}
void loop() { //....................................loop
  float Ta=getTemperature(0x06);  // address for ambient temp
  float To=getTemperature(0x07);  // address for object temp
  Serial.print(Ta); Serial.print("\t"); Serial.println(To);
  delay(1000);
}
```

First we have to include support for the I2C functionality by including the `Wire.h` header file, which also causes the compiler and linker to include the corresponding libraries. After we define the I2C address `0x5A` of the sensor, which we find in the datasheet, we encapsulate the I2C communication in a separate function called `getTemperature()`. It receives the register address of the I2C device as input parameter, reads values with the help of a number of

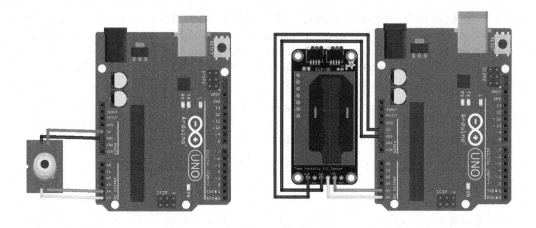

Figure 4.7 Connecting the MLX90614 contact-free thermometer (left) and an SCD30 CO_2 sensor (right) to an Arduino UNO[†].

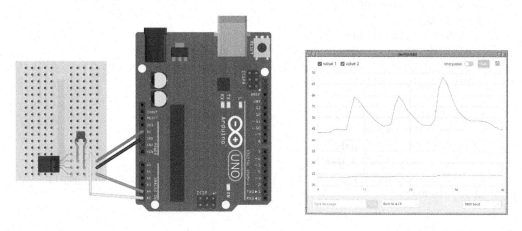

Figure 4.8 Connecting the HYT-221 humidity sensor to the Arduino (left[†]) and using the serial plotter to display the humidity and temperature measurements (right).

Wire.* function calls, and returns the associate temperature, properly scaled to degrees Celsius. The device also returns a byte that allows us to determine transmission errors, but we do not use that feature in this simple example. In the setup() function we only initialize Serial line and I2C communication. In the loop() function we use the getTemperature() function to retrieve the ambient temperature from address 0x06 and the object temperature from address 0x07, as described in the datasheet, and print both temperatures, separated by a tabulator \t, to the Serial line. The entire process is then repeated after waiting 1000 ms. We can use the Serial monitor or plotter built into the Arduino IDE in order to view or display the results.

Several other sensors, often related to environmental parameters such as humidity or barometric pressure, also report the temperature as a by-product, because it is internally needed to calibrate the reported primary measurement value. For the next sensor, the HYT-221, this is the humidity. Connecting the sensor to the Arduino is illustrated on the left in Figure 4.8. From the datasheet we know that the four pins of the sensor, when looking towards the front of the sensor, are SDA, GND, VDD, and SCL. The pins for GND and VDD are connected to the GND and 5 V pins on the Arduino, and SDA to A4 and SCL to A5 in the same way we did for the temperature sensor in the previous example. In the following sketch, we read the HYT sensor, and send the measurement data of humidity and temperature via the serial line to the host computer.

```
// HYT221 humidity sensor, V. Ziemann, 161203
#include <Wire.h>
const int HYT=0x28;  // I2C address
void setup() {  //...........................setup
  Serial.begin (9600); while (!Serial) {;}
  Wire.begin();
}
void loop() {  //............................loop
  int b1,b2,b3,b4,raw;
  double humidity,temp;
  Wire.beginTransmission(HYT);
  Wire.requestFrom(HYT,4);
  delay(50);
  if (Wire.available()==4) {
    b1=Wire.read(); b2=Wire.read();  // humidity rawdata
    b3=Wire.read(); b4=Wire.read();  // temperature rawdata
    Wire.endTransmission(true);
  }
  raw=(256*b1+b2) & 0x3FFF;          // humidity
  humidity=100.0*raw/16384.0;
  raw=((256*b3+b4) & 0xFFFC)/4;      // temperature
  temp=165.0*raw/16384.0-40.0;
  Serial.print(humidity); Serial.print("\t"); Serial.println(temp);
  delay(1000);
}
```

First we include the I2C functionality with Wire.h, and define a variable HYT that contains the I2C address 0x28. In the setup() function, we first initialize the serial communication and then the I2C functionality with the call to the Wire.begin() function. In the loop() function we first declare a number of auxiliary integer variables and double-precision float variables to hold the measurement data before calling the Wire.beginTransmission() function with the address HYT as argument, and tell the Arduino to fetch four bytes as response from the sensor in the following line. After a short delay, we check whether the four bytes have arrived, with the Wire.available() function, and in case they are available, we read them into the variables b1,..,b4 before closing the transmission. The first two bytes contain the raw data for the humidity, and the next two for the temperature measurement. In the Arduino sketch, we follow the instructions from the HYT-221 datasheet to convert the raw data to the physical quantities. For the humidity, the datasheet tells us that the first received byte b1 is the most significant byte of a two-byte word and b2 is the least significant byte. Therefore, we can reconstruct the word by 256*b1+b2. But we also have to throw away (mask) the two highest bits, which we do by a logical **and** (&) with a binary value that has ones everywhere except at the two highest bits, which is & 0x3FFF. Once we have the raw value, we again follow the datasheet and scale by 100/16384 to obtain the humidity in percent. For the temperature, we follow a similar procedure that is described in the HYT-221 datasheet. Finally, we send the humidity and temperature values to the host computer via the serial line and wait a while before repeating the measurement. On the right of Figure 4.8 we show the data using the Serial-plotter in the Arduino IDE. The upper trace shows the humidity, and the three positive excursions are due to exhaling on to the sensor.

Often the concentration of CO_2 is a good proxy to judge the quality of indoor air in congested rooms. This is of quite some concern, especially during the Covid pandemic. We measure the concentration with a SCD30 CO_2 sensor that is connected to an UNO via four cables for power, ground and the two I2C lines for clock and data, shown on the right-hand side on Figure 4.7. The circuit is brought to live with the code shown below. First we include support for I2C and define the address of the sensor as well as additional variables. Then we define the function **data_ready()** which reads the status of the conversion and returns 1 if data from a new conversion is available and 0 otherwise. The datasheet specifies that upon sending the two-byte command 0x02 0x02, the SCD30 returns three bytes: the first two contain the status and the last one is a checksum, which we do not use in this basic example. In the **setup()** function, we only initialize the serial and I2C communication.

```
// Sensirion SCD30 CO2 concentration, V. Ziemann, 221006
#include <Wire.h>
const int SCD30=0x61;
char line[30];
int running=0;
int data_ready() {
  byte b1,b2,b3;
  Wire.beginTransmission(SCD30);
  Wire.write(0x02); Wire.write(0x02);
  Wire.endTransmission();
  Wire.requestFrom(SCD30, 3);
  b1 = Wire.read(); b2 = Wire.read(); b3= Wire.read();
  return  (b1<<8) | b2;
}
void setup() {
  Serial.begin(9600); while(!Serial){;}
  Wire.begin();
}
void loop() {
  if (Serial.available()){
    Serial.readStringUntil('\n').toCharArray(line,30);
    if (strstr(line,"V?")) { // Get firmware version
      byte  b1,b2,b3;
      Wire.beginTransmission(SCD30);
      Wire.write(0xD1); Wire.write(0x00);
      Wire.endTransmission();
      Wire.requestFrom(SCD30,3);
      b1 = Wire.read(); b2 = Wire.read(); b3=Wire.read();
      Serial.print("V "); Serial.print(b1);
      Serial.print("."); Serial.println(b2);
    } else if (strstr(line,"Start")) {  // Start data taking
      Wire.beginTransmission(SCD30);
      Wire.write(0x00); Wire.write(0x10); // two command bytes
      Wire.write(0x00); Wire.write(0x00); // ambient pressure
      Wire.write(0x81); //CRC
      Wire.endTransmission();
```

```
        running=1;
      } else if (strstr(line,"Stop")) {
        Wire.beginTransmission(SCD30);
        Wire.write(0x01); Wire.write(0x04); // two command bytes
        Wire.endTransmission();
        running=0;
      }
    }
    if (running) {    // measure
      unsigned long  b1,b2,b3,b4,b5,b6,raw;
      float CO2,temp,hum;
      while (!data_ready()) {delay(10);}  // wait until data ready
      delay(20);
      Wire.beginTransmission(SCD30);
      Wire.write(0x03); Wire.write(0x00);  // command bytes
      Wire.endTransmission();
      Wire.requestFrom(SCD30,18);
      b1 = Wire.read(); b2 = Wire.read(); b3 = Wire.read();
      b4 = Wire.read(); b5 = Wire.read(); b6 = Wire.read();
      raw=(b1<<24) | (b2<<16) | (b4<<8) | b5;  // CO2
      CO2=*(float*)&raw;    % ppm
      b1 = Wire.read(); b2 = Wire.read(); b3 = Wire.read();
      b4 = Wire.read(); b5 = Wire.read(); b6 = Wire.read();
      raw=(b1<<24) | (b2<<16) | (b4<<8) | b5;  // temperature
      temp=*(float*)&raw;    % degree Celsius
      b1 = Wire.read(); b2 = Wire.read(); b3 = Wire.read();
      b4 = Wire.read(); b5 = Wire.read(); b6 = Wire.read();
      raw=(b1<<24) | (b2<<16) | (b4<<8) | b5;  // temperature
      hum=*(float*)&raw;    % percent
      Serial.print(CO2); Serial.print("\t");
      Serial.print(temp); Serial.print("\t"); Serial.println(hum);
      delay(2000);
    }
  }
```

The loop function implements our standard query-response protocol and dispatches two-byte commands to the SCD30 via the I2C bus that are given in the datasheet. The first command **V?** requests the firmware version of the SCD30 by sending the two-byte command 0xD1 0x00 to the device which returns three bytes of which the first two contain the major and minor revision numbers. Again the third byte contains a checksum. The command **Start** initiates data taking by sending the two-byte command 0x00 0x10 and the default ambient pressure to the device. At the same time the variable **running** is set to unity. Likewise, sending **Stop** transmits 0x01 0x04 to the device and stops data taking and also sets **running** to zero. After the section with the query-response protocol we check the variable **running** and if it is set we read data as soon as it is ready, which we check with the **data_ready()** function. Then we transmit two-byte command 0x03 0x00 to the device which requests new data. It arrives in three blocks of six bytes each. The first block contains the CO_2 concentration, the second contains the temperature, and the third contains the humidity. We follow the instructions in the datasheet to first convert the bytes to a four-byte **unsigned long** variable and then cast them to a **float** value before printing them via the serial line. Since the default acquisition time is 2 s, we just wait that long. The program thus continuously displays the CO_2 concentration, the temperature, and the relative humidity via the serial line such that we can watch it with the serial plotter of the Arduino IDE or any other program that controls the serial line.

The next environmental sensor – we use it later in a weather station – is the BME680. We already briefly encountered this chip in Section 2.3.3. It features sensors for barometric pressure, humidity, temperature, and volatile organic compounds, all on the same chip. Since each individual BME680 is factory-calibrated and has calibration constants for all sensors stored on-board, programming it is rather intricate. Have a look at the low-level drivers available from `https://github.com/BoschSensortec` to get an impression. Instead of using that library we use the high-level interface, provided by the manufacturers of the breadboard shown in Figure 2.36 that we incorporate into the Arduino IDE from the library manager where we search for "Adafruit BME680" and install the library.

With this library the sketch to read out the sensor is very short, because all the details of the I2C communication, sensor readout, and data calibration data are encapsulated in the high-level functions. We make the library accessible by including `Adafruit_BME680.h` header file and then creating an object `bme` to access the sensor functionality. In the function `setup()` we initialize the serial and I2C communication, make sure the sensor is actually connected, and start to configure it. The three commands to set the oversampling determine the accuracy of the returned temperature, humidity, and pressure. The IIR filter defines and filter to average over successive samples, which smoothes the data. Finally the small on-board heater for the MQ-x sensor, which measures volatile organic compounds in air is configured to heat up to $320^{\circ}C$ and wait for 150 ms before reading the resistance that quantifies the air quality, where higher resistance implies better air quality.

```
// BME680, V. Ziemann, 221006
#include <Wire.h>
#include <Adafruit_BME680.h>
Adafruit_BME680 bme;
void setup() {
  Serial.begin(9600); while (!Serial);
  Wire.begin();
  if (!bme.begin()) {Serial.println("Error: BME680 not found");}
  bme.setTemperatureOversampling(BME680_OS_8X);
  bme.setHumidityOversampling(BME680_OS_2X);
  bme.setPressureOversampling(BME680_OS_4X);
  bme.setIIRFilterSize(BME680_FILTER_SIZE_3);
  bme.setGasHeater(320, 150); // 320*C for 150 ms
}
void loop() {
  bme.performReading();          // here the sensor is read out
  Serial.print(bme.temperature); Serial.print("\t");
  Serial.print(bme.pressure/100.0); Serial.print("\t");  // hPa=mbar
  Serial.print(bme.humidity); Serial.print("\t");
  Serial.println(bme.gas_resistance/1000.0);             // kOhm
  delay(1000);
}
```

In the function `loop()` we only need to initiate an acquisition with `bme.performReading()` and then access one value at a time. Note that we convert the pressure to hPa, which is the same as mbar. Moreover the air quality is returned as the resistance of the MQ-x sensor.

We can use the next sensor to determine the acceleration and speed of rotation on carousel rides at a fair. Incidentally, similar sensors are used in smart phones to determine whether they are held upright or sideways. This device, the *MPU-6050*, contains separate sensors for acceleration in three dimensions and for angular velocity around three axes. It uses a standard I2C with default address 0x68. In the following sketch, we explain how to interface it to the Arduino. Since we will frequently read and write single bytes via the I2C lines we create a file named I2Crw.h that contains the functions I2Cwrite() and I2Cread(). Note that it must reside in the same directory as the sketch that includes it. Using this file in later chapters will save us quite a bit of writing.

```
// I2Crw.h, V. Ziemann, 221118
void I2Cwrite(int addr, uint8_t reg, uint8_t val) {
  Wire.beginTransmission(addr);
  Wire.write(reg);
  Wire.write(val);
  Wire.endTransmission(true);
}
uint8_t I2Cread(int addr, uint8_t reg) {
  Wire.beginTransmission(addr);
  Wire.write(reg);
  Wire.endTransmission(false);
  Wire.requestFrom(addr,1);
  return Wire.read();
}
```

With the basic functionality encapsulated in I2Crw.h we just need to include it in the following sketch

```
// MPU6050, V. Ziemann, 221118
#include <Wire.h>
#include "I2Crw.h"
const int MPU6050=0x68;  // address
void mpu6050_init() {  //.....................mpu6050_init
  Wire.begin();
  int whoami=I2Cread(MPU6050,0x75);
  if (whoami!=0x68) {
    Serial.println("No MPU6050 found!"); while (1) {delay(10);}
  }
  I2Cwrite(MPU6050,0x6B,0); // wake up device
  I2Cwrite(MPU6050,0x1B,0); // Gyro config, FS=0
  I2Cwrite(MPU6050,0x1C,0); // Acc config, AFS=0
}
  void mpu6050_read_float(int addr, float fdata[7]) { //...read_float
    int16_t intval;
    Wire.beginTransmission(addr);
    Wire.write(0x3B);
    Wire.endTransmission(false);
    Wire.requestFrom(addr,14);
    intval=Wire.read()<<8 | Wire.read(); fdata[0]=intval/16.384; //ax
    intval=Wire.read()<<8 | Wire.read(); fdata[1]=intval/16.384; //ay
    intval=Wire.read()<<8 | Wire.read(); fdata[2]=intval/16.384; //ay
    intval=Wire.read()<<8 | Wire.read(); fdata[3]=intval/340.0+36.53; //T
    intval=Wire.read()<<8 | Wire.read(); fdata[4]=intval/131.0; //gx
    intval=Wire.read()<<8 | Wire.read(); fdata[5]=intval/131.0; //gy
```

```
    intval=Wire.read()<<8 | Wire.read(); fdata[6]=intval/131.0; //gz
    Wire.endTransmission();
}
void setup() { //.................................setup
    Serial.begin(115200);
    while (!Serial) {yield();}
    mpu6050_init();
}
void loop() { //.................................loop
    float fdata[7];
    mpu6050_read_float(MPU6050,fdata);
    Serial.print(fdata[0],0); Serial.print("\t");
    Serial.print(fdata[1],0); Serial.print("\t");
    Serial.print(fdata[2],0); Serial.print("\t");
    Serial.print(fdata[4],0); Serial.print("\t");
    Serial.print(fdata[5],0); Serial.print("\t");
    Serial.print(fdata[6],0); Serial.print("\t");
    Serial.println(fdata[3]);
    delay(1000);
}
```

Here we first include I2C support by including `Wire.h` and `I2Crw.h` as well as define a symbolic name for the I2C-address of the device. In the `mpu6050_init()` function, we initialize the I2C library and check for the `whoami` byte at address `0x75`. According to the datasheet, a value of `0x68` indicates that the device is present. Then we need to write to address `0x6B` in order to wake up the device, and then configure the gyro and accelerometer to use the standard ranges of $\pm 250^o$/s and $\pm 2g$. The `mpu6050_read_float()` function requests 14 consecutive bytes from the device, which encodes the three gyro sensors, three accelerometers, and temperature with two bytes each. Here `0x3B` is the starting address of the range we intend to read. We then convert the corresponding bytes to floating point values. The first three values in the array `fdata[]` are the accelerations in the three spatial directions, converted to units of 10^{-3} g, where g is the gravitational acceleration on the surface of the earth. The fourth entry is the temperature that is converted from raw ADC readings to degrees Celsius by conversion constants found in the datasheet. The last three entries contain the rate of rotation around three axes, converted to degrees per second. Again the conversion constant is taken from the datasheet. The `setup()` function only initializes the serial line and the MPU6050. The `loop()` function retrieves the data from the device by using the function we defined earlier, and displays the values in a single line. We note in passing that there are libraries available that encapsulate the details of the communication and conversion, and therefore simplify the code significantly, albeit at the expense of reduced transparency and flexibility.

In the next example, however, we take full control of the color sensor TCS34725 that we briefly discussed toward the end of Section 2.3.3. In the following code, we first load support for the I2C functionality, define the address of the device, and define digital pin 2 as output to control the LED that is mounted on the small breadboard, shown in Figure 2.40. We then define the function `I2Cwrite()` that receives the address and the register, as well as the value that will be written as input. Note that we always have to set the most significant bit `0x80` when specifying the register, a requirement that is buried deep in the datasheet. In the function `setup()` we initialize I2C and serial communication, define the led-pin as output and turn the LED off, before we wait 10 ms and turn on the device by writing the

PON bit in the configuration register at address 0x00, wait 3 ms, and set the AEN bit. This enables continuous acquisition of color data.

```
// TCS34725.ino, V Ziemann, 221006
#include <Wire.h>    // include support for the I2C functionality
const int TCS34725=0x29;                    // I2C address
int led = 2;                                // LED on breakout board
void I2Cwrite(int addr, int reg, int val) { // I2C write
  Wire.beginTransmission(addr);             // address
  Wire.write(0xFF & (0x80 | reg));          // command bit set
  Wire.write(val);                          // write value
  Wire.endTransmission(true);
}
void setup() {
  Wire.begin();                             // initialize I2C
  Serial.begin(9600); while (!Serial) {;}
  pinMode(led,OUTPUT);                      // LED on breakout board
  digitalWrite(led,LOW);                    // initially it is off
  delay(10);                                // wait
  I2Cwrite(TCS34725,0x00,0x01);             // first PON
  delay(3);                                 // see note 2 on page 20
  I2Cwrite(TCS34725,0x00,0x03);             // then AEN
}
void loop() {
  int b1,b2,clear,red,green,blue;
  digitalWrite(led, HIGH); delay(100);
  Wire.beginTransmission(TCS34725);
  Wire.write(0xFF & (0x80 | 0x14));      // start register address
  Wire.endTransmission(false);
  Wire.requestFrom(TCS34725,8);          // request eight bytes in a row
  b1=Wire.read(); b2=Wire.read(); clear=b2*256+b1;
  b1=Wire.read(); b2=Wire.read(); red=b2*256+b1;
  b1=Wire.read(); b2=Wire.read(); green=b2*256+b1;
  b1=Wire.read(); b2=Wire.read(); blue=b2*256+b1;
  Serial.print(clear); Serial.print("\t");
  Serial.print(red); Serial.print("\t ");
  Serial.print(green); Serial.print("\t "); Serial.println(blue);
  digitalWrite(led, LOW);
  delay(1000);
}
```

In the function loop() we declare a number of variables and turn the LED on. If we only want to record the ambient light, we just leave the LED turned off. In the next step we start reading the color information by first specifying the register (0x14) from which we intend to start reading and, at the same time, setting bit 0x80. Then we request to receive the contents of eight consecutive registers. The datasheet tells us that the first two, 0x14 and 0x15, contain the information about the intensity. We combine these two bytes to the 16-bit word clear. Likewise we read and assemble integers for red, green, and blue before we write them to the serial line, such that we can use them further in a calling program. Finally, we turn the LED off and wait one second before repeating to read the color information.

Occasionally the number of input or output pins on the microcontroller is insufficient for a certain project, but luckily there are integrated circuits that help us to extend the

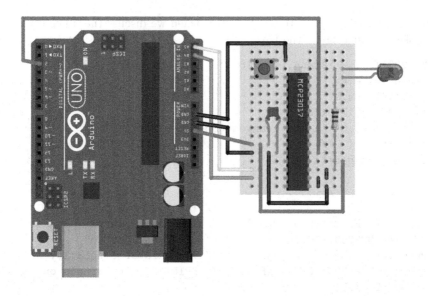

Figure 4.9 Connecting the MCP23017 IO-extender to the UNO[†].

number of IO pins. One of these is the MCP23017, a device that provides 16 IO pins, each of them freely configurable as either input or output pin. The device has an I2C interface, such that it is equally easy to address as the other examples in this section, and it can even share the same I2C lines, SCL and SDA, with other devices, because it has its own address: 0x20. Should an address-conflict with another device arise, it can be changed to any value in the range from 0x20 to 0x27 by three address pins that need to be set to either ground or supply voltage. This also implies that we can place eight copies of an MCP23017 on the same I2C bus and can use 128 extra IO pins. In the following example, we are less ambitious, and use one chip only. We configure it such that eight pins are output and eight pins are input. The chip has the attractive feature that it can generate an interrupt if an input pin changes its value. We will use that feature, despite having to use one extra pin on the microcontroller, because it relieves the controller of continuously checking whether any input pin on the MCP23017 has changed. The controller only needs to determine which pin has changed, once a change has occurred. In Figure 4.9 we show the circuit with an Arduino UNO interfaced to one MCP23017. On the extender, we connect one pin to an LED that we want to turn on and off. Another pin is connected to a button, and we want to determine whether it is pressed or not.

In the circuit shown in Figure 4.9 we see the UNO on the left and a small solderless breadboard with the extender on the right. Wires connect the breadboard to the 5 V and ground terminals of the UNO, and the I2C lines for SDA and for SCL are connected to pins A4 and A5, respectively, just as before in earlier examples. Moreover, we pull the three address pins at the lower right of the MCP23017 to ground, which causes the device to have the default address 0x20. We also pull the reset pin, the fourth from the bottom on the right-hand side, to 5 V, which is required for stable operation. There are eight IO pins on the top-right part of the chip, labeled A0 to A7, where A7 is connected to the anode of the LED, and eight more IO pins on the top left of the chip. They are labeled B0 to B7 from top to bottom, with B0 connected to the small button on the top left. An additional wire connects the pin called INTB on the MCP23017 to pin D2 on the UNO. We configure the extender in such a way that INTB indicates that one of the pins B0 to B7 has changed. INTB is normally HIGH, but will go LOW once a change occurs, and that triggers an interrupt on the UNO, but more on that later when we discuss the inner workings of the following sketch that runs on the UNO.

```
// MCP23017, V. Ziemann, 221118
#include <Wire.h>
#include "I2Crw.h"
#define MCP23017 0x20
volatile uint8_t pin_has_changed=0,portB=0,portA=0;
void action() { //.................................action
  pin_has_changed=1;
}
void setup() { //.................................setup
  Wire.begin();                    // Init I2C bus
  I2Cwrite(MCP23017,0x00,0x00);    // IODIRA=all output
  I2Cwrite(MCP23017,0x0D,0xFF);    // GPPUB Pullups on
  I2Cwrite(MCP23017,0x03,0xFF);    // IPOLB reverse polarity
  I2Cwrite(MCP23017,0x05,0xFF);    // GPINTENB, interrupts on
  pinMode(2,INPUT_PULLUP);
  attachInterrupt(digitalPinToInterrupt(2),action,FALLING);
  Serial.begin(9600); while (!Serial) {;}
}
void loop() { //.................................loop
  if (Serial.available()) {
    char line[30];
    Serial.readStringUntil('\n').toCharArray(line,30);
    if (strstr(line,"A7 ")==line) {
      int val=(int)atof(&line[3]);
      if (val==0) {portA &= B01111111;} else {portA |= B10000000;}
      I2Cwrite(MCP23017,0x012,portA);  // write GPIOA
    }
  }
  if (pin_has_changed) {
    portB=I2Cread(MCP23017,0x11);   // 0x11=INTCAPB, 0x13=GPIOB
    pin_has_changed=0;
    if (portB & B00000001) {
      Serial.println("Button B0 pressed");
    }
  }
  delay(1);
}
```

In the sketch we first include `Wire.h` and `I2Crw.h` from page 74 for the I2C communication and a few additional variables. Here we use `volatile` to indicate that at least one of them can change in an interrupt handler. The function `action()` is called once the interrupt is triggered, which happens as a consequence of a pin changing on the MCP23017 that is subsequently signaled via the INTB pin going low. In the interrupt handler we only set a variable to indicate that we need to read the pins, which we later do in the `loop()` function. In the `setup()` function, we initialize I2C communication by calling `Wire.begin()`, and configure the MCP23017. Setting register 0x00 to 0x00 configures all pins of port A (those on the right-hand side with the LED attached to pin A7) as output. We do not need to do that for port B, because we use the pins in their default configuration as input. Next we enable the internal pull-up resistors on all pins of port B, and reverse the polarity with which a change is reported. It is convenient to invert the state, which normally is low if a button is pressed, to pull the input pin low. We prefer, however, to have the state reported as high. Finally we write 0xFF to register 0x05, which enables interrupts to be generated whenever the state of an input pin of port B changes. This completes the configuration of the MCP23017, but we still need to configure the UNO to listen to a change of state on its pin 2. This pin is configured as an input pin with pull-up enabled, and we attach the function `action()` to be executed if pin 2 changes state from high to low; in other words, if it is `FALLING`. We complete the configuration by initializing the serial line. In the `loop()` function, we first handle commands received over the serial line, and test whether the command starts with `A7`. We access the rest of the line by `&line[3]`, because `line` is an array of characters, and using the ampersand indicates the address of the array element number 3. Here we expect an integer, rather than a float value, but it is easy to use the construction `(int) fval` to cast the float value `fval` to an integer. This construction of parsing the rest of the `line` with `atof()` turns out to be a compact and robust way to decipher the numbers, and we will use it repeatedly. The result is placed in the variable `val` and if it is zero, we toggle bit 7 of the variable `portA` off and write the variable to the output register 0x12, which is responsible for the state of the output pins of port A, those on the right-hand side. Note that we use the prepended letter "B" to specify a number in binary representation. In particular, the LED is connected to the uppermost pin, labeled A7. By toggling the appropriate bit, we switch it on and off. After handling the Serial communication, we check whether the variable `pin_has_changed` is set, and if that is the case, we read a register with the present state of port B. We can directly read the present state of port B by reading from the `GPIOB` register 0x13. But the MCP23017 has a special feature that internally captures the state of port B immediately after the interrupt is triggered, and we retrieve that value by reading from register 0x11, which is called `INTCAPB` in the datasheet. We then check whether bit 0 is set and address that by sending some text to the serial line. Given this template and some reading of the datasheet, it should be straightforward to adapt the software to build a user interface with many buttons and LEDs.

To summarize this section, we point out that we can communicate with the sensors using basic I2C communication functions such as `Wire.read()`. In that case, we have to convert the retrieved raw data to physical quantities ourselves, which requires careful reading of the datasheets. If a library for the particular sensor is available, we can include it in the Arduino IDE from the library manager or by directly copying it to the `Arduino/libraries` subdirectory. We can use the new libraries after restarting the IDE. In any case, we strongly advise careful inspection of the examples that are normally included.

A second communication bus, similar to I2C, is the SPI bus. A number of sensors, but also other devices, such as analog-to-digital converters, use it, as we shall see in the next section.

4.4.4 SPI

In this section we discuss interfacing the MCP3304 ADC. It supports an SPI interface, and we use it to expand the number of ADC channels of a NodeMCU microcontroller, which otherwise is equipped with a single ADC channel only. In Figure 4.10 we show how to connect the ADC to the NodeMCU on a small solderless breadboard, with the NodeMCU visible on the right and the ADC on the left. From left to right, the eight input pins A0 to A7 of the ADC are on the side facing downwards. A0 and A1 are used together to form differential input channel 0. On the upper side, we connect the supply voltage to the top left (pin 16) of the MCP3304 and ground to pins 9 and 14. The four SPI lines of the chip are connected to pins D5, D6, D7, and D8 on the NodeMCU: SCL connects to D5, MISO

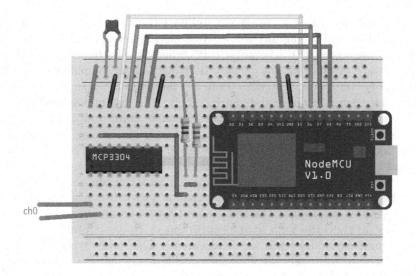

Figure 4.10 Connecting the MCP3304 ADC to the NodeMCU[†].

to D6, MOSI to D7, and chip-select to D8. We use a voltage divider with an 8.2 kΩ and a 68 kΩ resistor to lower the reference voltage to 0.4 V. It is connected with a wire to the Aref pin, pin 15 on the MCP3304. This increases the resolution of the ADC to 13 bits in the range ±0.4 V.

In the following we discuss two ways to read the ADC values from the MCP3304. Later we program the interface routine using the built-in SPI interface libraries, but first we use a method called bit-banging, which means that we write and read the respective pins in the correct order, quasi *by hand*. The following sketch is adapted from the example code on the Arduino playground.

```
// MCP3304 bit-banged, V. Ziemann, 170726
#define CS   15      // D8, ChipSelect
#define MOSI 13      // D7, MasterOutSlaveIn
#define MISO 12      // D6, MasterInSlaveOut
#define CLK  14      // D5, Clock
void mcp3304_init_bb() {  //.....................mcp3304_init
  pinMode(CS,OUTPUT);
  pinMode(MOSI,OUTPUT);
  pinMode(MISO,INPUT);
  pinMode(CLK,OUTPUT);
  digitalWrite(CS,HIGH);
  digitalWrite(MOSI,LOW);
  digitalWrite(CLK,LOW);
}
int mcp3304_read_bb(int channel) { //..........mcp3304_read_bb
  int adcvalue=0, sign=0;
  byte commandbits = B10000000;
  commandbits|=(channel & 0x03) << 4; // 5 config bits, MSB first
  digitalWrite(CS,LOW);     // chip select
  for (int i=7; i>0; i--){  // clock bits to device
    digitalWrite(MOSI,commandbits&1<<i);
```

```
      digitalWrite(CLK,HIGH); // including two null bits
      digitalWrite(CLK,LOW);
  }
  sign=digitalRead(MISO);    // first read the sign bit
  digitalWrite(CLK,HIGH);
  digitalWrite(CLK,LOW);
  for (int i=11; i>=0; i--){
    adcvalue+=digitalRead(MISO)<<i;
    digitalWrite(CLK,HIGH);
    digitalWrite(CLK,LOW);
  }
  digitalWrite(CS, HIGH);
  if (sign) {adcvalue = adcvalue-4096;  }
  return adcvalue;
}
void setup() {  //...................................setup
  mcp3304_init_bb();
  Serial.begin(115200);
  while (!Serial) { delay(10);}
}
void loop() {  //....................................loop
  int val=mcp3304_read_bb(0);
  Serial.print("CH0 = "); Serial.println(val);
  delay(1000);
}
```

In the sketch we first give meaningful names to the pins we use on the NodeMCU, and declare the function mcp3304_init_bb() where we collect the code needed to initialize the bit-banged SPI communication. We declare the respective pins to be INPUT or OUTPUT and initialize their state with a call to digitalWrite(). The mcp3304_read_bb() function takes the channel number as input and returns the digitized word. In this routine we assume that we use two input pins of the MCP3304 in differential mode. Inside the routine we first declare some variables and initialize the commandbits to start with a most significant bit 7 that has the value 1. The next bit 6 is 0, indicating differential mode, and then we add the channel information as bits 5 and 4 and set bit 3 to zero, followed by three more zeros, following the description from the datasheet of how to configure the ADC. Now we can pull the CS line LOW to indicate that a transaction is about to start, and then clock the five configuration bits plus two extra bits to allow some time for the conversion by setting the MOSI pin to the respective value and toggling the CLK pin. Once we complete this we can read the sign bit from the MISO line and toggle the CLK pin. Then we repeat to read MISO and toggle CLK 12 times to fill the appropriate bits of the integer adcvalue, and finally conclude the transaction by pulling the CS pin HIGH. Before returning the adcvalue we ensure that the sign bit is properly folded into the reading. In the setup() function we only initialize the serial line and the MCP3304, and in the loop() function we read channel 0 from the ADC and display its value on the serial line. The latter is repeated once a second.

Instead of bit banging the pins, we can also utilize the SPI library that comes with the Arduino IDE. The sketch, based on the <SPI.h> library, but otherwise equivalent to the previous one, is the following.

```
// MCP3304, V. Ziemann, 170726
#include <SPI.h>
#define CS 15  // D8
```

```
int mcp3304_read_adc(int channel) {  //...0 to 3.......read_adc
  int adcvalue=0, b1=0, hi=0, lo=0, sign=0;;
  digitalWrite (CS, LOW);
  byte commandbits = B00001000; // Startbit+(diff=0)
  commandbits |= channel & 0x03;
  SPI.transfer(commandbits);
  b1 = SPI.transfer(0x00);      // always D0=0
  sign = b1 & B00010000;
  hi = b1 & B00001111;
  lo = SPI.transfer(0x00);      // input is don't care
  digitalWrite(CS, HIGH);
  adcvalue = (hi << 8) + lo;
  if (sign) {adcvalue = adcvalue-4096;}
  return adcvalue;
}
void setup() {//..............................setup
  pinMode(CS,OUTPUT);
  SPI.begin();
  SPI.setFrequency(2100000);
  SPI.setBitOrder(MSBFIRST);
  SPI.setDataMode(SPI_MODE0);
  Serial.begin(115200);
  while (!Serial) {delay(10);}
}
void loop() {    //.........................loop
  int val=mcp3304_read_adc(0);
  Serial.print("CH0 = "); Serial.println(val);
  delay(1000);
}
```

Here we first include the `<SPI.h>` functionality and define the CS pin before defining the mcp3304_read_adc function to read a `channel` from the ADC. In this function we first declare a number of variables and pull the CS pin LOW in order to start the transaction. Then we build the `commandbits`; this time they are aligned such that the start bit is bit 3 and the channel information is stored in bits 1 and 0, and we use the SPI.transfer function to send it to the ADC. Note that the first bit recognized by the ADC is the first non-zero bit, which is the start bit 3. We then send 0x00 in a second call to SPI.transfer and receive the readings from the MISO pin in b1. Since we have a number of idle clock toggles, the sign bit is bit 4 and the next four bits are the four most significant bits of the ADC reading, which we store in the variable hi. The next call to SPI.transfer() returns the subsequent bits, which are the lower eight bits from the MISO pin, and we store them in the variable lo. After pulling the CS pin HIGH to conclude the transaction, we build the ADC word adcvalue and add the sign information. Finally we return the reading to the calling program. In the setup() function we initialize the SPI functionality by calling the SPI.begin() function, setting the clock frequency, byte order, and MODE, and then initialize the serial line. The loop() function is a straight copy of the previous example, where we only read channel 0 and print the value to the serial line.

This external ADC greatly expands the analog input capabilities of the NodeMCU, which only has a single ADC channel with unipolar 10-bit resolution on board. Now we have four extra channels at 12-bit resolution, including an additional polarity bit. We note in passing that the MCP3304 has a close relative, the MCP3208, which has eight unipolar channels and can be programmed in a similar way. Consult the datasheets for the details. Sensors that support SPI communication can be controlled in a similar way to the ADCs, either by bit banging the respective pins or by using the SPI.transfer() function. Again, reading the detailed specifications in the datasheets is mandatory.

So far, we discussed the standard communication channels, RS-232, I2C, and SPI, but there are several others that we briefly address in the next section.

4.4.5 Other protocols

The DHT11 and DHT22/AM2302 humidity sensors use a nonstandard communication protocol, but this is reasonably well explained in the datasheet. The protocol is based on using a single IO-pin, which requires a pull-up resistor of about 4.7 kΩ. On breakout boards, the resistor is likely already mounted. Initially the microcontroller configures the pin as OUTPUT and pulls it to LOW for at least 18 ms, before releasing control to the sensor by reconfiguring the pin to INPUT. The sensor then acknowledges the transaction by sending a pulse of $50\,\mu s$ LOW followed by $50\,\mu s$ HIGH before clocking 40 data bits to the pin using the following convention. A zerobit is defined by a LOW state for $50\,\mu s$ and a subsequent HIGH state for about $27\,\mu s$. A onebit is defined by a $50\,\mu s$ LOW state followed by a $70\,\mu s$ long HIGH state. The 40 data bits contain the information about the humidity and the temperature. In the following code we implement this protocol.

```
// DHT11, V. Ziemann, 170804
#define DHT 2    // sensor pin
float read_dht11(float *T) {
  bool p[500]= { 0 };
  uint8_t data[5]={0},checksum;
  int ic=0,goes_up=0;
  pinMode(DHT,OUTPUT);
  digitalWrite(DHT,LOW);    // 20 ms low pulse
  delay(20);
  digitalWrite(DHT,HIGH);
  noInterrupts();  // interrupts off for accurate timing
  delayMicroseconds(20);  // make sure we start at low
  pinMode(DHT,INPUT);
  for (int i=0;i<500;i++) {  // read 5 ms worth of data
    p[i]=digitalRead(DHT);
    delayMicroseconds(10);    // one every 10 us
  }
  interrupts();     // interrupts on again
  while (p[ic++] == 0); // next HIGH, acknowledge bit
  while (p[ic++] == 1); // next LOW, first data bit
  for (int i=0;i<5;i++) {        // loop over the
    for (int j=7;j>=0;j--) {      // 40 data bits
      while (p[ic++] == 0) {goes_up=ic;}
      while (p[ic++] == 1) ;
      (ic-goes_up > 4) ? data[i] |= (1<<j) : 0;
    }
  }
  checksum=((data[0]+data[1]+data[2]+data[3]) & 0xFF);
  if (checksum==data[4]) {
    *T=(float) data[2];        // temperature
```

```
    return (float) data[0];   // humidity
  }
  *T=-100;    // only get here if bad reading
  return -1;
}
void setup() {  //...........................setup
  Serial.begin(9600);
  while (!Serial) {;}
  pinMode(DHT,INPUT);
  digitalWrite(DHT,HIGH);
  delay(2000);       // wait for DHT to wake up
}
void loop() {  //...............................loop
  float temperature,humidity;
  humidity=read_dht11(&temperature);
  Serial.print(humidity); Serial.print("\t");
  Serial.println(temperature);
  delay(2000);
}
```

In this sketch we first define the pin to which the DHT11 is connected and the function read_dht11() that encapsulates the communication with the sensor. There we first declare a number of variables and then directly progress to configuring the pin to which the sensor is connected as OUTPUT, and produce a 20-ms-long LOW pulse before pulling the pin HIGH and reconfiguring it as INPUT. Since the timing of the response from the DHT11 is critical, we briefly turn interrupts off, which disables background processing of serial or WLAN activities. Then we read the pin every $10\,\mu s$ 500 times for a total duration of 5 ms, which is sufficient to read 40 bits, each having a maximum duration of $120\,\mu s$, into the boolean array p[]. Once the acquisition is complete, we turn interrupts on again. Now that we have the entire data set in memory we can decode it. The variable ic steps through the received array p[] and searches for LOW to HIGH transitions and HIGH to LOW transitions. First we take care of the acknowledge pulse before looping over 5 bytes of 8 bits each. For each bit we measure the separation from the time the signal goes up until it goes down again. If the difference of the indices is more than four, which means that the temporal separation is longer than $40\,\mu s$, we set the corresponding bit in the respective data byte. Once 40 bits are decoded we calculate the checksum, which is the sum of the first four data bytes, truncated to eight bits. If this equals the fifth byte, the transmission is correct and the function returns data[0] as the humidity and data[2] as the temperature, as specified in the datasheet. If the checksum is incorrect, the function returns values that are obviously wrong to allow checking in the calling program. Note that the DHT22/AM2302 sensors have a slightly different interpretation of the data bytes. For these sensors, the humidity and temperature are instead given by

```
humidity=(0.1*(256*data[0]+data[1]));
*T=(float) (0.1*(256*data[2]+data[3]));
```

In our version of the read_dht11() function, we first read the entire time series of 500 points with a spacing of $10\,\mu s$ before decoding. This wastes a little memory, but is more robust than waiting for transitions of the pin while decoding, as is done in other implementations of the DHT libraries. Note also that sampling a pin in this way constitutes a simple logic analyzer. This may come handy in other contexts as well.

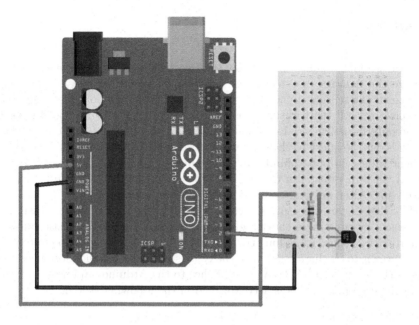

Figure 4.11 Connecting the DS18b20 sensor to the Arduino UNO†.

Another data bus, similar to the one for the DHT sensors, but with additional error handling, is the Dallas 1-wire protocol that is used by the popular DS18b20 temperature sensor, but also by other devices, such as humidity sensors, memory circuits, or autonomous data loggers. We will, however, focus on the DS18b20 and connect it to the Arduino using the schematics shown in Figure 4.11 where we connect ground and supply voltage and the single data pin to pin D2 on the UNO. We also added a 4.7 kΩ pull-up resistor from the data pin to the supply voltage. This is needed for a bare sensor only. If we use a breakout-board version of the DS18b20, the pull-up can likely be omitted, because it is already mounted on the breakout board. For interfacing to the Arduino, we use ready-made libraries that encapsulate the low-level interaction. Before using the libraries, we need to install the *OneWire* and *DallasTemperature* libraries. An easy way is to use the library manager of the Arduino IDE, which is accessible from the *Sketch→Include Library→Manage Libraries*, where we enter the keywords *OneWire* and *DallasTemperature*. This displays a list of supported libraries that can be installed directly from the menu. After installation, we enter the following code, which shows a simple example.

```
// DS18b20 1-wire temperature sensor, V. Ziemann, 170120
#include <OneWire.h>
#include <DallasTemperature.h>
OneWire oneWire(2); //use pin D2
DallasTemperature sensors(&oneWire);
void setup() {
  Serial.begin(9600); while (!Serial) {;}
  sensors.begin();
}
void loop() {
  sensors.requestTemperatures();
  float temp=sensors.getTempCByIndex(0);
  Serial.println(temp);
```

```
    delay(1000);
}
```

In the sketch, we first include support for the `Onewire` and `DallasTemperature` devices, and initialize the `OneWire` bus using pin number 2, followed by the initialization of the `DallasTemperature` sensors, to use the bus initialized in the previous statement. Note that we can connect several temperature sensors in parallel on the same bus. In the `setup()` function we only initialize serial communication and the sensor. The code in the `loop()` function sends out the query to the connected sensors, which causes them to send back their measurement values. In our example we only connect a single DS18b20 such that we only can read out the temperature in Celsius from the device number zero, and print the result to the serial line. After waiting a second we repeat the procedure.

The HC-SR04 distance sensor, shown in Figure 2.35, determines the distance to the nearest obstacle by emitting a short high-frequency (non-audible) acoustic pulse when its trigger pin receives a short 10-μs-long trigger pulse. Then it pulls the echo pin to low voltage and only returns it to the high state once the echo is received. We therefore need to connect the sensor with four wires (GND, 5V, Trig, Echo) to the Arduino, and need to measure the duration of the echo pulse. This is accomplished in the following sketch.

```
// HC-SR04 distance logger, V. Ziemann, 161204
const int trig=2, echo=3;
void setup() {
  Serial.begin(9600); while (!Serial) {;}
  pinMode(trig,OUTPUT);
  pinMode(echo,INPUT);
}
void loop() {
  float duration,distance;
  digitalWrite(trig,LOW); // make a 10 us trigger pulse
  delayMicroseconds(2);
  digitalWrite(trig,HIGH);
  delayMicroseconds(10);
  digitalWrite(trig,LOW);
  duration=pulseIn(echo,HIGH);      // wait for echo
  distance=100*duration*340e-6/2;   // in cm
  Serial.println(distance);
  delay(1000);
}
```

In the sketch, we first declare the constants with the pins used in this sketch. The trigger pin of the sensor is connected to pin 2 on the Arduino and the echo pin on the sensor to pin 3. In the `setup()` function, we declare the serial line and whether the respective pins are used as input or output. In the `loop()` function, we first declare the variables for distance and duration and then create a 10-μs-long trigger pulse by writing HIGH and LOW to the trigger pin, and wait a short time in between. Once the trigger pulse is dispatched, we use the `pulseIn()` function to wait for the echo pin to return to the HIGH state. The `pulseIn()` function returns the elapsed time in microseconds such that 1000 μs correspond to a distance of about 0.17 m, or 17 cm if we assume a value for speed of sound of 340 m/s and note that the sound has to go back and forth from the sensor to the first obstacle. Then the measured value is sent over the serial line to the host computer, and the whole process is repeated after some delay. We note in passing that we can use the `pulseIn()` function to measure the duration of short pulses in general.

Output pins of a rotary encoder carry a sine-like and a cosine-like signal, respectively. We can determine whether the shaft is turned clockwise or counterclockwise depending on whether the two pins are equal or not while one of them changes on a falling edge, going from a HIGH to a LOW state. The following code implements this method using interrupts. Using interrupts has the advantage that the state of the pins does not have to be measured continuously and compared, but an interrupt_handler is registered with a certain action, in our case on a falling edge on the pin connected to the interrupt. The Arduino has two pins, numbers 2 and 3, that have the interrupt functionality. But let us look at the sketch first.

```
// Rotary encoder, V. Ziemann, 161205
const int pinA=2,pinB=4;
volatile int pos=0;
void setup() {
  Serial.begin(9600); while (!Serial) {;}
  pinMode(pinA,INPUT_PULLUP);
  pinMode(pinB,INPUT_PULLUP);
  attachInterrupt(0,interrupt_handler,FALLING);//0=pin2,1=pin3
}
void loop() {
  Serial.println(pos);
  delay(1000);
}
void interrupt_handler() {
  if (digitalRead(pinA)==digitalRead(pinB)){pos++;}else{pos--;}
}
```

The sketch starts by declaring the used pins and a variable pos that contains the encoder position. The variable is declared volatile, which instructs the compiler that it can change asynchronously within the interrupt routine, and prevents the compiler from optimizing it away, because it does not change in the main program. In the setup() function we declare the serial line parameters and the input pins with pull-up resistors enabled. One of the pins *must* be pin 2 or 3, those with interrupt capability. The attachInterrupt() function is used to connect a so-called callback function, here interrupt_handler. It is defined at the end of the sketch. In our sketch we chose interrupt 0 (connected to pin 2) to trigger the callback function interrupt_handler. We also specify that the callback function is called on a FALLING edge on the interrupt pin. Instead of hard coding the value 0 as the source of the interrupt, we can use the service function digitalPinToInterrupt() that takes the pin number, 2 in our case, as argument, such that the line reads

```
attachInterrupt(digitalPinToInterrupt(2),interrupt_handler,FALLING);
```

instead. Using digitalPinToInterrupt() is the preferred way, because it relieves the user of remembering which interrupt number is connected to which pin. The loop() function is very simple and only writes the position of the encoder shaft to the serial line once a second. Finally, we define the callback function interrupt_handler, which increments the pos variable by one step if the two encoder pins are equal and decrements otherwise. In this way, the sense of rotation of the shaft is taken care of. We need to point out that we had to use separate external pull-up resistors ($10\,\mathrm{k\Omega}$) to guarantee stable operation, and in the sketch we may encounter a so-called race condition when the variable pos changes while it is transmitted on the serial line.

After the discussion of interfacing the sensors, we now move on to investigate how to work with actuators, such as switches, motors, or analog voltages.

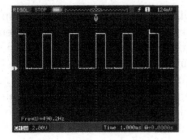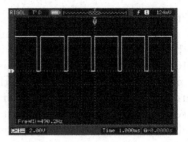

Figure 4.12 Oscilloscope trace of a pulse-width modulated signal on pin D9 of an Arduino UNO with values of 88 (left) and 220 (right).

4.5 INTERFACING ACTUATORS

In this section we discuss how to interface the actuators discussed in Chapter 3 to the Arduino, and start with the simplest actuator, the LED.

4.5.1 Switching devices

In Section 4.3 we used the `digitalWrite()` function to turn an LED on and off. If we connect a transistor, Darlington driver, or H bridge to an output pin, we can use the same function to handle larger current than the 20 mA the Arduino can drive. Consult the circuits shown in Section 3.1.2 for details.

Dynamically adjusting the brightness of an LED is achieved by pulse-width modulation, and that feature is available on pins D3, D5, D6, D9, D10, and D11 on the UNO, which indicated by a tilde written next to the pin on the board. We use the `analogWrite(pin,value)` function that takes the pin number and a value between 0 and 255 (0 to 1023 on the ESP8266) to adjust the duty cycle of the pulse-width modulated signal from completely off to completely on. The switching frequency is between 500 and 1000 Hz, and in Figure 4.12, we show an oscilloscope trace of the result of `analogWrite()` to pin D9 of a UNO with values 88 and 220, respectively. The frequency of the signal is about 490 Hz, and we see the length of the signal differing from about 1/3 of the time at 5 V on the left to 5 V almost all the time on the right.

In order to be able to manipulate actuators in the same way as the sensors, we use the same query-response protocol to communicate the required actions to the microcontroller. We want to use the convention that `DWx 0` turns digital pin number x off and `DWx 1` turns it on, and that sending `AWx nnn` sets pulse-width modulation on pin number x to value `nnn`. The code that achieves this is the following:

```
// Switching_and_pwm, V. Ziemann, 170614
char line[30];
void setup() {
  pinMode(2,OUTPUT);
  Serial.begin (9600);
  while (!Serial) {;}
}
void loop() {
  if (Serial.available()) {
    Serial.readStringUntil('\n').toCharArray(line,30);
    if (strstr(line,"DW2 ")==line) {
```

```
        int val=(int)atof(&line[4]);
        digitalWrite(2,val);
    } else if (strstr(line,"AW9 ")==line) {
        int val=(int)atof(&line[4]);
        analogWrite(9,val);
    } else {
        Serial.println("unknown");
    }
  }
}
```

where we first declare the variable `line` that contains the characters received on the serial line. In the `setup()` function, we only declare pin 2 as output, and initialize the serial communication, exactly as before. In the `loop()` function, we use the previously seen construction to parse the message received on the serial line and perform some action, depending on the command received. If the command starts with `DW2` we read the characters following the first 4 into the variable `val` and pass `val` to the `digitalWrite()` function. Here we employ the same method of using the `atof()` function to decode the rest of the character array `line[]` that we first used in the sketch to interface the MCP23017 IO extender. We immediately encounter the `atof` construction again when parsing the value received after `AW9`. We then use it to set pin D9 to the desired value with the `analogWrite()` function. In this way we have a simple mechanism to instruct the Arduino to turn pins on and off and adjust those that have pulse width capability to the desired duty factor.

We note that the code above does not contain graceful handling of errors. For example, passing any value `val` received from the Serial line to the `digitalWrite()` function when handling the `DW2` command can result in weird behavior if `val` is not `0` or `1`. Even a negative number could be written to the pin, which is meaningless. Replacing that line by the following construction will only write meaningful values to the pin. Implicitly, we use the convention that `0` turns the pin off, and anything else turns it on.

```
// digitalWrite(2,(int)val);
if ((int)val == 0) {
  digitalWrite(2,LOW);
} else {
  digitalWrite(2,HIGH);
}
```

Of course, other ways of catching errors can be implemented, but that goes beyond the scope of our presentation.

Testing the above program with a UNO is quickly done. In Figure 4.13 we show an implementation on a small breadboard. Two cables connect the power lines to the circuit, where the ground signal connects to the emitter of the NPN transistor. Here we use a BC547, but any small-signal NPN transistor will work. The collector of the transistor is directly connected to the cathode of the LED, and the anode of the LED is connected via a $220\,\Omega$ to the positive supply voltage. The base of the transistor is connected via a $1\,k\Omega$ resistor to the controlling pin, here pin D9, on the Arduino.

This example covers the basic functionality of turning on and off, as well as controlling the power delivered to a load, the LED. In the next example we use very similar circuitry to control speed and direction of motors.

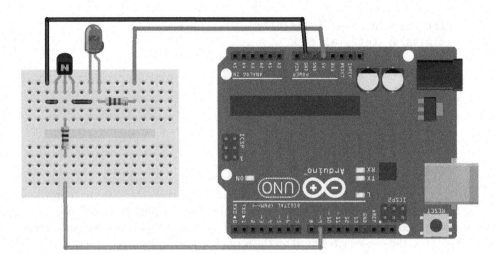

Figure 4.13 Using pulse-width modulation and a transistor to adjust the brightness of an LED[†].

4.5.2 DC motors

If we only want to control the speed of a very small motor, we can replace the LED and the 220 Ω resistor in the previous example by the motor. We also need to add flyback diodes to prevent the back-emf from damaging the transistor. For slightly larger motors, we need to use a transistor with a higher power rating, such as a TIP-120 Darlington power transistor. The TIP-120 has a flyback diode from emitter to collector already built in. Note that the performance of larger transistors normally degrades at higher frequency. Inspection of the datasheet, however, shows that this only affects frequencies well above 10 kHz.

In Figure 4.14 we show the setup with a UNO controlling a small motor. The terminals of the transistor are base, collector, and emitter, from left to right, and the motor is connected between the collector and the supply voltage. The emitter is directly connected to ground, and we include an external flyback diode (horizontally mounted) between the motor leads, with the cathode pointing towards the right. The vertically mounted diode illustrates the connection of the built-in diode. We control the speed of the motor by pulse width modulating the base of the transistor that is connected to pin D9 on the UNO via a 1 kΩ resistor.

Controlling the speed is useful, but often we also need to control the direction of rotation; for example, to drive a vehicle not only forward but also backward. The standard way to accomplish this feature is by using an H bridge, as discussed in Chapter 3. Here we use an L293D H bridge driver, which is called a *quadruple half-bridge driver*, a term that becomes clear by considering the pin layout of the 16-pin chip, shown in Figure 4.15. There are ground pins and two supply voltages, one for logic levels and one to supply power to the motor. Then there are four (quadruple) inputs with corresponding outputs. The circuit works in such a way as to translate the logic level on the input pins to the corresponding output pins, which in turn deliver the voltage level from the motor power supply to the motor. Thus, setting input 1 to HIGH and input 2 to LOW connects one motor lead to the motor supply voltage and the other to ground, which causes the motor to turn in one direction. Setting input 1 to LOW and input 2 to HIGH will cause the converse, and the motor will turn in the other direction. Two inputs and outputs therefore provide the functionality of a single H bridge. Since there are four inputs and outputs, we can use the L293D to implement two H

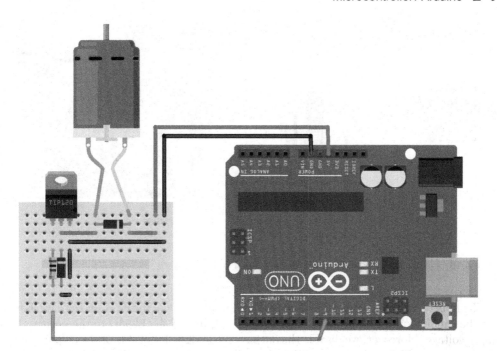

Figure 4.14 Controlling the speed of a motor with pulse width modulating the base of a TIP-120 Darlington transistor. See the text for a discussion of the diodes[†].

bridges. Moreover, pin 1 on the L293D can be used to enable the outputs of input 1 and 2. Applying a pulse-width modulated signal to this pin will therefore work as speed control for the motor. Pin 9 provides the same functionality for inputs 3 and 4. Finally, we remark that the character D in L293D indicates that flyback protection diodes are already built into the integrated circuit.

With the theoretical background covered we are ready to build a motor controller with an L293D on a breadboard, and control it with an Arduino UNO. We show the circuit in Figure 4.16. The DC motor is shown on the left and the L293D is the only component on the small breadboard. The orientation of the chip is the same as in Figure 4.15, making the functionality easy to understand. The pin numbering goes from top left (pin 1) to bottom left (pin 8), continuing via bottom right (pin 9) to the top right (pin 16), which is the standard for most integrated circuits.

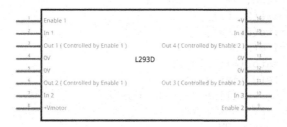

Figure 4.15 The pin assignment of the L293D H-bridge driver[†].

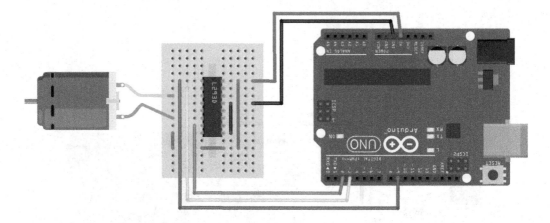

Figure 4.16 Controlling speed and direction of a DC motor[†].

The upper two wires connect ground and the supply voltage to the chip. Here we assume that the supply voltage is the same, and we connect pin 8 of the L293D to the positive logic supply voltage. Moreover, we disable the output on the right-hand side of the chip by pulling pin 9 to ground. Two wires connect digital outputs D2 and D3 on the Arduino to the input of two half-bridges on pins 2 and 7 on the L293D. The corresponding outputs from pins 3 and 6 are connected to the motor cables. Finally, pin 1 on the L293D is connected to the pulse-width modulated output pin D9 on the UNO.

We want control the motor from the Arduino by sending commands via the serial line according to the protocol that FW **nnn** sets the speed in the forward direction and BW **nnn** sets the speed in the reverse direction. Anything else will stop the motors. This is easy to do with the following sketch.

```
// H bridge DC motor Controller, V. Ziemann, 170614
char line[30];
void setup() {
  pinMode(2,OUTPUT);
  pinMode(3,OUTPUT);
  Serial.begin (9600);
  while (!Serial) {;}
}
void loop() {
  if (Serial.available()) {
   Serial.readStringUntil('\n').toCharArray(line,30);
   if (strstr(line,"FW ")==line) {
     digitalWrite(2,LOW);
     digitalWrite(3,LOW);
     digitalWrite(3,HIGH);
     float val=atof(&line[3]);
     analogWrite(9,(int)val);
   } else if (strstr(line,"BW ")==line) {
     digitalWrite(2,LOW);
     digitalWrite(3,LOW);
     digitalWrite(2,HIGH);
     float val=atof(&line[3]);
```

```
      analogWrite(9,(int)val);
   } else {    // STOP in all other cases
      digitalWrite(2,LOW);
      digitalWrite(3,LOW);
      analogWrite(9,(int)0);
   }
 }
}
```

The `setup()` function only declares the pins used to control the H bridge as outputs and initializes the serial line. In the `loop()` function we employ the standard query-response mechanism. A command on the serial line that starts with FW first turns the motor off as a precaution against having both outputs positive, which might short the motor supply voltage. Then pin D3 is pulled high, resulting in the motor turning one way. Finally we parse the rest of the command line and write the value to set the motor speed by pulse-width modulating the enable pin of the motor-driver. Any command received, apart from FW n and BW n, will disable both outputs and sets the speed to zero. Of course, we can easily implement other commands, such as STOP, that will cause the motor to stop.

Even more elaborate commands can be programmed into the controller, such as turning one way at some time for a given time, then stopping and turning backward for another time. The following code fragment may serve as an example.

```
} else if (strstr(line,"BACKANDFORTH")==line) {
  digitalWrite(2,LOW);  // turn all off
  digitalWrite(3,LOW);
  digitalWrite(2,HIGH); // chose one direction
  analogWrite(9,255);   // full speed
  delay(500);
  digitalWrite(2,LOW);  // stop
  delay(1000);
  digitalWrite(3,HIGH); // other direction
  delay(500);
  digtalwrite(3,LOW);   // stop
  analogWrite(9,0);     // stop PWM
} ...
```

Other examples are to move for a short time, do some measurements, using code examples from the previous section, and continue doing the same for a number of steps. Yet another example is moving until a limit switch is engaged, which will cause the motors to turn off, maybe do some measurement, and later return to the parking position at a second limit switch.

Or, what about continuously reading a sensor and stopping once a certain value is achieved? This is actually what a model-servo does. So, let us look at it more closely.

4.5.3 Servos

As we discussed in Chapter 3, servos are used to carefully change the position or orientation of some device, such as the rudder of a boat or the steering system of a radio-controlled car. Servos only need three wires to be connected: ground and the supply voltage as well as one wire that carries the control information according to the timing shown in Figure 3.9. In Figure 4.17 we show how to connect a small servo to an Arduino UNO. If the servo requires more current or other voltages than the UNO provides, one can use a suitable external power

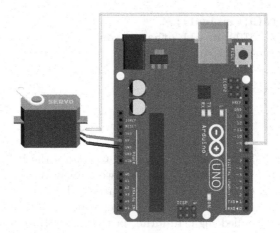

Figure 4.17 Connecting a model-servo to the Arduino UNO[†].

supply. The ground cables of the external supply and the UNO need to be connected, and one wire from the servo, which is often red, needs to be connected to the positive terminal of the external supply instead of the 5 V power pin on the Arduino. Controlling a servo from the Arduino is rather simple, and we use the following code to do that.

```
// Servo controller, V. Ziemann, 170614
#include <Servo.h>
Servo myServo;
char line[30];
void setup() {
  myServo.attach(9);    // pin D9
  Serial.begin (9600);
  while (!Serial) {;}
}
void loop() {
  if (Serial.available()) {
    Serial.readStringUntil('\n').toCharArray(line,30);
    if (strstr(line,"SERVO ")==line) {
      float val=atof(&line[6]);
      myServo.write((int) val);  // 0 to 180
    }
  }
}
```

The code follows the normal template with opening serial line in the setup() function and the query-response construction in the loop() function. Servo-specific means the inclusion of the <Servo.h> file and the declaration of a Servo object that we call myServo. In the setup() function, we attach the servo functionality to pin D9 on the Arduino and write new values to the servo with the call to myServo.attach(). The latter function takes a value between 0 and 180 as argument and moves the arm of the servo to the desired position. Note that <Servo.h> is part of the standard Arduino distribution. Note also that we can use any pin on the Arduino to control servos, even multiple servos, but that using the servo functionality disables the pulse-width modulation feature of the analogWrite() function on pin D9 and D10 because it uses the same hardware timer.

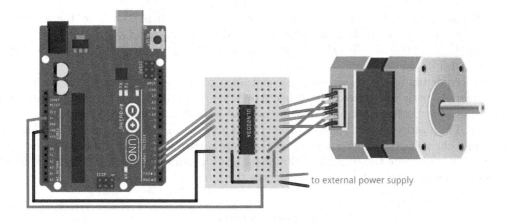

Figure 4.18 Connecting a unipolar stepper motor to the UNO with a ULN2003 Darlington driver[†].

Using servos in more advanced ways, such as scanning, is also easy to implement. The following code fragment scans back and forth through the entire range

```
} else  if (strstr(line,"SCAN")==line) {
  for (int pos=0;pos<180;pos+=1) {
    myServo.write(pos); delay(10); // and do a measurement
  }
  for (int pos=180;pos>=0;pos-=1) {
    myServo.write(pos); delay(10);
  }
```

and it is easy to envision including some measurement inside the loops that perform the scanning. Imagine an HR-SR04 distance sensor attached to the servo. Scanning in a semi-circular motion and continuously measuring the distance to the closest object mimics the functionality of radar by using a sonar-like device instead. This could serve as a collision detection system for moving vehicles.

The last type of motors that we consider are stepper motors, and they come next.

4.5.4   Stepper motors

As we discussed before, stepper motors can be very accurately controlled by a given pulse sequence that turns the shaft by a desired number of discrete steps. The coils inside the motors need to be excited in certain patterns, either by a unipolar or bipolar power supply. We start with the unipolar type.

In Figure 4.18 we show the connection of a unipolar stepper motor with the help of a ULN2003 Darlington driver that is mounted on the small breadboard between the Arduino UNO and the motor. The inner connection of the ULN2003 is very simple. Pins 1–7 on the left-hand side are the bases of seven Darlington transistors with resistors built into the chip. The facing pins on the right-hand side are the collectors of the respective transistors. All emitters are wired to pin 8 on the lower left-hand side of the chip, which is also ground. Pin 9 on the lower right-hand side is the terminal for the motor power supply and is often connected to an external power supply that provides adequate current and voltage to power the motor. In the figure we assume that the stepper motor is small, and we connect pin 9 to the 5 V pin on the Arduino with one wire connected to the left-hand side on the Arduino. Ground is connected with the other wire connected to the left-hand side. The bases of the Darlington transistors are directly connected to pins D2 to D5 of the Arduino and need to be controlled by the program.

The center tap of the motor coils are connected to the positive terminal of the motor power supply with two wires. The other four wires are hooked up to the collectors of the upper four Darlington transistors in the ULN2003. Since the center tap of a coil is connected to positive supply voltage, placing a positive voltage on the base of a Darlington will cause the collector–emitter link to conduct and a current to flow through the coil. In this way we can excite the coils in a suitable pattern to rotate the shaft either clockwise or counterclockwise. The sketch that implements this functionality using the query-response protocol is the following.

```
// Stepper controller, V. Ziemann, 170616
#include <Stepper.h>
Stepper myStepper(200, 2, 4, 3, 5);
char line[30];
void setup() {
  myStepper.setSpeed(60);
  Serial.begin (9600);
  while (!Serial) {;}
}
void loop() {
  if (Serial.available()) {
    Serial.readStringUntil('\n').toCharArray(line,30);
    if (strstr(line,"MOVE ")==line) {
      int val=(int)atof(&line[5]);
      myStepper.step(val);
    } else {
      Serial.println("unknown");
    }
  }
}
```

Here we first include the `<Stepper.h>` library that comes with the Arduino IDE and declare a `Stepper` object called `myStepper`. The first argument is the number of steps per revolution and the next four integers are the pins to which the four coils are attached. This may differ from motor to motor, and normally this information is provided in the datasheet. Otherwise, a little experimenting with swapping numbers may lead to a moving motor. In the `setup()` function, we use the `setSpeed()` method to declare how fast we want the motor to rotate, and we also initialize the serial line. In the `loop()` function we employ our standard query-response protocol and make the motor respond to the command `MOVE`, and read the characters behind the command as the number of steps to move. This number can be either positive or negative, depending on the desired direction to rotate.

The advantage of using a bipolar stepper motor is increased torque for the same supply voltage, because there is one coil pulling the permanent magnet of the rotor and the other, which has opposite polarity, is pushing. This makes it worthwhile to describe how to drive a bipolar stepper motor.

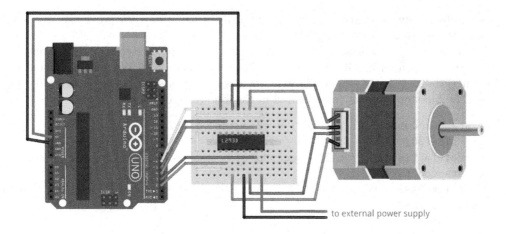

to external power supply

Figure 4.19 Connecting a bipolar stepper motor to the UNO with an L293D H-bridge driver[†].

We connect a bipolar stepper motor in the way shown in Figure 4.19. Here we use an L293D H-bridge driver, the same integrated circuit we use to provide direction control for DC motors. Based on the pin-description for the circuit shown in Figure 4.15, we immediately see that two supply wires from the left-hand side of the UNO connect to the pins for logic supply voltage and ground, and the enable pins are connected with the light-colored wire to pin D6 on the UNO. The wires connected to pins D2, D3, D4, and D5 on the Arduino are connected to the inputs of the four half-bridge drivers on the L293D. The four leads of the bipolar stepper motor are wired to the four output terminals of the corresponding drivers. In this circuit the terminal for the positive motor supply voltage on pin 8 of the L293D is connected to an external power supply. Since internally all four ground pins of the chip are connected, we can add a wire to connect the ground terminal on the other side of the L293D of the motor to the ground terminal of the power supply.

It is remarkable that we can use the same code we used for the unipolar stepper motor for the bipolar motor as well. The same patterns, sent to the driver input pins, causes the bipolar motor to operate in full step mode. We can therefore use the same Arduino sketch as before to operate the motor, with the small reservation that we may need to swap the assignment of pins in the declaration of `myStepper`.

We need to point out that at least one coil is always excited, and that causes the motor to get warm over time, even when it is not moving. This problem can be alleviated by adding a command to turn off all coils, albeit at the expense of losing some accuracy, because the holding torque from the excited coil is absent. If we can accept the small loss in precision, we can use the pin of the Arduino connected to the yellow wire to turn on and off the outputs of the driver chip. It must, however, be pulled high with `digitalWrite(6,HIGH)` for normal operation.

Using the built-in libraries to control stepper motors is the quickest way to get started, and using well-tested libraries normally leads to robust code, but, at any rate, it is rather instructive to code the step sequence for the stepper motors *by hand* so to speak. And that is what we do in the following example, which implements the same functionality as the previous one, but provides insight into the inner workings of the stepper library. Moreover, the built-in library only implements full-step operation of the motor, where the permanent magnet is directly facing a coil, such as permanent magnet 1 in Figure 3.10, which directly faces coil A. However, by exciting coil pairs simultaneously, we can also fix the permanent magnet halfway between the coils, and move the shaft by half a step. We previously discussed this half-step mode of operation in Section 3.2.3.

For the following exercises, we assume that we have a bipolar stepper motor connected to the Arduino with an L293D H bridge, as shown in Figure 4.19. We first implement the same functionality that the stepper library provides, and later show how to easily add other modes. The full-step mode is implemented in the following code:

```
// Stepper controller by hand, V. Ziemann, 170624
char line[30];
int settle_time=2; // milli-seconds
int stepcounter=0;
const int PA=2;
const int PB=3;
const int PC=4;
const int PD=5;
const int ENABLE=6;
void set_coils(int istep) {  //........full-step mode
  bool patA[]={1,1,0,0};  // or {1,0,0,0}
  int pat_length=4;
  int ii;
  istep=istep % pat_length;
  if (istep < 0) istep+=pat_length;
  digitalWrite(PA,patA[istep]);
  ii=(istep+2) % pat_length;
  digitalWrite(PB,patA[ii]);
  ii=(istep+3) % pat_length;
  digitalWrite(PC,patA[ii]);
  ii=(istep+1) % pat_length;
  digitalWrite(PD,patA[ii]);
  delay(settle_time);
}
void setup() {    //.........................setup
  Serial.begin (9600);
  while (!Serial) {;}
  pinMode(PA,OUTPUT);
  pinMode(PB,OUTPUT);
  pinMode(PC,OUTPUT);
  pinMode(PD,OUTPUT);
  pinMode(ENABLE,OUTPUT);
  digitalWrite(ENABLE,HIGH);
}
void loop() {  //...........................loop
  if (Serial.available()) {
    Serial.readStringUntil('\n').toCharArray(line,30);
    if (strstr(line,"MOVE ")==line) {
      int steps=(int)atof(&line[5]);
      if (steps > 0) {
        for (int i=0;i<steps;i++) set_coils(stepcounter++);
      } else {
        for (int i=0;i<abs(steps);i++) set_coils(stepcounter--);
      }
```

```
      } else if (strstr(line,"STEPS?")==line) {
        Serial.print("STEPS "); Serial.println(stepcounter);
      } else if (strstr(line,"STEPS ")==line) {
        stepcounter=(int)atof(&line[6]);
      } else if (strstr(line,"DISABLE")==line) {
        digitalWrite(ENABLE,LOW);
      } else if (strstr(line,"ENABLE")==line) {
        digitalWrite(ENABLE,HIGH);
      }
   }
}
```

First we declare some global variables, such as settle_time, the time to wait between changing coil excitations, and the stepcounter, which keeps track of the distance traveled. PA through PD are the pins to which the coil terminals are connected, and ENABLE connects to the enable pin of the H-bridge driver and can be used to de-excite all coils to prevent them from overheating. Then the set_coils() function, which will excite the coils in the correct pattern, is declared. Inside the function, we first define the excitation pattern patA for coil A. In this case it is 1100, the pattern in which two coils are always excited, and and provide the larger torque. Alternatively we can also declare the pattern 1000 which would result in the single-coil excitation pattern. See Section 3.2.3 for a discussion. The variable pat_length is the number of different steps, four in this case. Next we calculate the remainder with respect to pat_length of the input variable istep to determine what the new state of the pattern excitation is, and write the corresponding entry in the array patA to the output that is connected to terminal PA. Next we calculate the entry in patA that is shifted by two time slots, also modulo the pat_length, and write the corresponding entry to output pin PB. In the same fashion, we set the remaining two output pins, PC and PD, to the entry in patA that is shifted by one or three time slots, respectively. Finally, we wait a short time, given by the variable settle_time. This function implements a single step; in order to move a larger number of steps, we simply call it repeatedly.

The remainder of the sketch consists as usual for Arduino sketches of a setup() function, where we initialize serial communication and declare the output pins to control the motor as OUTPUT. Finally we enable the motor driver, by setting the ENABLE pin HIGH. In the loop() function we employ the often-used construction to read from the serial line, and decode the command with the strstr() function. If the command MOVE is received, we decode the rest of the line as the integer steps, the number of steps we want to move the stepper motor. This number can be positive or negative, depending on the desired direction in which to move. If it is positive, we call the set_coils() function the required number of times while incrementing the variable stepcounter, in order to keep track of the currently applied step as well as the accumulated number of steps. If steps is negative, we call set_coils() the necessary number of times while decrementing the variable stepcounter. The other commands read and set the stepcounter variable with STEPS? or STEPS nnn, respectively. The commands ENABLE and DISABLE turn the driver stage of the L293D on and off, which may be convenient to prevent overheating of the coils during long times of idleness. This program moves the motor back and forth. Implementing speed control by adjusting settle_time is left as an exercise.

The previous sketch implements full-step mode, but changing it to half-step mode only requires changing the set_coil() function to produce the pattern for half stepping, which, according to Section 3.2.3, is given by 11100000. The replacement for set_coil() implementing half-step mode is

```
void set_coils(int istep) { //............half-step mode
```

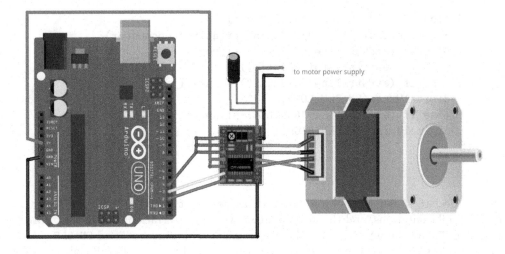

Figure 4.20 Connecting a bipolar stepper motor to the UNO with a DRV8825 stepper motor driver with microstep capability[†].

```
bool patA[]={1,1,1,0,0,0,0,0};
int pat_length=8;
int ii;
istep=istep % pat_length;
if (istep < 0) istep+=pat_length;
digitalWrite(PA,patA[istep]);
ii=(istep+4) % pat_length;
digitalWrite(PB,patA[ii]);
ii=(istep+6) % pat_length;
digitalWrite(PC,patA[ii]);
ii=(istep+2) % pat_length;
digitalWrite(PD,patA[ii]);
delay(settle_time);
}
```

which is very similar to the previously discussed full-step version. The only difference is the array patA, which now contains the excitation pattern for the half-step mode. The variable pat_length is increased correspondingly to eight, and the steps to excite the other pins are 4, 2, and 6. Remember, this corresponds to 180, 90, and 270 degrees of the eight time slots. Depending on the sense of wiring for the coils, we may have to swap the cables, or, as an alternative, exchange the step increment in the set_coils() function. Now we can full step and half step the motors, even do so by hand, but for the micro-stepping modes we need additional hardware to control the excitation of the coils more accurately.

A driver circuit that implements these microstepping modes is the DRV8825. Interfacing a bipolar stepper motor with an Arduino using this driver is shown in Figure 4.20. Before connecting the motor to the circuit, we need to match the driver to the motor by adjusting the maximum current by which the coils are excited. So we start from the configuration shown in Figure 4.20, but *without* the four leads to the motor in place. The motor power supply must provide between 8.2 and 40 V, and we add a 100 μF capacitor to stabilize the supply voltage. We then connect the, normally black, ground lead of a multimeter to digital ground on the motor driver and connect the other, normally red, lead of the multimeter to

Figure 4.21 Adjusting the maximum current on the DRV8825 breadboard.

the tip of a screwdriver. We use the latter to adjust the potentiometer on the top left of the driver breadboard until the voltage shown on the multimeter is half the desired current limit for the motor, as specified in the datasheet of the DRV8825. We illustrate this is Figure 4.21, where we see the multimeter on the bottom left and the DRV8825 placed on a small breadboard with the screwdriver touching the potentiometer on the driver board. The multimeter in this case shows a voltage of 0.56 V, which implies that the current limit for the driver is 1.12 A.

Once we have set the maximum current, we connect the motor to the terminals labeled 1A, 1B and 2A, 2B on the driver circuit. Here we chose to connect the first and third wires on the motor to the B-pins, and the second and fourth to the A-pins, which accounts for the crossed wires on the lower motor coils. Ground and positive voltage of the motor power supply are connected to the two terminals on the top right, and the digital ground is connected to the Arduino ground connector with the black wire. One wire connects UNO pin D2 to the direction pin at the lower left of the DRV8825, and another wire connects D3 to the step input. A wire, connected to pin D4, is used to select the microstepping mode. Several modes such as full-step, half-step, 1/4, 1/8, 1/16, and 1/32 stepping are available. See the datasheet for the details. Here we only implement two modes. If the three mode pins M0, M1, and M2 are pulled to ground, the driver operates in full-step mode. If the three mode pins are pulled to the logic supply voltage, the microstepping mode with 32 microsteps is selected. Thus by toggling UNO pin D4 we can switch between full- and micro-stepping mode. The enable pin on the top left of the driver is left unconnected, because it is internally pulled low to enable the driver by default. Moreover, one wire, connected to the 5 V supply of the UNO, pulls the reset and sleep pins of the driver high, thus permanently enabling the driver.

In order to move the motor, we use a sketch for the Arduino UNO that suitably changes the states of the direction, step, and mode pin. This is done with the following code.

```
// Stepper controller with DRV8825, V. Ziemann, 170626
char line[30];
int settle_time=30,stepcounter=0;
const int DIR=2;  // direction pin
```

```
const int STEP=3; // step pin
const int MODE=4; // mode pin, LOW=FULLSTEP, HIGH=MICROSTEP
void setup() { //.............................setup
  Serial.begin (9600);
  while (!Serial) {;}
  Serial.println("starting");
  pinMode(DIR,OUTPUT); digitalWrite(DIR,LOW);
  pinMode(STEP,OUTPUT); digitalWrite(STEP,LOW);
  pinMode(MODE,OUTPUT); digitalWrite(MODE,HIGH);
}
void loop() { //.................................loop
  if (Serial.available()) {
    Serial.readStringUntil('\n').toCharArray(line,30);
    if (strstr(line,"MOVE ")) {
      int steps=(int)atof(&line[5]);
      if (steps > 0) {
        digitalWrite(DIR,LOW);
        for (int i=0;i<steps;i++) {
          stepcounter++;
          digitalWrite(STEP,HIGH);
          delayMicroseconds(settle_time);
          digitalWrite(STEP,LOW);
          delayMicroseconds(settle_time);
        }
      } else {
        digitalWrite(DIR,HIGH);
        for (int i=0;i<abs(steps);i++) {
          stepcounter--;
          digitalWrite(STEP,HIGH);
          delayMicroseconds(settle_time);
          digitalWrite(STEP,LOW);
          delayMicroseconds(settle_time);
        }
      }
    } else if (strstr(line,"STEPS?")) {
      Serial.print("STEPS "); Serial.println(stepcounter);
    } else if (strstr(line,"STEPS ")) {
      stepcounter=(int)atof(&line[6]);
    } else if (strstr(line,"WAIT?")) {
      Serial.print("WAIT "); Serial.println(settle_time);
    } else if (strstr(line,"WAIT ")) {
      settle_time=(int)atof(&line[5]);
    } else if (strstr(line,"MICROSTEP")) {
      settle_time=30;
      digitalWrite(MODE,HIGH);
    } else if (strstr(line,"FULLSTEP")) {
      settle_time=2000;
      digitalWrite(MODE,LOW);
    }
  }
}
```

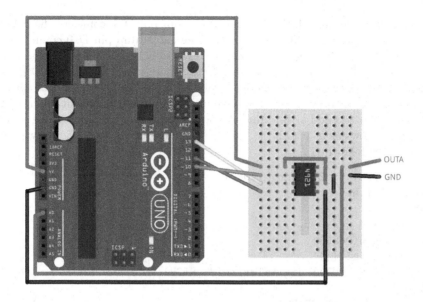

Figure 4.22 Connecting an MCP4921 12-bit DAC to the Arduino UNO[†].

At the top of the sketch we declare a number of variables, such as the settle_time and the stepcounter, but also the used pins. In the setup() function we initialize serial communication, declare the control pins for the stepper driver as output, and initialize their state. By pulling the mode pin high, we select the microstepping mode. The loop() function uses the query-response construction in which the first command deciphers the MOVE command and interprets the characters following it as the number of steps. Then the code branches, depending on the sign of steps. If it is positive, we pull the direction pin HIGH and then increment the stepcounter. Then the STEP pin is pulled high and low, with a small delay in between. Note that we use the delayMicroseconds() function, because the rotor moves very little between the microsteps, and we can afford to reduce the waiting time between steps to achieve smooth motion of the motor. But in any case, the values for the waiting time should be adapted to the actual motor connected. If steps is negative, we pull the direction pin LOW and decrement the stepcounter before toggling the STEP pin first HIGH and then LOW. After handling the MOVE command, we add the bookkeeping commands to read and set both stepcounter and settle_time, and finally, select the operation mode. If the command is MICROSTEP, the settle_time is set to 30 μs and the MODE pin is pulled HIGH. If the command is FULLSTEP, we choose a larger settle_time and pull the MODE pin LOW.

After being able to control various types of motors, we move on to controlling analog voltages.

4.5.5 Analog voltages

In this section we connect an MCP4921 12-bit digital-to-analog converter, which supports SPI-communication, to an Arduino UNO, and discuss a sketch to set the output voltage of the DAC using our query-response protocol. This could, for example, be used to set the control voltage of a power supply or any other device that requires an analog voltage level as input. The circuit that achieves this is shown in Figure 4.22, where we see the UNO on the left and the supply and ground connections to the MCP4921. The bright wire from pin

D13 connects the respective clock pins, the wire from pin D11 connects the MOSI pins. It carries the data from the UNO to the DAC. The cable from pin D10 connects the chip select pins. In this case there is no information read back from the DAC to the UNO, and we do not need a MISO wire. As a matter of fact, there is not even a MISO pin on this DAC. On the DAC we connect the reference voltage pin to the 5 V supply voltage and pull the LDAC pin low to transfer the new voltage immediately to the output at the end of the SPI transaction. We also add a wire from the output pin of the DAC to analog input A0 of the UNO, which allows us to read back and verify the output voltage. The software running on the UNO is the following.

```
// MCP4921 DAC, V. Ziemann, 170801
#include <SPI.h>
#define CS 10
void setup() { //.............................setup
  Serial.begin(9600);
  while (!Serial){;}
  pinMode(CS,OUTPUT); digitalWrite(CS,HIGH);
  SPI.begin();
  SPI.setBitOrder(MSBFIRST);
}
void loop() { //.................................loop
  char line[30];
  if (Serial.available()) {
    Serial.readStringUntil('\n').toCharArray(line,30);
    if (strstr(line,"DAC ")==line) {
      uint16_t val=(int)atof(&line[3]);
      val|=(B0011 << 12);   // 0=chA,0=unBuf,1=x1,1=ON
      digitalWrite(CS,LOW);
      SPI.transfer(highByte(val));
      SPI.transfer(lowByte(val));
      digitalWrite(CS,HIGH);
    } else if (strstr(line,"A0?")==line) {
    Serial.print("A0 ");
    Serial.println(analogRead(0)*5.0/1023.0);
    }
  }
}
```

The sketch starts by including the header file SPI.h for the SPI functionality, and declares the chip-select pin CS. In the setup() function, we initialize serial and SPI communication, as well as declaring the CS pin as output, and set its value to the SPI idle-state, which is HIGH. In the loop() function we use the query-response protocol and respond to the command DAC n to set the DAC. First we read the desired 12-bit word from the serial line into the variable val, and then add four configuration bits B0011 to the most significant end, where the first bit corresponds to the channel number, which is zero for the single-channel DAC MSP4921, and the second bit declares that we use unbuffered voltages. With the third bit we choose the internal amplification level to be unity, and the fourth bit enables output. Once the 16 bits in the variable val are assembled, we can pull CS low to initiate communication and then transfer the 16 bits in two chunks of 8 bits each to the chip with two calls to SPI.transfer() before pulling CS high again to end the communication and internally transfer the voltage to the output buffer of the DAC. The command A0? reads back the voltage, already converted to volts. This simple sketch will allow us to set any device that requires analog control voltages as input, and also read back the voltage to verify correct operation.

In other cases we do not need to set the voltage, but rather generate periodic signals.

4.5.6   Periodic signals

The following sketch turns an Arduino UNO into a simple pulse generator that repeatedly toggles an output pin on and off with microsecond precision. This is accomplished by including the header file for the `TimerOne` library that gives us control over the built-in hardware timer on the UNO. Note, however, that this library only has access to either pin D9 or pin D10. After defining the period and pulse width in microseconds, we initialize `Timer1` with the period specified and start the pwm (pulse-width modulation) with the time the pin is high scaled to within the range between zero and 1023. At this point pin D9 repeatedly turns on for $10\,\mu s$ and off for $90\,\mu s$.

```
// Pulse Generator using PWM, V. Ziemann, 221021
#include <TimerOne.h>
const int outpin = 9;
float period=100, width=10; // in microseconds
void setup() {
  Serial.begin(9600); delay(2000);
  Timer1.initialize((long) period);
  Timer1.pwm(outpin, (long)(width*1023/period));
}
void loop() {
  if (Serial.available()) {
    char line[30];
    Serial.readStringUntil('\n').toCharArray(line,30);
    if (strstr(line,"PERIOD ")) {
      period=atof(&line[7]);
      Timer1.setPeriod((long)period);
      Timer1.pwm(outpin, (long)(width*1023/period));
    } else if (strstr(line,"PERIOD?")) {
      Serial.print("PERIOD "); Serial.println(period);
    } else if (strstr(line,"WIDTH ")) {
      width=max(1,min(period,atof(&line[6])));
      Timer1.pwm(outpin, (long)(width*1023/period));
    } else if (strstr(line,"WIDTH?")) {
      Serial.print("WIDTH "); Serial.println(width);
    } else if (strstr(line,"START")) {
      Timer1.start();
    } else if (strstr(line,"STOP")) {
      Timer1.stop();
    }
  }
  delay(10);
}
```

In the `loop()` function we can adjust the parameters of `Timer1` with our standard query-response construction to specify the `PERIOD` and the `WIDTH`. After changing these parameters, we immediately call `Timer1.setPeriod()` and `Timer1.pwm()` as needed. The output pin immediately outputs a signal with the new parameters. The commands `STOP` and `START`

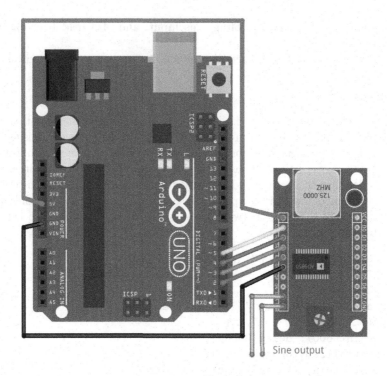

Figure 4.23 AD9850 on a breadboard connected to a UNO[†].

disable and restart the output. Note that the output signal has a rectangular shape. It is either on or off. To obtain, for example, a sinusoidal output, we use the AD9850 DDS, already shown on the right-hand side in Figure 3.14.

Figure 4.23 shows a AD9850 controlled from an UNO. Apart from wires for power and ground, four additional wires are required to set the frequency. W_CLK on pin D5 on the UNO and DATA on D3 are used transmit the control word that sets the frequency, FQ_UD on D4 updates the output to the new frequency and RESET on D2 is needed to initialize the device. A sinusoidal output signal is available from the two wires labeled *sine output*. The following program running on the UNO first defines the initial frequency and the frequency of the crystal on the breadboard and the four pins on the UNO. The macro pulse_high simply pulses the specified pin. It is used in the function set_frequency(), which receives the frequency in Hz as input and initializes the device. It follows the datasheet to calculate the value of the phase-increment register freq1, before clocking it, one bit at a time, into the device. Note that there are only 32 bits in freq1, that we successively shift to the right. Therefore the last eight bits are all zeros and set the initial phase of the sine wave to zero. Finally the new frequency is made active by pulsing FQ_UD.

```
// AD9850, V. Ziemann, 221024
double freq=100000;  // initial frequency
#define AD9850_CLOCK 125000000
const int W_CLK=5,FQ_UD=4,DATA=3,RESET=2;
#define pulseHigh(pin) {digitalWrite(pin,HIGH); digitalWrite(pin,LOW);}
void set_frequency(double frequency) {
  pulseHigh(RESET); pulseHigh(W_CLK); pulseHigh(FQ_UD); // Reset
  int32_t freq1=frequency * 4294967295/AD9850_CLOCK;
```

```
  for (int i=0; i<40; i++){
    digitalWrite(DATA,(freq1 >> i) & 0x01);
    pulseHigh(W_CLK);
  }
  pulseHigh(FQ_UD);
}
void setup() {
  Serial.begin(9600);
  pinMode(FQ_UD,OUTPUT); pinMode(W_CLK,OUTPUT);
  pinMode(DATA,OUTPUT); pinMode(RESET,OUTPUT);
  set_frequency(freq);
}
void loop() {
  if (Serial.available()) {
    char line[30];
    Serial.readStringUntil('\n').toCharArray(line,30);
    if (strstr(line,"FREQ ")) {
      freq=atof(&line[5]);
      set_frequency(freq);
    } else if (strstr(line,"FREQ?")) {
      Serial.print("FREQ "); Serial.println(freq);
    }
  }
}
```

The `setup()` function only initializes the serial line, declares the four control lines at output, and sets the starting frequency. In the `loop()` function, we provide the ability to set a new frequency and to request information about the presently set frequency.

So far we have a number of methods to control the amplitude and timing of voltages, switches, and motors. Now we will wrap up this section by discussing a few means to attract the attention of a user, which is useful in case of malfunction or another event that requires human intervention.

4.5.7 Human attention actuators

NeoPixel strips are a bit like LED on steroids. Units of a red, a green, and a blue LED, together with a WS2811 controller, are assembled in strips of eight or more individually addressable such units. Before using them, we need to install the *Adafruit_NeoPixel* library from the *Library manager* in the *Tools* menu. This library provides functions to set the intensity of all individual LEDs. We illustrate its use by displaying the temperature, measured with a DS18b20 as shown in Figure 4.11 and again in Figure 4.24 between 21 and 28 degrees on an eight-unit NeoPixel strip. The DS18b20 is connected to 5 V and ground on the UNO and is read out via pin D2. Likewise, the NeoPixel strip is connected to the power rails and is controlled via pin D6. Since the strip can suddenly draw large currents, we add capacitor with a few μF to the power rails. The following sketch is based on the example from page 85; it only adds code to initialize and use the NeoPixels. First we specify the number of pixels and the pin the strip is connected to before creating an object `pixel` with standard parameters. In the `setup()` function we only need to call `pixels.begin()` to initialize the interface. In the `loop()` function, we first measure the temperature and convert the real value to the temperature above 20 degrees, the integer `imax`.

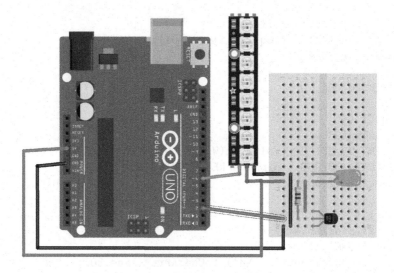

Figure 4.24 Measuring the temperature with a DS18b20 and displaying it on a NeoPixel strip[†].

```
//Neopixel thermometer, V. Ziemann, 221115
#include <OneWire.h>
#include <DallasTemperature.h>
#define ONE_WIRE_BUS 2  // pin 2
OneWire oneWire(ONE_WIRE_BUS);
DallasTemperature sensors(&oneWire);
#include <Adafruit_NeoPixel.h>
#define NUMPIXELS 8
#define PIN 6
Adafruit_NeoPixel pixels(NUMPIXELS, PIN, NEO_GRB + NEO_KHZ800);
void setup() {
  sensors.begin(); pixels.begin();
}
void loop() {
  sensors.requestTemperatures();
  float temp=sensors.getTempCByIndex(0);
  int imax=(int)temp-20;  // temperature above 20 degrees
  pixels.clear();
  for (int i=0;i<imax;i++) {
    pixels.setPixelColor(i, pixels.Color(20+10*i,20, 100-10*i));
  }
  pixels.show();
  delay(1000);
  }
}
```

Then we illuminate the units up to imax, which mimics the behavior of a conventional thermometer with a mercury column. Note that we give the units different colors; the lowest is almost blue, but the upper units become progressively more red.

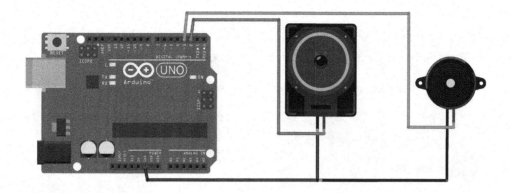

Figure 4.25   Connecting a speaker and a piezo buzzer to the Arduino[†].

LED or NeoPixel are not the only way to attract attention. Whereas they obviously address the eye we can also address our ears with acoustic signals. The simplest example is a small loudspeaker that we can either use as a simple buzzer, or even code with simple acoustic patterns to identify certain things. For instance, if a moving robot gets stuck in a corner, we may want it to sound like a siren.

We interface a speaker to the UNO by connecting one terminal to ground on the Arduino and the other to a digital output pin, say pin 2, as shown in Figure 4.25. In order to produce a tone of 440 Hz for 1000 ms, we use the Arduino command

```
tone(2,440,1000);
```

placed anywhere in the Arduino sketch. Initialization of the pin as digital output is done automatically. Replacing the 2 by a 3 will sound the piezo buzzer shown in Figure 4.25.

After being able to interface a number of sensors and actuators over the serial line, we need to take a closer look at the different ways to communicate to the host computer and how to receive the data there.

4.6   COMMUNICATION TO HOST

So far, we used only the serial monitor in the Arduino IDE to communicate with the Arduino, but normally we also want to communicate from other programs, and that is the topic of this section.

4.6.1   RS-232 and USB

The serial monitor in the Arduino IDE uses an RS-232 protocol that is transported over a USB line. The translation happens automatically, and plugging the USB cable into the host computer causes the latter to automatically create a device file (/dev/ttyUSBx or /dev/ttyACMx on Linux, COMx on Windows, and /dev/tty.usbserial-xxx on a Mac) that represents the other end of the communication channel from the UNO to the host computer. We can then use any terminal program on the host computer to connect to that device file, but have to keep in mind to use the same baud rate that we specify in the setup() function on the Arduino. To find out the device file to which the Arduino is connected, we can use the Arduino IDE and inspect the Tools→Port menu, and write the name of the device file down. In my case it was /dev/ttyACM0. After we close the Arduino IDE, we can use any terminal-emulator program, such as putty on Windows, or the very basic program screen on Linux or Mac, to connect to the UNO. In my case the command

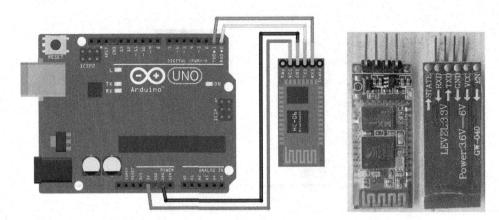

Figure 4.26 Connecting an HC-06 Bluetooth dongle to the Arduino (left†) as well as front and back side of the dongle (right).

```
screen /dev/ttyACM0 9600
```

does the trick, and we can send commands to the Arduino and receive the response in the terminal window. The `screen` program can be exited by pressing `Ctrl-a-k` and confirming the question of whether one really wants to exit with `y`. The number `9600` is the baud rate used for the communication and must match the number in the `Serial.begin()` statement in the Arduino sketch. Note that `9600` is the default baud rate and could be omitted in the `screen` command.

4.6.2   Bluetooth

Bluetooth functionality can be added by attaching an HC-06 Bluetooth dongle to ground, power, and the Arduino pins 0 and 1. On the right-hand side in Figure 4.26 we see four exposed pins labeled `RXD, TXD, GND,` and `VCC`. The latter two are connected to the respective power supply connections, and `RXD` on the HC-06 is connected to pin 1, labeled `TX` on the Arduino, which is illustrated on the left-hand side in Figure 4.26. On this connection the information flows from the Arduino to the HC-06. The pin labeled `TXD` on the HC-06 must be connected to pin 0, labeled `RX` on the Arduino, and the information flows from the HC-06 to the Arduino on this line. This crossed connection from `TXD` to `RX` and vice-versa is equivalent to a null-modem cable. After connecting the HC-06 in this way, all communication is sent via the HC-06 and the USB connection in parallel. For reliable operation, the USB link should not be used while Bluetooth is in operation.

On the host computer, we have to *pair* a new Bluetooth device with the host computer. On Windows, this is done in the Bluetooth administration program. On some Linuxes, similar user interfaces exist, but we can always pair Bluetooth devices using a number of command-line programs. First we need to find out whether the host computer has Bluetooth capabilities, by the command `hcitool dev`, which should report at least one device, normally called `hci0`. Then we scan the surroundings for Bluetooth devices with the command `hcitool scan`. If the HC-06 is powered we should see at least one device with a line `xx:xx:xx:xx:xx:xx HC-06` where the six-byte string is the MAC address of the Bluetooth device. To establish pairing, we need to use `sudo` and call the `sudo bluetoothctl`. At the `[bluetooth]#` prompt we initiate a search for new devices with `scan on`, which reports all known devices. The following two commands establish the pairing

```
trust xx:xx:xx:xx:xx:xx
pair xx:xx:xx:xx:xx:xx
```

where xx:xx:xx:xx:xx:xx is the MAC address of the HC-06 dongle we want to pair. During the previous actions we are prompted for a PIN number, and unless we have changed the default on the HC-06, we use 1234. Then we exit the bluetoothctl program and create the device file for the serial communication by issuing

```
sudo rfcomm bind 0 xx:xx:xx:xx:xx:xx
```

which creates a device file /dev/rfcomm0 that has the same functionality as the /dev/ttyACM0 device file we used earlier to communicate with the Arduino. Therefore we can again use the screen command to communicate with the Arduino, but using /dev/rfcomm0 as first argument to the screen command instead. Note that the HC-06 is configured to communicate with 9600 baud. This can be changed using AT commands, but we will not discuss this further, and assume that all Bluetooth communication using the HC-06 is done at 9600 baud. Once we are done using the Bluetooth serial link, we should take it down by issuing the command rfcomm release 0, which will remove the /dev/rfcomm0 device file, and we will no longer be able to use it. The bottom line is that we can communicate using Bluetooth in much the same way as using native RS-232 or USB. We only need to pair the device and the host computer once, and then create the device file using the sudo rfcomm bind command before using the serial line, and delete it with the sudo rfcomm release command once we are finished.

The ESP32 has bluetooth already built in and we use it by including support for the BluetoothSerial library, declare SerialBT device and use it henceforth in the same way as the Serial device to communicate. Note that we give the device the name ESP32BT to identify itself. The loop() function follows the query-response scheme, but uses SerialBT instead of Serial. Here is a boiled-down version of a sketch that reads the analog voltage from pin 33 of a ESP32.

```
// SerialBT, V. Ziemann, 221101
#include <BluetoothSerial.h>
BluetoothSerial SerialBT;
void setup() {SerialBT.begin("ESP32BT");}
void loop() {
  if (SerialBT.available()) {
    char line[30];
    SerialBT.readStringUntil('\n').toCharArray(line,30);
    if (strstr(line,"A0?")) {
      int val=analogRead(33);
      SerialBT.print("A0 "); SerialBT.println(3.3*val/4095);
    }
  }
}
```

Once this sketch is compiled and downloaded to the ESP32, calling hcitool scan will show a device ESP32BT and its MAC address. The setup with trusting and pairing on the Raspi works in the same way as for the HC-06 dongle. When asked to confirm some pin, you can just accept it.

Apart from the channels based on serial communication, some microcontrollers, such as the ESP8266 or ESP32, offer WLAN-based communication using network sockets. We will discuss this nifty feature in the next section.

4.6.3 WiFi

For this section, we use a NodeMCU system that is connected to the host computer with a USB cable for programming. Moreover, the device support package as described in Section 4.2 must be installed in the Arduino IDE. Our task is to connect the NodeMCU to a local wireless network; we assume it is called `MyHomeNet` and that it can be reached by other computers connected to the same network. All these computers will be able to query the measurement values of sensors connected to the NodeMCU. To better understand the setup, let us briefly discuss some general features of computer networks.

Everybody is probably familiar with the fact that computers on the Internet are identified by IP numbers, such as `192.168.10.200`. Because the numbers are difficult to remember, there are also aliases such as `www.cnn.com`, and the translation is done by so-called *domain name resolution* or DNS servers. On our simple network, we assume that we can keep track of the numbers, and all sensors and host computers are connected to the (class-C) network `192.168.10.nn` where `nn` is a number between 2 and 254. The numbers 0 and 255 are reserved for special purposes, and we assume that the router that connects our network to the outside world ("the Internet") has the IP number `192.168.10.1`. Note that the addresses starting with `192.168` are private numbers and can be used by anyone, provided that a router separates this network from the Internet. But we confine all our communication to within the `192.168.10` network, and identify each computer by its IP number. Each computer can potentially provide different services, such as running a web server, a measurement server, a mail server, or a server to allow logging onto the computer with standard protocols such as `telnet, ftp,` or `ssh`. The different services that a computer provides are identified by a *port number*. As an analogy, it may help to think of the IP number as a street address of an apartment building and the port numbers as the apartment numbers. The communication with a server that provides a service on a computer therefore requires the specification of the IP number and the port number.

But how does a computer know its own IP number? There are two ways to specify this: by configuring the network setup manually and assigning the IP number explicitly. We use the second way and dynamically acquire the IP number via the *dynamic host configuration protocol*, DHCP, which is much more convenient, provided such functionality is available on a given network. In most networks with a wireless router, the router provides this service and one only has to tell a computer to use DHCP. In that case, the computer sends a request for an IP number at power-up. The DHCP server responds with an IP number that is then assigned to the newly connected computer. We assume that a DHCP server is running on our `192.168.10` network and the NodeMCU is also configured by default to use DHCP.

Wireless networks are normally protected from unauthorized use by encrypting the communication. Most networks use an encryption standard called WPA that requires entering a password to connect to the network. We assume that the "MyHomeNet" WLAN is of that type. Having covered the networking basics, we are ready to connect our Arduino clone, the NodeMCU, to the WLAN.

First we discuss the running of a simple web server on the NodeMCU, that uses the standard `http` protocol to communicate with other computers on the network. The following program shows how to set up a web server that listens on the default `http` port number 80. It provides access to temperature measurements and allows one to control the brightness of an LED. In our example we query the temperature measured by the NodeMCU, by directing a browser such as Mozilla Firefox to the address of the NodeMCU at `http://192.168.10.nn/temperature`, and the NodeMCU returns a web page with the temperature measured by an LM35 temperature sensor. The brightness of the built-in LED is controlled by adding a value to the address after a question mark, such that

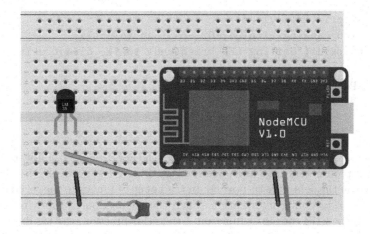

Figure 4.27 The NodeMCU with an LM35 temperature sensor[†].

`http://192.168.10.nn/led?b=1023` sets the brightness to its maximum value. The circuit schematic is shown in Figure 4.27, and the code that brings it to life is the following.

```
// TemperatureWebServerBrightness, V. Ziemann, 221214
#include <ESP8266WiFi.h>
const char* ssid     = "MyHomeNet";
const char* password = "........";
WiFiServer server(80);  // server listens on html port 80
void setup() {
  pinMode(D0,OUTPUT);    // D0=GPIO16=LED_BUILTIN on NodeMCU
  Serial.begin(115200); delay(10);
  Serial.print("Connecting to "); Serial.println(ssid);
  WiFi.begin(ssid, password);
  while (WiFi.status() != WL_CONNECTED) {delay(500); Serial.print(".");}
  Serial.print("\nWiFi connected and "); server.begin();
  Serial.print("server started at "); Serial.println(WiFi.localIP());
}
void loop() {
  WiFiClient client = server.available();
  while (client) {
    Serial.println("new client");
    while(!client.available()){delay(1);}
    String req = client.readStringUntil('\r');
    Serial.println(req);
    client.flush();
    client.print("HTTP/1.1 200 OK\r\nContent-Type: text/html");
    client.print("\r\n\r\n\r\n<!DOCTYPE HTML>\r\n<html>\r\n");
    if (req.indexOf("/temperature") != -1) {
      float temp=100*3.3*analogRead(0)/1023;
      client.print("Temperature="); client.print(temp,2);
    } else if (req.indexOf("/led") != -1) {
      int i1=req.indexOf("?b="); int i2=req.indexOf("HTTP");
      String payload=req.substring(i1+3,i2-1);
```

```
        Serial.println(payload);
        if (i1>0) analogWrite(D4,1023-payload.toInt());
        client.print("Setting LED brightness to "); client.print(payload);
    }
    client.println("</html>"); delay(1);
    client.stop();
  }
}
```

In this example we first include the header file with the `ESP8266WiFi.h` library information, define the access parameters `ssid` and `password` for the wireless network, and instantiate the `server` to listen on the default http-port 80. In the `setup()` function, the serial line is opened to be able to listen to debug information. Then the connection to the wireless network is established with the `WiFi.begin()` function. Inside this function, the log-on to the wireless network and the communication with the DHCP server is handled. Once `WiFi.status` reports that the connection is established, we print information about the connection, such as the acquired IP number, to the serial line. Then we start the server with the call to the `server.begin()` function. In the `loop()` function a similar construction as before receives a request from a client, stores the request in the variable `req`, returns a standard http header with `Content-type` to the client and then checks whether the request `req` contains the string `/temperature`. If it does, the sketch determines the temperature with a call to the `analogRead()` function, and writes the value back to the client. Note that this time we use the `indexOf` method of a `String` to determine whether a substring is present. If `req` contains `/led`, we determine the position of the question mark and trailing characters `HTTP` and extract everything in between as the `payload`. After conversion to an integer, we use the `payload` to set the brightness of the built-in LED. We can inspect how `req` is constructed and why we have to find the trailing `HTTP` by printing with `Serial.println(req)`. If an unknown request arrives, we are notified on the serial line. Finally, we add the concluding `</html>` tag and close the connection to the client. We point out that in this simple example we do not check the validity of the `payload`, which poses a potential security risk, if the system is reachable from the Internet.

A few words are needed about the cryptic HTTP header that the NodeMCU returns to the calling browser. When a browser, such as Firefox or Chromium, connects to a web server on the default port 80, it first sends the name of the requested web page, such as `GET /temperature` in the above example, as an HTTP-GET request. Before returning the requested web page, the server sends some meta-information, such as a status code, and what type of information comes next. The status code for a properly understood request is 200, and for a missing page it is `404`, a number probably everyone has seen as a response to a typo in the specification of a web page. The type may be an image, a media file such as a video, or HTML-formatted text, which is what we specify in the sketch as `Content-Type`. After an empty line the normal HTML header with the `DOCTYPE` follows, and the information embedded in `<html>` tags. We will discuss HTML and the structure of web pages in more detail in Section 5.7. Browsers do not render the meta-information, which therefore remains invisible, but we can eavesdrop on the communication with the `netcat` or `telnet` commands (more on those commands later in Section 5.3) by pointing them to port 80 on the NodeMCU at IP address `192.68.20.184` and issuing `GET /temperature` by hand. *All* output from the server sub-sequentially appears in the same window. Figure 4.28 illustrates the exchange.

Setting the brightness of the LED uses the same mechanism that is used to enter credit-card numbers when shopping online. It is based on the HTTP-GET method as well, but adds a *query-string* that follows the question mark after the address. The query-string has the form `name=value`. This mechanism to send information to a web server is often used

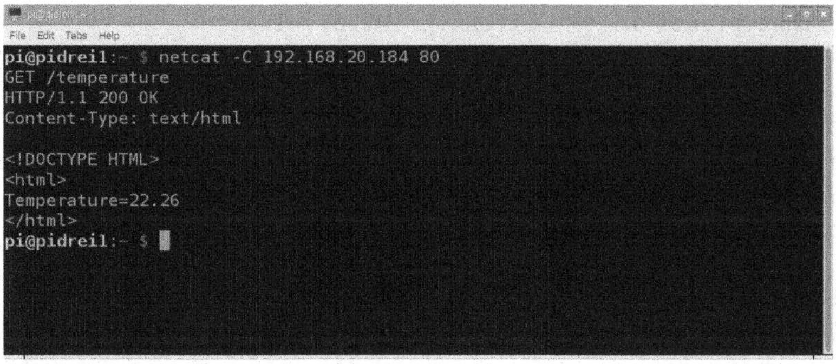

Figure 4.28 The communication with a web server showing the HTTP header.

with HTML forms. It allows the user to enter text or select things from a list and then click a submit button to send it to the server, where it causes some reaction, such as sending the desired purchase – or to set the brightness of an LED.

In a second example, we configure a NodeMCU to run a server listening for connections on port number 1137, and once a connection is established, it waits for commands and then replies appropriately using the previously discussed query-response protocol. The hardware is the same as in the previous example, already shown in Figure 4.27. The code that runs on the NodeMCU is the following.

```
// Socket-based measurement server, V. Ziemann, 161211
const char* ssid    = "MyHomeNet";
const char* password = ".........";
const int port = 1137;
#include <ESP8266WiFi.h>
WiFiServer server(port);
void setup() { //.......................................setup
  Serial.begin(115200);
  WiFi.begin(ssid, password);
  while (WiFi.status() != WL_CONNECTED) {
    delay(500); Serial.print(".");
  }
  Serial.println("");
  Serial.print("WiFi connected to "); Serial.println(ssid);
  Serial.print("Server IP address: "); Serial.println(WiFi.localIP());
  server.begin();
  Serial.print("Server started on port "); Serial.println(port);
}
void loop() { //.......................................loop
  char line[30];
  float volt,temp;
  WiFiClient client = server.available();
  while (client) {
    while(!client.available()) {delay(1);}
    client.readStringUntil('\n').toCharArray(line,30);
    Serial.print("Request: "); Serial.println(line);
    if (strstr(line,"A0?")) {
      volt=3.3*analogRead(0)/1023;
```

```
        client.print("A0 "); client.println(volt);
      } else if (strstr(line,"T?")) {
        temp=100*3.3*analogRead(0)/1023;
        client.print("T "); client.println(temp,1);
      } else {
        Serial.println("unknown command, disconnecting");
        client.stop();
      }
      client.flush();
//    client.stop();
    }
  }
```

In this sketch we first define two character strings, `ssid` and `password`, for the name and passphrase of the wireless network. We also specify the port number, here 1137, in the following line before including the functionality for WiFi support on the ESP8266 device. Once that is included, we define a WiFiServer named `server` that listens on the specified port. In the `setup()` function, we start up the serial communication via the USB line that we use for logging, and then connect to the wireless network with the `WiFi.begin()` function. Once the NodeMCU is connected to the wireless network, we start the server with the `server.begin()` command. In the `loop()` function we use the same structure that we used earlier for communication via the serial line, but this time over the socket that is dynamically opened once a client computer connects to our server. The connection to the requesting computer stays open until an unknown command closes the connection. But while the connection is open, values can be requested repeatedly. If we want the connection to close immediately after the reply is sent, we can uncomment the line with `client.stop();` immediately following the line with `client.flush()`. Note that this server can only handle one request at a time. If one computer is already connected and a second computer tries to establish contact, the latter is put on hold until the first computer disconnects. This program on the NodeMCU allows us to query via the custom socket on port 1137 any parameters or measurement values that the NodeMCU has available. How to connect to it from another computer we defer to the next chapter.

Beyond the standard communication protocols there are several others available. We discuss those only briefly in the following section.

4.6.4   Other communication

In the previous sections, we used standard and easy-to-use protocols (RS-232, USB, Bluetooth, and WiFi) to communicate with the microcontroller. They have the advantage that they are readily available on many host computers. There are, however, other communication channels, of which we name a few in the following paragraphs. All are supported by Arduino libraries.

One other wireless communication channel is *ZigBEE*, which is related to Bluetooth but uses a simpler protocol overhead and is optimized to consume as little power as possible, to make long-lived battery operation possible. Another option is using inexpensive *433 MHz transceivers*, which can be used to connect slave microcontrollers to their master using a simple wireless link.

MIDI is a protocol developed to interface electronic musical instruments. The physical interface is based on current loops that are galvanically decoupled through optocouplers at the input of each device. Logically the communication is very similar to RS-232 at a baud rate of 31250. Each MIDI message consists of three (or in some cases, two) bytes that have standardized interpretation, such as channel number, tone (corresponding to a specific key on a piano keyboard), and velocity (the intensity with which the key is struck).

It is possible to interface nearby hardware using infrared (IR) light, which is used in TV remote controls to send information from the remote to the TV to change channel or volume. Incidentally, even the first generation Lego^{TM} Mindstorm microcontrollers communicated in this way. The communication is based on an IR diode using a wavelength of 940 nm and modulating the light at a rate of 38 kHz. Sending bursts of modulated light represents either a LOW or a HIGH signal level and is used to emulate an RS-232-like protocol at low baud rates such as 2400 baud. Another protocol is RC-5, which is commonly used in TV remote controls. Receivers such as the TSOP2238 have optical filters built in and are sensitive only to a narrow band of wavelengths around 940 nm. Moreover, they demodulate the 38 kHz carrier frequency and only deliver a 3.3 or 5 V signal on their output pin, making them very easy to interface.

Note the generic structure: A communication channel is based on the low-level hardware and on a protocol stack that often consists of several layers on top of the hardware implementation. In our examples we use the convention to send a string, terminated with a question mark to signify a request. The reply then consists of the same string followed by a value. This convention defines a simple protocol for the communication. We can easily come up with other protocols, and add features such as checksums to test the integrity of the transmission. Also, I2C and MIDI communication are based on a standardized protocol where a number of bytes is transmitted and each byte signifies some particular information, such as the register to be addressed or the value that is written to the register. Of course, it is mandatory that all participants in a transaction agree on the interpretation of the bits and bytes that are transmitted, otherwise confusion will reign supreme.

So far we always assumed that some unspecified host computer, for example, a desktop computer, is available and serves as the communication partner for the microcontroller. In the next chapter we consider one specific host computer that is widely available, well documented, and inexpensive – the Raspberry Pi.

QUESTIONS AND PROJECT IDEAS

1. Convert the hex number 0x5CA3 to a) binary and b) decimal representation.

2. The ADC that reads pin A0 on the NodeMCU is not very reliable. For example, connecting A0 to GND will not necessarily give a zero ADC reading. Make an attempt to calibrate it.

3. Can you connect an MCP23017 IO-extender and a BME680 environmental sensor to the same I2C bus?

4. Under what circumstances is it advantageous to use interrupts?

5. You need up to 128 digital input pins. Use multiple copies of the MCP23017 and explain how to connect them.

6. Inspect the datasheet of the MCP3304 and find out how to configure it to read eight unipolar input voltages. What do you need to change in the software?

7. If you need 128 unipolar analog input signals, how do you connect the MCP3304 to an Arduino? Is it possible at all, and if so, how?

8. Connect a three-color LED to an Arduino and independently control the brightness of the three colors via the serial line. Add features to directly set standard mix-colors such as orange or magenta.

9. Explain why we use (1023-value) to set the brightness in the example with the web server on page 113.

10. What is the difference between *HTTP* and *HTML*?

11. Learn about HTML forms and implement a web page with a user interface to set the brightness of the LED on the NodeMCU.

12. Mount an HC-SR04 distance sensor on top of a model-servo and scan the neighborhood for the distance to obstacles. Produce a beep, if it is closer than 20 cm.

13. Cause the NeoPixels on the thermometer from Figure 4.24 to flicker by rapidly reprogramming them while adding a random number to the value that specifies red, green, and blue.

14. Use the `tone()` function to make the Arduino sound like a *siren.*

15. Build a *magnetic field sensor* with a Hall probe.

16. Build a device that senses the tilt angle with an MPU6050.

17. Build a *carousel comparator* and use an MPU6050 to measure the acceleration you experience in a merry-go-round. Use a battery-powered NodeMCU to publish a web page with the data. Enhance the system by measuring and displaying the angular velocity as well. Also display the maximum and minimum values during the past 5 minutes.

18. Build a contact-free *wireless thermometer* with an MLX90614 FIR sensors and a NodeMCU that makes the measured values available on a web page or via a socket-based server.

19. Measure the wind speed with a propeller that breaks the light path in a *slotted optical switch* or between a discrete LED and a phototransistor.

20. Investigate materials for their phosphorescence by briefly flashing an ultraviolet LED and record the response of the material with phototransistors or diodes sensitive to different wavelengths.

21. Investigate how to interface the dust sensors from Section 2.3.6 to the Arduino.

22. Connect an MQ-x gas sensor to the UNO and determine the air quality.

23. Investigate how to interface the GPS sensor from Section 2.3.6 to the Arduino. Consider using the `SoftSerial` library to add additional RS-232 ports to the microcontrollers.

24. Investigate the Arduino NANO. In what way does it differ from the Arduino UNO? Discuss circumstances where you would use a NANO.

25. Connect your favorite sensor to an ESP32 and communicate with it via Bluetooth.

## CHAPTER 5

Host Computer: Raspberry Pi

The Raspberry Pi [11] is a small single-board computer that first appeared in 2012, with the intention of providing an inexpensive platform to introduce students and other interested parties to computers in general and to programming in particular.

5.1 HARDWARE

Since its first appearance, the Raspi went through several hardware revisions. The smaller board on the bottom left in Figure 5.1 is a Pi Zero with 512 MB of RAM and no Ethernet port. This makes it too limited for our purpose. Instead we use the most recent incarnation, model 4, which appeared in February 2019. It is available in two form factors, the classical version 4B, shown on the bottom right in Figure 5.1 and the newer version 400, which has a built-in keyboard, shown on the top. Both feature quad-core ARM central-processing units (CPU) operating at speeds up to 1.8 GHz and have between 2 and 8 GB RAM memory on board. This hardware is sufficiently powerful to run a full-fledged Linux system. An external and replaceable micro-SDHC card (at least 16 GB and speed class 10) serves as a hard disk to hold the operating system and user files. For our purposes 2 or 4 GB RAM are adequate; also the previous version 3B works fine.

But beyond the CPU, the Raspberry Pi 4B and 400 sport a video processor that can display videos at 4k resolution (up to 3840×2160) via two built-in micro HDMI-connectors. Audio output is available either via the HDMI connectors or via a 3.5-mm headphone jack. Moreover, there are two USB-2 and two USB-3 ports on board to connect peripheral components, such as USB sticks, keyboard, mice, or web cameras. Communication with the outside world is feasible via a built-in wired Ethernet port, and since version 3, the Raspi has built-in Bluetooth and WiFi in the 2.4 GHz and on the 4B even in 5 GHz frequency bands (802.11ac).

The Raspis are very attractive due to their built-in low-level peripherals. There are 17 general-purpose input–output (GPIO) pins exposed on the board, some of which support I2C, SPI, and UART (RS-232-like) communication. Moreover, a specific audio bus (I2S) is available, as well as a high-speed CSI interface to connect the tailor-made Raspberry Pi camera, and a DSI interface to connect LCD panels. Some of these features can be used to implement the same functionality as on the Arduino, but we will not use that here, but rather use the Raspi as a standardized host computer system.

DOI: 10.1201/9781003341703-5

Figure 5.1 Raspberry Pi 400 (top), version 4B (bottom right), and version Pi Zero (bottom left).

5.2 GETTING STARTED

The CPU and peripherals are adequate for using a Raspi as media center via the *LibreElec* distribution, which runs the *KODI* media center software and turns a dumb TV into a smart TV. There is the *OpenWRT* distribution that turns the Raspi into an Internet router including firewall functionality. Other software packages turn it into a retro-gaming console or into a network-addressed storage (NAS) server. Apart from these special-purpose uses, we can also install a general-purpose Linux operating system on the SDHC card from which we boot the Raspi. Different flavors of Linux are available, such as Ubuntu, Fedora, or OpenSuse, but the standard system that we use in the remainder of this book is based on the Debian distribution and is called *Raspberry Pi OS* or *Raspios*. Over the years, a large number of books, such as [24], were published with recipes of how to use the Raspi in many circumstances. Use those and others to complement the discussion in this book.

We find download instructions for the Raspberry Pi OS operating system on the `https://www.raspberrypi.org/software` web page, from where the `rpi-imager` program can be downloaded for all common desktop operating systems. A link on the page points to a short video tutorial about using the imager. Follow the instructions and have a little patience while the imager downloads the selected image of the Raspi operating system and then copies it to the SD card. In most cases the 32-bit image with all recommended software is a good choice, though with a size of 2.7 GB the download is rather large. Optionally, on Linux systems, it is also possible to manually download the operating system image from a link further down on the web page, use the `unxz` program to unpack it, and then copy the image file to the SD card by issuing the command

```
dd bs=4M if=2022-09-22-raspios-bullseye-armhf-full.img of=/dev/sdX
```

as user `root`. We may have to prepend the `dd` (disk-duplicator) command by `sudo` depending on the flavor of Linux on the desktop computer. In the `dd` command, the destination, the output file `/dev/sdX`, must be the device file of the SD card. We can easily determine it by inspecting the system log file after inserting the SD card in the writer. Just before the end of the log file a recently inserted card is typically referenced as /dev/sdX, with X being c or d or e. We must make sure to use the correct drive! Picking the wrong one could erase the harddisk of our desktop system. If unsure, we carefully follow the instructions in the installation help, which are more extensive than those given here. Once the dd command completes, which can take up to 30 minutes, depending on the SD card writer, we insert the SD card into the slot on the Raspi, connect a monitor to the HDMI connector, mouse and keyboard to USB connectors, and apply power via the USB-C connector to boot the Raspi.

When booting the first time the Raspi automatically expands the operating system image on the SD card to use all available space and it generates secret keys that are later used in encrypted communication before it reboots. During the second boot it asks about the localization, especially country, timezone, and keyboard layout as well as a username for the standard user. Here, in order to be consistent with the first edition of this book, I choose the username pi and a secret password. In later steps the WLAN is configured, the software is brought up to date, and finally it reboots.

Once the Raspi comes up again we are directly logged into the desktop system, without a password being asked. We can change this behavior by starting the *Raspberry Pi Configuration* program which resides behind the Raspberry icon in the Menu list on the top of the screen and under *Preferences*. In the *System* tab we can change the login behavior though I usually keep the default if the Raspi is in a safe location. I usually change the *hostname* to something memorable, such as sensorpi. In the *Interface* tab I recommend to enable *ssh*. We later use this functionality to connect to our Raspi from other computers.

At this point we have a running Raspios Linux system on our Raspi, and it is time to explore it by selecting the Menu button with the raspberry image on it. There we find all available programs, sorted according to groups such as *Programming*, with links to Java, Python, and Mathematica. The group *Office* contains links to the Libreoffice programs for word processing, spreadsheet, and presentation creation. Behind the *Internet* menu item, there are links to the Internet browser and mail client. Moreover, there are menu items for *Games, Accessories* and *Preferences*. The latter, shown in Figure 5.2, contains links to the *Raspberry Pi Configuration* and other configuration options. The *Accessory* menu contains a link to the *Terminal* program that provides a console to enter commands. We will use it extensively in later sections, but here are a few basic commands that are useful to know. When we open the terminal window, we are placed in the home directory of user pi, namely /home/pi, which we can verify by entering the command pwd to "print the working directory." We can list its contents by entering the ls command, or its long form ls -l, which displays all files and subdirectories in the present directory. In order to enter a subdirectory we use the cd <dirnam> command to change the working directory to <dirnam>. The command cd .. takes us back to the directory where we started. If we want to navigate to a directory for which we only know the absolute name that starts with a slash, such as /usr/local, we use cd /usr/local. We should play around with the programs and test them to familiarize ourselves with the new Raspi system. Either type the commands on the command line or click on the programs in the menus and explore. Finally, there is the *Preferences* menu item with programs to customize the Raspi. It contains, among other programs, the *Raspberry Pi Configuration* we used earlier to enable *ssh*.

In the menu list, next to the Menu button with the raspberry on it, there are quick-start buttons for several programs, such as an Internet browser, file manager, and a *Terminal*. Programs can be added to this area by right-clicking on the menu list and selecting *Add/Remove Panel Items*. In the window that appears, we highlight Application Launch Bar and click on Preferences. On the left-hand side of the window, the programs already present in the quick-launch area are listed. From the right-hand side we can add programs from a list of all applications installed on the computer. If the *Terminal* program is missing from the Quick launch area, we add it from the *Accessory* group, because we will use it a

Figure 5.2 The menu behind the raspberry icon with the *Preferences* submenu showing programs to configure the Raspi in great detail.

lot later on. On the rightmost end of the menu list, widgets for network, audio, and other features are located. We add or remove such widgets from the *Add/Remove Panel Items* window that we launch by right-clicking the menu list and pressing the *Add* button, whence a list of available widgets appears. From it we select whatever we want to add. Finally, we move the entire menu-list to our preferred location on the desktop by right-clicking on it and selecting *Panel Settings* where we choose the location of the menu list. The default location is on the top, but we can also move it to the bottom of the desktop.

In order to log on to the Raspi from our desktop computer, we need to know its IP address, which is found either by hovering the mouse over the network applet at the right of the menu list or by executing `ifconfig` in a terminal window on the Raspi and finding the point labeled `inet` for the interface `eth0`. Once we know it, we log on to the Raspi via the secure shell program `ssh` from our desktop computer by typing `ssh -X pi@192.168.10.nn`, where `192.168.10.nn` is the IP number of the Raspi. The `-X` option allows us to display additional windows on the desktop computer. If this is the first time we connect, a warning message appears that the Raspi is unknown to your desktop and we should answer *yes* that all is well. Then we are presented with a request to enter the password for user `pi` on the Raspi. If we enter that correctly, we get a command prompt from the Raspi at which we enter commands, such as `ls -l` or any other. From a Windows computer, you may use any ssh client program, such as `putty`, to log on to the Raspi command line. This way of logging on can be simplified significantly by closing the connection, and back on the desktop computer we enter

```
ssh-copy-id pi@192.168.10.nn
```

which presents a prompt to enter the password for user `pi` on the Raspi for the last time, and henceforth we log on to the Raspi from our desktop by typing `ssh -X pi@192.168.10.nn` and are at the command prompt of the Raspi without being asked for a password. Note that this only works from our account on the desktop computer. The `ssh-copy-id` exchanges a secret with the Raspi that is used to authenticate our account on this desktop computer and no other. Thus, logging onto the Raspi is just as safe as logging onto our desktop computer. We could now remove screen, mouse, and keyboard and always use the `ssh` command to log onto the "naked" Raspi, but for convenience we keep the peripheral devices attached a little longer until we have installed software to allow running a virtual desktop via the network.

The default system already comes loaded with a good selection of programs, but we can add more programs, and the process to do so is extremely simple. We discuss this in the next section.

5.3   INSTALLING AND USING NEW SOFTWARE

Before installing new software we first update the current system, which is based on an image file prepared some time ago. To do this we start a terminal program and enter the following two commands:

```
sudo apt-get update
sudo apt-get upgrade
```

where the first command updates the database of installed programs and finds out whether newer versions are available. The second command installs the upgrades after a question to approve its selection. First, the newer versions are downloaded and then installed. This may take quite a while – about 30 minutes if the time since the last update or the creation of the initial installation media is long ago. Note that the system-administration command is called `apt-get`, and we have to run it with superuser privileges by prepending `sudo`. After the upgrade is complete and the program returns to the command line prompt, we have an up-to-date system and are ready to install additional software.

Since we want to test the interface with the Arduino, we need to install the simple terminal emulator program `screen` by issuing the command

```
sudo apt-get install screen
```

and after confirming that some kilo-bytes are downloaded, the program is installed in a few seconds. We assume that we have an Arduino UNO programmed with the query-response sketch from page 66 and connect the USB cable from the UNO to a USB port on the Raspi. We then check the device name of the serial line by issuing `dmesg` in a terminal window, and inspect the output near the end. In my case, one line contains the string `ttyACM0`, but `ttyUSBn` with any number `n` can also occur, depending on the serial-to-USB converter installed on the UNO. Since the Arduino sketch uses 9600 baud for the serial line, we connect to it with

```
screen /dev/ttyACM0 9600
```

on the command line of the Raspi. Then we query the UNO by typing A0? followed by `Enter` in the screen terminal-window. The UNO should respond with A0 `nnn` where `nnn` is the value measured by the ADC on the UNO. Congratulations, we have established a communication link from the Raspi to the Arduino UNO. Once we are done with getting data from the UNO, we close the connection by pressing `Ctrl-a-k` in the screen window, confirm the question of whether we really want to quit by y, and return back to the command prompt.

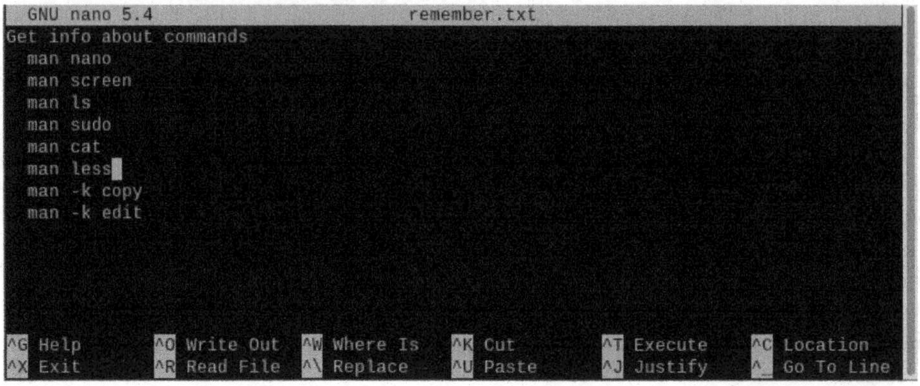

```
 GNU nano 5.4                            remember.txt
Get info about commands
 man nano
 man screen
 man ls
 man sudo
 man cat
 man less█
 man -k copy
 man -k edit




^G Help        ^O Write Out    ^W Where Is    ^K Cut      ^T Execute    ^C Location
^X Exit        ^R Read File    ^\ Replace     ^U Paste    ^J Justify    ^_ Go To Line
```

Figure 5.3   The nano text-editor with available commands listed on the bottom.

The **screen** program and in fact most programs on a Unix or Linux system bring their own help system, called manual pages, which can be accessed from the command line by typing **man <program name>** – for example, **man screen** in the present case. This displays information on how to use the program in the terminal window. In particular, all command line switches are explained. We exit from the **man** program by typing "q" in the window. Using **man** only works if the program name is known, but we can use the command **man -k <keyword>** to search the system for manual pages that have to do with **<keyword>**. The latter command works if we initialize a database once with **sudo mandb**. Check with **man mandb** to see what it really does!

Later on, we will often have to edit configuration files or to write program code. The edited files are usually plain text files, so-called ASCII files, and we create and edit them with an editor program such as **nano**, **vi**, or **emacs**. The first two are already installed on the system, and since **vi** has a steep learning curve, **nano** is a good choice to start, unless we have another preference. We start **nano** from the command line by typing **nano <filename>**, and after pressing **Enter**, the terminal window opens with the contents of the file, if it exists, or an empty file, if it does not exist. Figure 5.3 shows **nano** after we added some text. At the bottom of the window the most important commands are listed, especially **Ctrl-x** to exit the program and **Ctrl-g** to open the built-in help system with more information about using **nano**. Here, both small and capital letters work in conjunction with the **Ctrl** key to execute commands, such as **Ctrl-k** and **Ctrl-u** to cut and paste the line with the cursor. Note that once a text file is written to disk, we can also view it using the **less** command or with **cat**. Please check the respective manual pages for additional information.

We immediately put **nano** to good use and replace the default desktop manager, called **mutter**, which has an annoying bug that sometimes prevents windows from being resized or moved. We need to edit the configuration file **/etc/xdg/lxsession/LXDE-pi/desktop.conf** with

```
sudo nano /etc/xdg/lxsession/LXDE-pi/desktop.conf
```

and change the second line in the file from **window_manager=mutter** to **window_manager= openbox**, save the file and reboot. Note that editing file requires administrator privileges and we therefore need to prepend the command with **sudo**.

Any program, library, or other package that is available in the official Raspberry Pi repositories can be installed by first updating the database with **sudo apt-get update**, which is only necessary once per session and then installing the package with **sudo apt-get** install <package-name>. We use this newly won information to install the MATLAB-clone octave [25] by issuing

    sudo apt-get install octave

on the command line. After a lengthy installation that takes a few minutes, depending on the download speed, the program is ready to use. We start it from the command line by the command octave, followed by Enter. Octave greets the user with its version number and some copyright information before the prompt octave:1> appears. There we can enter MATLAB-compatible commands like A=[1,2;3,4]; B=inv(A), which defines a 2×2 matrix A, inverts it, and stores the result in the variable B. Later we will also connect to the Arduinos from within octave.

In order to be able to display and convert graphic files from one format to another, the imagemagick package is invaluable. We install it by executing

    sudo apt-get install imagemagick

from the command prompt. Once it is installed, we can display almost any graphics file with display graph.png, where graph.png is just an example of any graphics file. Converting an existing file graph.png to JPEG format is done by typing convert graph.png graph.jpg on the command line. Use man display and man convert to find out more about what the programs can do.

Installing with apt-get is already rather convenient because dependencies to external libraries are automatically resolved, and any missing libraries are installed in order to guarantee a reliably working system. There is, however, an even more convenient way, namely by using the package manager synaptic. We install it by issuing

    sudo apt-get install synaptic

on the command line. Once the installation is complete, it appears as *Synaptic Package Manager* in the *Preferences* section on the menu, behind the raspberry logo. Starting it from there first requests the login password before a window appears, shown in Figure 5.4. We use the *Search* button on the top to find programs. Just click on it, enter a keyword and peruse the list, install any interesting program, and try it out.

We use that feature immediately, and install programs to access remote computers, such as the NodeMCU that provides server capabilities via sockets. For this we need one or both of the netcat and telnet (client) programs, and install them via synaptic by searching for the program name; we pick the required package from the list by right-clicking it and choosing *Mark for Installation*. Once all programs are selected, we press the *Apply* button near the top left of the synaptic user interface. Upon completion of the installation, we are ready to use the programs to connect to the NodeMCU running the socket-based server, which we discuss in detail later. Here we briefly illustrate the capabilities of the netcat program by opening an ad-hoc server in one terminal window. Running netcat -l 11111 starts a server that listens (-l) on port 11111. Note that normal users can only use port numbers above 1023. We connect to that server by running netcat localhost 11111 from a second terminal window, where localhost is the default name for *this computer* and the port number on which the server listens. Now anything written in one terminal automatically appears on the other. In this example, we only connected two running versions of netcat on the same computer. One is started as a server, the other as a client, but the same functionality works between any computer with a working network connection. Find out more about netcat by consulting its manual page with man netcat. Should we wish to remove a previously installed program, we select it in synaptic by right-clicking the program in the list and selecting *Mark for Removal*. Pressing *Apply* promptly removes it.

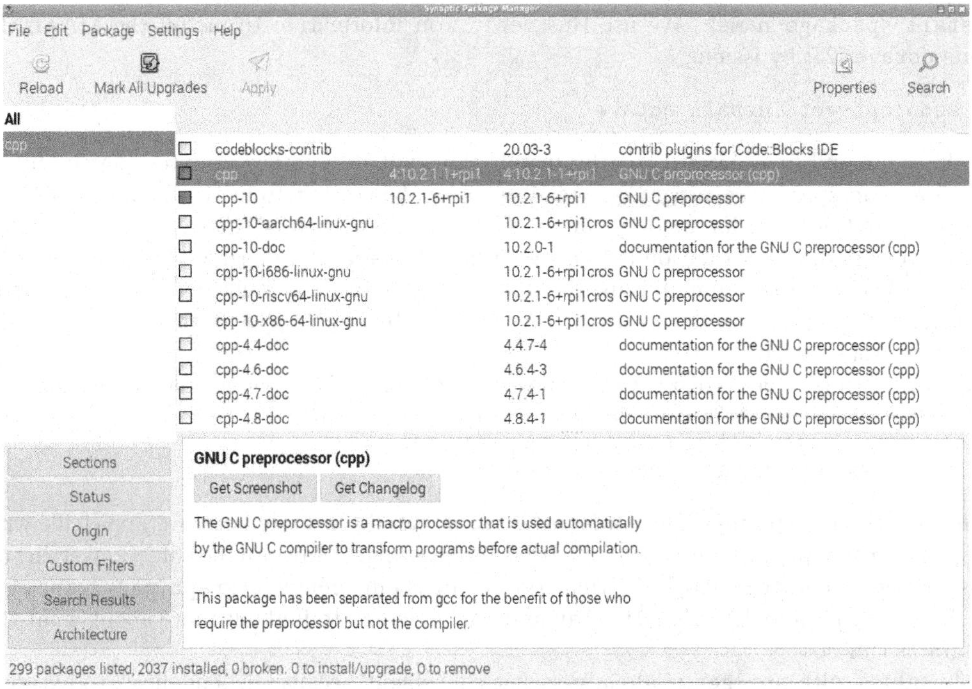

Figure 5.4 The Synaptic package manager.

Nowadays many software packages are available from so-called `git` repositories. This includes the software accompanying this book, which is hosted on `https://github.com` and can be downloaded with

```
git clone https://github.com/volkziem/HandsOnSensors2ed.git
```

where the command `git clone` creates a subdirectory `HandsOnSensors2ed` into which the contents of the repository is copied. Should the `git` program be missing, we install it with `sudo apt-get install git` or with `synaptic`. Again, the command `man git` provides a wealth of information about the many options `git` and `man git clone` about the specific options when downloading – also called cloning – a repository.

So far, we have used the Raspi as a standalone computer with screen, keyboard, and mouse directly attached to the Raspi, but that ties up the peripheral hardware. So, now is the time to install software to allow logging onto the Raspi via the network, and receiving a virtual copy of the Raspi desktop on our regular desktop computer. Then we have a Raspi-desktop in a window that behaves just like the one on a regular monitor. There already is such a server from *RealVNC* installed on the raspi, which can be enabled in the *Raspberry Pi Configuration* program, but in order to use it you need to install the proprietary *RealVNC viewer* on your Desktop. If you are reluctant to do so, you can achieve the same functionality by installing the `tightvncserver` on the Raspi with the command

```
sudo apt-get install tightvncserver
```

or, alternatively, via `synaptic`. This command removes the pre-installed *realvnc-server* and replaces it by one that is supported by many open source `vncviewer` programs. Once the installation has finished, we run `vncpasswd` from the command line as user `pi`, answer the questions about passwords, and remember them.

On the desktop computer from which we want to log onto the Raspi via the network, we install the vncviewer software using the package manager of our Linux distribution. During the installation, the vncpasswd program should be installed as well, and we need to run it from the command prompt on our desktop computer. It prompts us to enter the same password for the VNC connection that we selected on the Raspi when running the vncserver program for the first time. Once that is done, the basic infrastructure to log onto the Raspi using VNC is in place. To simplify the entire process, we create a file named raspi.sh on the desktop computer with the following contents

```
#!/bin/bash
RASPI=192.168.10.nn
ssh -X pi@${RASPI} vncserver -geometry 1280x960 &
sleep 10
vncviewer ${RASPI}:5901 -passwd ${HOME}/.vnc/passwd &
```

and place that file in the ~/bin directory. This little program remotely logs onto the Raspi with ssh, starts the vncserver program on the Raspi with the specified desktop geometry (we may use a smaller one such as 1024x768), then waits a few seconds and starts the vncviewer on the desktop computer using the credentials in the .vnc/passwd file in the home directory. After a few further seconds the desktop of the Raspi appears in a window on the desktop computer, and we seamlessly move the mouse and keyboard focus to the Raspi desktop, and execute commands on the Raspi. This is a very convenient way to use the Raspi, because it only requires a network connection and no other peripheral devices. In passing we mention that starting graphical programs that require administrator privileges, for example by sudo synaptic, on some systems do not have access to the display. In such cases we may grant authorization to all local users by first executing xhost localhost as normal user in a terminal window.

At this point, we have a convenient setup to work on the Raspi, and are basically ready to communicate with all the microcontrollers with attached sensors. But using the hardware for both wired and wireless Ethernet on the Raspi is just too tempting to neglect, and in the next section we briefly describe how to use the wireless interface on the Raspi to span a private WLAN network to which we later connect the sensor nodes.

5.4   RASPI AS A ROUTER

In this setup we assume that the Raspi is connected to our normal network with its wired Ethernet RJ45 port, and we have access to it from our desktop computer to log on using ssh or vncviewer, as previously discussed. On the Raspi we need to install the following packages

```
sudo apt-get update
sudo apt-get install hostapd dnsmasq
```

where we first update the repositories, and then download and install the hostapd and dnsmasq package. The former program is responsible for spanning the private wireless network, and turns the Raspi into a WLAN access point. The latter provides IP numbers on the private network via the DHCP protocol, and translates website names to IP numbers.

As a first step we need to re-configure the network setup. In the default configuration there is a background process, a daemon called dhcpd, that tries to obtain IP numbers for every network interface on the Raspi, including the wireless interface called wlan0. Instead, in order to operate the Raspi as an access point, we use for example nano to configure the interface ourselves by adding the following lines

```
interface wlan0
  static ip_address=192.168.20.1/24
  nohook wpa_supplicant
```

at the end of the configuration file /etc/dhcpcd.conf. The first line indicates that the following two lines relate to the built-in wireless interface. The second line defines a static IP number to be 192.168.20.1 and the type of network (class C) with the shorthand notation /24. The third line prevents the dhcp daemon to take control of the interface.

We now enable *network address translation*, also called *masquerading*, between the wireless network on wlan0 and the wired network on interface eth0 by executing the following two commands in a terminal window

```
sudo iptables -t nat -A  POSTROUTING -o eth0 -j MASQUERADE
sudo sh -c "iptables-save > /etc/iptables.ipv4.nat"
```

The first line uses the built-in command iptables to establish the rules to apply and the second line saves them to the file /etc/iptables.ipv4.nat such that they can be restored after a reboot. During the next boot the system loads this configuration from the automatically executed file /etc/rc.local. We simply add the following iptables-restore command just before exit 0 at the end of /etc/rc.local.

```
iptables-restore < /etc/iptables.ipv4.nat
exit 0
```

This ensures that the network translation is properly initiated after rebooting the system. Finally we allow network packages to pass between the interfaces, which is disabled by default as a security measure. We enable *forwarding* by removing the hash # from the line

```
net.ipv4.ip_forward=1
```

in the file /etc/sysctl.conf. We omit this step if all sensor nodes should only communicate with the Raspi, but no other computer on the local network should be able to surf the outside Internet beyond the Raspi, which might not be needed and poses a potential security risk. This last step completes the basic network setup, and we turn to the configuration of the access point software.

The configuration file for the hostapd daemon is /etc/hostapd/hostapd.conf. It contains the following lines:

```
# /etc/hostapd/hostapd.conf
country_code=SE
interface=wlan0
ssid=messnetz
channel=7
macaddr_acl=0
auth_algs=1
ignore_broadcast_ssid=0
wpa=2
wpa_passphrase=zxcvZXCV
wpa_key_mgmt=WPA-PSK
wpa_pairwise=TKIP
rsn_pairwise=CCMP
hw_mode=g
```

The file must be made readable for the owner only by executing `sudo chmod 600 /etc/hostapd/hostapd.conf`. It specifies the country in which we live – this is needed to determine the permissible WLAN frequencies, which differ from country to country – and other definitions, such as the name of the network, the details of the encryption, and the passphrase. Here we choose a secret phrase to prevent others from misusing your wireless network. Moreover, we need to ensure that `hostapd` finds its configuration file by verifying the line

```
DAEMON_CONF=/etc/hostapd/hostapd.conf
```

appears near the top of the file `/etc/init.d/hostapd`. Add it unless it is already there. Then we instruct the system to start the `hostapd` daemon at boot time with the command

```
sudo systemctl unmask hostapd
sudo systemctl enable hostapd
```

After rebooting the system, we should be able to see a new WLAN named `messnetz` from a computer with a wireless network interface, but before we can connect to it we need to start the `dnsmasq` daemon that provides DHCP services on the `192.168.20.xx` network and distributes IP numbers. The configuration file is called `/etc/dnsmasq.conf` and it should contain the following lines:

```
domain-needed
interface=wlan0
dhcp-range=192.168.20.100,192.168.20.200,12h
listen-address=192.168.20.1
```

We save the originally installed version under a new name because it contains explanations of all parameters and is useful as a reference. For our system, however, we only need those in the above example, and they are almost self-explanatory. Finally, we register the `dnsmasq` daemon to start at boot time with the commands

```
sudo systemctl unmask dnsmasq
sudo systemctl enable dnsmasq
```

and reboot the Raspi to ensure that all parts of the system are synchronized.

In order to figure out which computers are connected to our WLAN `messnetz` and which services they offer on open ports, we install the program `nmap` with `sudo apt-get install nmap`. The following call

```
nmap -sn 192.168.20.1/24
```

lists the IP numbers of all computers connected to `messnetz`. The option `-sn` specifies the type of scan and `192.168.20.1/24` the range of IP numbers to probe, where `/24` is a shorthand notation for the netmask `255.255.255.0`. This implies that all IP numbers of the form `192.168.20.nn` are probed. If a device with, for example, `192.168.20.56` appears, the command

```
nmap 192.168.20.56
```

lists its open ports. This is a convenient way to ensure that a server started on the device is actually operational. Explore the many options of `nmap` with `man nmap`.

At this point we have turned the Raspi into a WLAN access point that spans the `messnetz` WLAN. If the IP forwarding is enabled, we can even surf the web from any computer on `messnetz`. But we are mostly interested in communicating with sensor nodes that are connected to `messnetz` and will discuss how to use the Raspi as the spider in the center of a network of sensor nodes connected by serial or WLAN links.

5.5 COMMUNICATION WITH THE ARDUINO

After the basic network infrastructure is set up, we connect the Arduino to the Raspi. Most programs on the Raspi work in a similar way as on desktop computers, which is no surprise, because a modern Raspi is about as powerful as a desktop computer from about a decade ago. We start by communicating via the Arduino IDE first, but later we will also use the Python programming language and octave.

5.5.1 Arduino IDE

We can install the Arduino IDE on the Raspi with `sudo apt-get install arduino`, but the version `1.8.13` in the repositories is moderately old. We can install a more recent version from `https://www.arduino.cc`. To do so, we open a browser on the Raspi, direct it to the Arduino download website, choose the *Linux ARM 32 bit* version, and download directly to the Raspi's download directory `/home/pi/Downloads/`. At the time of writing the most recent version was `1.8.19`. We suggest to use the legacy version until the newer IDE version 2 supports all features such as data upload that we later need.

If we chose to use the most recent version, we install it, as is customary, under the `/usr/local` tree by executing the following commands:

```
cd /usr/local
sudo tar xvJf /home/pi/Downloads/arduino-1.8.19-linuxarm.tar.xz
cd /usr/local/bin
sudo ln -s /usr/local/arduino-1.8.19/arduino arduino19
```

where the `tar` command unpacks the downloaded archive into the subdirectory named `/usr/local/arduino-1.8.19/`, and the `ln -s` creates a so-called soft link with the alias `arduino19` for the `arduino` executable in the `/usr/local/bin/` directory. In this way it is automatically found when typing `arduino19` at the command line in a terminal window. Optionally, we can create a link in the *Programming menu* by starting the *Main Menu Editor* in the *Preference* section behind the *Menu* button with the image of a raspberry. In the left panel of the *Main Menu Editor* we select *Programming*, and then click on *New Item* on the right-side panel. In the opening window, we enter the name of the new program, *Arduino,* and browse to the `arduino19` executable in `/usr/local/bin` directory. Once these steps are completed, there is an *Arduino* entry in the *Programming* section of the main menu. Selecting it opens the Arduino IDE on the Raspi, and we could start programming the Arduino directly from the Raspi.

Instead of retyping all the programs for the Arduino on the Raspi, we copy the files from our desktop computer to the Raspi with `scp`, which is part of the secure-shell program suite if it is a Unix-based system such as Linux or Mac. On a Windows system we use *WinSCP*. On a Linux Desktop computer we execute

```
scp -r Arduino pi@192.168.10.nn:
```

which copies the Arduino subdirectory and everything in it to the Raspi, including all hardware descriptions from the `Arduino/hardware` directory. Now we need to restart the Arduino IDE, and all our previously prepared sketches are available under the *File→Sketchbook* menu in the Arduino IDE on the Raspi. Before starting to compile our examples we add support for ESP8266 and ESP32-based microcontrollers by following the steps outlined at the end of Section 4.2 on page 60. On the Raspi, the procedure is just the same. Now we are ready to connect an Arduino UNO or NodeMCU to a USB port on the Raspi and program them directly from the Raspi.

Of course, we need to select the type of microcontroller and the port to which it is connected the same way as on the desktop computer. In order to test that all is working properly, we download the `query-response` sketch to an Arduino UNO and open the *Tools→Serial Monitor* in the Arduino IDE. This should allow us to communicate with the UNO via the USB serial line. In case it does not work, ensure that the *Port:* in the *Tools* menu points to the correct serial port with the Arduino UNO and that the UNO is actually selected in the *Board:* menu.

5.5.2 From the command line

At this point we have established the Raspi as a development system for Arduino software and ensure it communicates via the USB-based serial line with the Arduino IDE. But normally we want to access the Arduino from self-written programs rather than always using the Arduino IDE. The first option is to use the `screen` program or any other terminal program to connect to the UNO. After closing the Arduino IDE, we therefore start the following program on the command line on the Raspi:

```
screen /dev/ttyACM0 9600
```

just as we did earlier when the UNO was connected to the Desktop computer. Bluetooth communication works exactly the same way as on the Desktop computer. Just follow the details described in Section 4.6.2 on page 110.

In the next step, we use a NodeMCU and assume that the Router software from Section 5.4 is installed on the Raspi. We change the variables `ssid` and `password` in the NodeMCU sketches from Section 4.6.3 to those used on `messnetz`. Two lines near the top of the sketches now need to read

```
const char* ssid     = "messnetz";
const char* password = "zxcvZXCV";
```

where we need to make sure to enter the same password as the `wpa_passphrase` in the `hostapd.conf` file. We then connect the NodeMCU to a USB port on the Raspi and download the updated socket-based server to the NodeMCU. We can observe progress of the NodeMCU boot process with the *Serial Monitor* while it connects to the `messnetz` WLAN. After a while it should report the acquired IP number that the `hostapd` on the Raspi dished out. It lies in the `192.168.20.nn` network, which corresponds to the IP range we specified in Section 5.4, in the configuration files of `dnsmasq`. In my case the IP number received was `192.168.20.135`, but yours may be different. At this point we can connect to NodeMCU via WLAN by issuing the following command in a Raspi terminal window:

```
telnet 192.168.20.135 1137
```

where `1137` is the port number that is specified in the sketch that runs on the NodeMCU. The `telnet` program provides a prompt at which we can type commands to send to the NodeMCU. So, typing `A0?` on the telnet prompt should result in the reply `A0 nnn`, also displayed in the telnet window, as shown in Figure 5.5. We close the session by using the command `Ctrl-]` and then type `quit` at the telnet prompt.

Instead of using `telnet`, we can also use `netcat` and type the following command in a terminal window:

```
netcat -C 192.168.20.135 1137
```

where the `-C` option ensures that CR-LF characters are sent after each line, which is required by the NodeMCU to recognize the end of a command. Otherwise the communication works in the same way as with `telnet`.

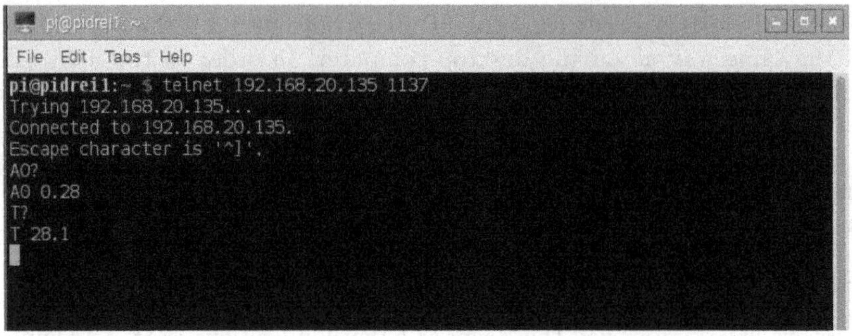

Figure 5.5 Telnet session connected wirelessly to the NodeMCU.

The previous two paragraphs only verify that the communication works properly and that we have our system under control. Now we proceed and use the Python language to achieve the same feat.

5.5.3 Python

Python [26] in its current version 3 is a modern programming language that is installed on any Raspi by default. Actually, the *Pi* in Raspberry Pi stands for *Python Interpreter.* Unless you have some experience with Python, I suggest you have a look at some of the tutorials at

https://wiki.python.org/moin/BeginnersGuide/Programmers

or search the web for detailed explanations of some of the simple, often even self-explanatory, commands we use here. The purpose to access an Arduino or NodeMCU from Python is to show how to write customized programs on the Raspi that work hand-in-hand with sketches that run on the Arduino.

We use the following Python script to query the UNO with the query-response sketch for a single value from analog pin 0.

```
# query_arduino.py, V Ziemann, 220930
import serial, time
query="A0?\n".encode('utf-8')
ser=serial.Serial("/dev/ttyACM0",9600,timeout=1)
time.sleep(3)     # wait for serial to be ready
ser.write(query)
time.sleep(0.1)
reply=ser.readline().decode('utf-8')
print(reply.strip())
ser.close()
```

Python is rather lightweight, and if the `import` statement reports an error, we have to import extra functionality, here for serial communication, by executing

```
sudo apt-get install python3-serial
```

at the command prompt. Now we can import the support for handling the serial line and basic time-handling. We need the latter to implement delays in the script. In the next line we define a variable `query` that contains the query string we send to the Arduino. Note that we explicitly add the carriage return \n character. Moreover, in Python3 we need to

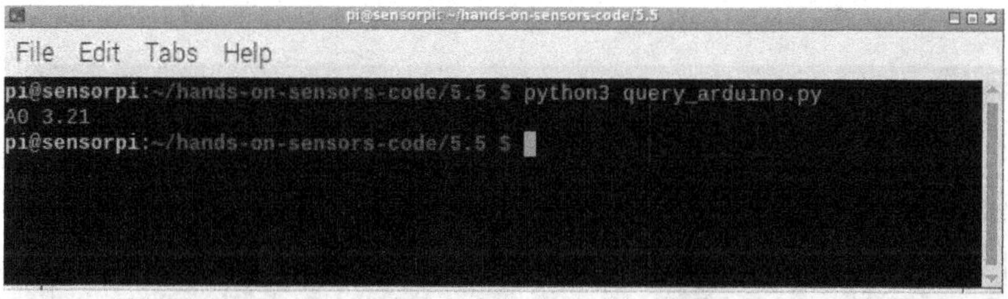

Figure 5.6 Using Python on the Raspi to communicate with the `query_response` sketch running on the Arduino UNO.

explicitly specify the encoding of strings by appending `.encode('utf-8')` to the definition. Then we open the serial port `ser` on `/dev/ttyACM0` with a baud rate of 9600 and a timeout of 1 second, which prevents unsatisfied read attempts from blocking the program. We wait for three seconds to allow the operating system to finish opening the serial port, and then submit the query string with the `ser.write` command. Note that we use the method `write` on the serial device `ser`. Then we wait for 0.1 seconds and read characters up to the CR-LF character with `ser.readline().decode('utf-8')` into the variable `reply`. Here we read the line and decode the characters coming from the Arduino. We display the reply with the `print command` after stripping off leading and trailing white-space characters, such as normal spaces or CR-LF characters, before closing the serial line. We run the Python script that we give the name `query_arduino.py` by entering

```
python3 query_arduino.py
```

on the command line of the Raspi and show this example in Figure 5.6.

The previous example script shows how to request a single measurement value. We easily expand the script to query the UNO repeatedly, and even provide simple ASCII graphics that show the measurement values as a function of time. This is accomplished by the Python script below. Example output is shown in Figure 5.7, where we see that we call the program by executing

```
python3 ask_arduino_repeat_plot.py
```

and the program then shows the elapsed time since it started, the measurement value, and a graphical representation of the value between the expected minimum and maximum values. This functionality resembles a simplified version of the Serial plotter built into the Arduino IDE. The Python script is the following:

```
# ask_arduino_repeat_plot.py, V. Ziemann, 220930
import serial, time, atexit
def cleanup():        # ensure serial line is closed after CTRL-c
   ser.close()
atexit.register(cleanup)  # register the cleanup function
query="A0?\n".encode('utf-8')   # the query string
amin=0                          # minimum expected value
amax=5                          # maximum expected value
width=70                        # number of character used for plot
ser=serial.Serial("/dev/ttyUSB0",9600,timeout=1)
lll=len(query)-1    # needed to remove 'A0' from reply
time.sleep(3)       # wait for serial to be ready
t0=time.time()      # starting time
```

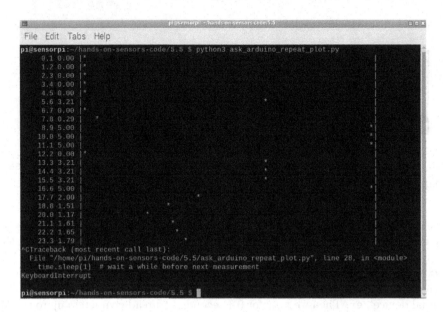

Figure 5.7 Simple ASCII graphics from querying the UNO repeatedly.

```
while 1:                          # repeat forever
    ser.write(query)              # send query
    time.sleep(0.1)               # wait a bit
    reply=ser.readline().decode('utf-8')      # read response
    value=reply[lll:].strip()                 # make numeric
    k=int((width-2)*float(value)/(amax-amin)) # where to place *
    p='%8.1f %4s |' % (time.time()-t0,value)
    for j in range(0,width-1):
        if j==k:
            p+='*'
        else:
            p+=' '
    p+='|'
    print(p)
    time.sleep(1)  # wait a while before next measurement
```

First we import support for serial communication, time, and the atexit functionality, which allows us to register a function that does some cleaning-up activities when the program terminates. Since we will employ an infinite loop to read the UNO, which we intend to stop asynchronously with Ctrl-C, this ensures that the serial line is properly closed. Note that Python uses indentation instead of brackets to define the scope of functionality, and the function body for the cleanup() function is just ser.close(), but indented by a few characters. After specifying and encoding the query string we define several variables, the serial line is opened, and we determine the expected length 111 of the characters preceding the value of the reply, wait a short time, and record the present time in the variable t0. The while 1: statement initiates a loop that runs forever. Inside the while loop (note the indentation to indicate the scope), we write the query to the UNO, wait a short while, read the response into the string reply and decode it. In the following line, we remove the first few characters such as A0, and remove white-space characters with the strip() method. The variable k scales the measurement value such that it lies within the range specified in the width variable, and then we start building the string p by first placing the seconds elapsed since t0 and the measurement value followed by a vertical bar denoting the start of the ASCII graphic. Then we loop over all subsequent places in the string p and place a * at the location corresponding to the scaled measurement value k, and a space otherwise. Finally, we add another vertical bar | and print the string p to standard output, before waiting one second for the next iteration to start. Running this Python script results in the rudimentary ASCII graphics shown in Figure 5.7.

The previous script works well with any serial line, including a serial port behind which a Bluetooth link is hidden. The only change we need to implement in the previous script is to replace the serial port /dev/ttyACM0 with the Bluetooth port that is typically called /dev/rfcomm0 or /dev/rfcomm1.

But what about connecting to the NodeMCU running the socket server that listens on port 1137? We only show the most basic network client that sends the query and displays the reply on standard output. More elaborate examples such as the simple ASCII graphics from above can be built quite easily once the basic network communication setup is under control. We illustrate these basics in the following example:

```
# socket_client.py, V. Ziemann, 220930
import socket, atexit, time
def cleanup():
    sock.send(b"quit\n")
    sock.close()
atexit.register(cleanup)
sock=socket.socket(socket.AF_INET,socket.SOCK_STREAM)
sock.connect(("192.168.20.184",1137))
sock.send(b"T?\n")
time.sleep(0.1)
reply=sock.recv(1000).decode('utf-8');
print(reply.strip())
```

The structure of the program is very similar to the one for serial communication. First, the necessary functionality is imported, and then we define a cleanup() function that executes before the program closes. This is the safest way to ensure that the server closes the connection properly. Inside the cleanup() function, we instruct the python script to send the string quit, which is unknown to the NodeMCU and causes it to close the connection, before closing the socket on the client side. Note that we use the shorthand notation for encoding literal text strings by prepending the letter b to the string. The function cleanup() is then registered in the call to atexit.register(). In the next line the socket sock is created with the specifications of a normal TCP/IP socket, before connecting to the server running on the NodeMCU at IP number 192.168.20.184 and port 1137. Then we send the query string T? and wait a short while before we receive the reply that we print to standard output after stripping white-space characters.

In this section, we cover how to interface the sensor nodes with Python, and even present them as rudimentary ASCII graphics. This might suffice for quick-and-dirty fixes to rapidly observe how some value changes with time, but for more professional-looking results, we turn to octave in order to interface the sensor nodes and prepare much nicer plots for presentations and reports.

5.5.4 Octave

We already installed octave in Section 5.3, but in order to use serial communication and network sockets, we need to install the octave-instrument-control package with

```
sudo apt-get update
sudo apt-get install octave-instrument-control
```

After the installation finishes, we add the following line:

```
pkg load instrument-control
```

to the octave startup file named .octaverc in our home directory, which in most circumstances on the Raspi is located in the /home/pi/ directory. This will cause the toolbox to be loaded every time we start octave. While we edit that file, we also add the line

```
graphics_toolkit('gnuplot')
```

to the .octaverc file to avoid a bug when using the octave plot command. But this may not be needed in all circumstances.

Once the installation completes, we are ready to use it to communicate with our sensor nodes. First, we try to communicate with the UNO connected to the USB port of the Raspi, which is accessible as /dev/ttyACM0 as before. In order to read one measurement from the UNO, we open the serial line from the octave prompt, submit the query, wait for the response, and display the result. The following program achieves this.

```
% read_serial.m
s=serialport("/dev/ttyUSB0",9600);   % open serial line
pause(3)                             % wait for this to complete
reply=queryResponse(s,"A0?\n")       % send query and receive reply
clear s                              % close serial port
```

It implements the simple query-response communication protocol we used earlier. Here we encapsulate the details of the query-response interaction in the function queryResponse(), because the installed version (0.6) of the instrument-control toolbox does not support reading from the serial device up to a termination character. We therefore implement that feature ourselves.

```
% send query string and return response up to termination character.
function out=queryResponse(dev,query,term_char)
if (nargin==2) term_char=10; end    % defaults to LF=0x0A
write(dev,query);                   % send query to device
i=1;
int_array=uint8(1);
while true                          % loop forever
   val=read(dev,1);                 % and read one byte
   if (val==term_char) break; end   % until term_char appears
   int_array(i)=val;                % stuff byte in output
   i=i+1;
end
out=char(int_array);                % convert to characters
```

The three arguments are the serial device dev, the query string, and an optional termination character, which defaults the line-feed character (hex=0x0A, dec=10). In the function, we first send the query string in the write() function call, and after initializing some variables, we repeatedly read one character at a time with the read() function and append the character to the int_array unless it equals the termination character. Once the termination character is received, the while true loop exits, and the int_array is converted to characters in the last line. The converted array is returned to the calling program. Needless to say, we can also use any Bluetooth device with a serial line interface, only the device file is /dev/rfcomm0 instead of the normal serial device files /dev/ttyACM0 or /dev/ttyUSB0.

Reading via WLAN by connecting to the socket on the NodeMCU is done in much the same way as reading from the serial line. Instead of opening a serial line we open a TCP connection with the tcpclient() function. The basic code snippet that accomplishes this is shown here:

```
% read_tcp.m
s=tcpclient("192.168.20.184",1137);   % open connection
pause(0.1)                             % wait, not really needed
reply=queryResponse(s,"A0?\n")         % send query and get reply
write(s,"quit\n");                     % close remote socket
clear s                                % close local socket
```

which follows the logic from the serial communication example above. The function queryResponse() is the same as the previous example for the serial line. Note also that we need to explicitly close the remote socket on the NodeMCU by sending quit.

To illustrate the usefulness of the octave interface to serial line or networked devices, we write a simple temperature logger that connects to the NodeMCU and produces a nice plot of the temperature as a function of time. The following octave program achieves that.

```
% temperature logger, V. Ziemann, 220930
clear all
s=tcpclient("192.168.20.184",1137);
pause(0.01)
running=0;
while running<1000
  running=running+1;
  reply=queryResponse(s,"T?\n");
  val(running)=str2double(reply(2:end));
  x(running)=now;
  plot(x,val,'*')
  ylim([22,28])
  datetick('x','ddd/HH:MM:SS')
  ylabel('Temperature [C]')
  xlabel('Time')
  pause(1);
end
write(s,"quit\n");
clear s
```

First, we clear all variables before we establish the TCP connection to the server on the NodeMCU, on port 1137 and IP number 192.168.20.135. Then we wait a short while and initialize a variable running to zero. We use this variable as an iterator in the loop and check whether the limit has been reached. In this simple example, we only iterate 10 times. In the loop, we increment the running variable and then send the query to the NodeMCU and receive the measurement in the string reply. Note that reply starts with a T, which we remove in the following line by converting reply from position 2 to the end to a double variable. The result we copy to the variable val at index position running. In this way we use the iterator not only to count the loop iterations, but also as an index of where to put the measurement values. In the variable x we copy the current time, which is returned by the built-in function now(). Then we plot the value versus the time using the plot() function, specify the vertical temperature range with the ylim() function, and specify the axis labels

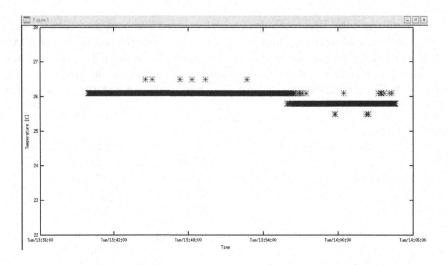

Figure 5.8 Plot from the temperature logger.

for vertical and horizontal axes. Finally we use the `datetick()` function to specify the type of tick marks we want. In this case we specify the day of the week and the time in hours, minutes, and seconds. The `datetick()` function uses the special format in which the `now()` function returns the time to extract the desired format of the tick mark. The full set of options is explained in the help text of the `datestr()` function that can be accessed by typing `help datestr` at the octave prompt. Before repeating the measurement, we wait a specified time, one second in the example. Finally, once the desired number of iterations is completed we ensure that the socket is closed on both the server and in the octave client. We show the resulting plot in Figure 5.8.

By now we have managed to record data from sensor nodes via serial and WLAN and display them online, even with reasonably attractive graphics, but for this privilege we need to have octave running while taking data. This may be feasible for a period of a few hours, but is hardly attractive for longer periods. In that case we need a separate repository to store the data, and only create plots when needed. We therefore need to separate the presentation from the data storage process, and this is the topic of the next section.

5.6 DATA STORAGE

In this section, we address several ways to store our measurement data, and the archetypical data storage repository is a *database*. We introduce a number of databases and use octave as well as pythonfor the presentation of the data.

5.6.1 Flatfile

The simplest database is certainly a file containing a time stamp and one or more measurement values, possibly even in human-readable form. This is called a flatfile database, and we create one by using the following Python script:

```
# ask_repeat.py, V. Ziemann, 220930
import serial, time, sys, atexit
def cleanup():
    ser.close()
```

```
atexit.register(cleanup)
query="A0?\n".encode('utf-8')
ser=serial.Serial("/dev/ttyUSB0",9600,timeout=1)
time.sleep(3)      # wait for serial to be ready
while 1:
    ser.write(query)
    time.sleep(0.1)
    reply=ser.readline().decode('utf-8')
    print(int(time.time()), reply[3:].strip())
    sys.stdout.flush()
    time.sleep(1)
```

and store it in a file called **ask_repeat.py**. We recognize the same organization as before, when we discussed Python scripts with importing functionality, registering the **cleanup()** function, opening the serial line, and repeatedly sending the query string and receiving the reply. The only difference is that we print the *Unix time*, which is the number of seconds since January 1, 1970, also called the *epoch*, before the measurement value. In the output, we therefore see the following text

```
     :
1670955286 3.29
1670955287 3.30
1670955288 3.29
     :
```

scrolling by. If we redirect this output into a file using the following line of code:

```
python3 ask_repeat.py > db.dat
```

that we run in a terminal window on the Raspi, we create a flatfile database **db.dat** with the time stamps and the measurement values. Later we can retrieve and convert the data to a nice plot.

Running the data acquisition program in a dedicated terminal window may be useful for short measurement sessions, but is hardly useful if we want to log, for example, the temperature in a building for an extended period of time, say a few months. In that case it is much preferable to have a dedicated process that wakes up once every few minutes, reads the temperature, and stores the value together with a timestamp in a file before going back to sleep for a few minutes. As it turns out, Unix systems normally have a system that takes care of these repeated tasks. It is called **cron**, and it is a background process that wakes up once a minute, checks whether there is a task to do, does it, and sleeps for a minute before checking again.

Let us start by creating a Python script that we want to execute at regular intervals. In this particular case we place it in the **/home/pi/bin** directory and call the file **single_request.py**. It contains the following lines:

```
# single_request.py, V. Ziemann, 220930
import serial, time
ser=serial.Serial("/dev/ttyUSB0",9600,timeout=1)
time.sleep(2)          # wait for serial to be ready
ser.write(b"A0?\n")    # shorthand notation for encoding
time.sleep(0.1)
reply=ser.readline().decode('utf-8')
print(int(time.time()), reply[3:].strip())
ser.close()
```

We see that it is rather similar to the previous scripts to read from the Arduino. It just acquires one measurement and prints the value with the timestamp to standard output. In order to make this program accessible for the cron software, it is convenient to encapsulate it in a shell script file /home/pi/bin/readA0.sh that contains the following lines:

```
#!/bin/bash
/usr/bin/python /home/pi/bin/single_request.py >> /home/pi/A0.dat
```

We make the file executable with the chmod program by executing chmod +x readA0.sh in the /home/pi/bin directory. The first line instructs the operating system to interpret the following lines using the bash shell. The next line starts the single_request.py script using the /usr/bin/python3 program, and redirects the output to the file /home/pi/A0.dat. Here >> implies that new data is appended to an existing file. Note that absolute file names must be given, including the absolute path. We test the readA0.sh script to ensure that it creates the /home/pi/A0.dat file or appends a reasonable measurement value with timestamp to the file. Once we are satisfied, we register readA0.sh with the cron software and edit the configuration file for the cron program with the crontab program. We execute it from the command prompt by typing

```
crontab -e
```

The first time crontab -e is called it asks for an editor. We pick our favorite or follow the suggestion. Once the editor opens with the configuration file, we append the following line at the end of the file:

```
* * * * * /home/pi/bin/readA0.sh
```

and save the file. This automatically registers the readA0.sh script to execute once every minute whenever the Raspi is running. We check that the file is registered with the crontab -l command. It lists the contents of the crontab file for user pi provided we are logged on as user pi. The meaning of the six columns in the crontab file are *minute, hour, day of month, month, day of week,* and *program to execute.* Placing asterisks in the first five columns instructs cron to execute the program every time it wakes up. The output of the crontab -l command has some basic explanations, and more is available by executing man 5 crontab at the command prompt. So, now we have a background process that records a measurement once a minute and fills the file /home/pi/A0.dat until we remove the entry in the crontab file.

Our next task is to produce plots with python3 and two packages. The first package, numpy, contains many numerical functions, but also a powerful function to load data files. The second package, matplotlib, contains functions to plot data. Unless already installed, we add the packages with

```
sudo apt-get install python3-numpy python3-matplotlib
```

and are ready to read and display the data recorded in the file /home/pi/A0.dat with the following script.

```
# matplotlib_plotter.py, V. Ziemann, 221004
import numpy as np
import matplotlib.pyplot as plt
d=np.loadtxt('/home/pi/A0.dat',delimiter=' ')
plt.plot(d[:,0]-d[0,0],d[:,1])
plt.xlabel('Time [s]')          # axis labels
plt.ylabel('A0')
plt.show()                      # make plot visible
```

In the first lines we import the functionalities and assign short names (np,plt) to the packages, which saves us quite a bit of writing. Note that we use the pyplot features, which contains functions that make plotting particularly simple. The loadtxt() function in the following line loads the data from the file into the array d. The next lines simply plot the data, add axis labels, and display the plot on the screen.

A second option to display contents of the same data file, even with the timestamps taken into account nicely is to use octave, which has powerful capabilities to handle timestamps. We extract the data stored in the flatfile database and display it with the following script.

```
% flatfile_reader.m, V. Ziemann, 221004
d=importdata('/home/pi/A0.dat');
TZ=1; t=719529+(d(:,1)+TZ*3600)/86400.0;
plot(t,d(:,2))
datetick('x','HH:MM:SS')
```

Most of the work is done by the importdata() function. It figures out the format of the A0.dat file and loads its contents into the variable d, which is a matrix with two columns, one with the time information and the other with the measurement values. In the next line we convert the time into the standard format that the datetick() function expects. It will properly format the displayed time on the horizontal axis. Note that we use the variable TZ to denote the time zone. Since I live in Sweden, the local clock is one hour ahead of UTC standard time, to which the epoch refers. The constant 719529 is the number of days from January 1, 0000 until January 1, 1970, and 86400 is the number of seconds per day. The plot looks similar to the one shown in Figure 5.8.

After storing data in a flatfile database, retrieving it, and displaying the data, we now progress to using a more mature database, MariaDB. Earlier it was known as MySQL, but the copyright for that name is reserved, which caused the name change.

5.6.2 MariaDB

Among several databases available for the Raspi, we choose *MariaDB* [27] because internally it works just like MySQL and we keep referring to it as such. Moreover, MySQL-like databases are widely used and can be accessed reasonably easily from most programming languages. These include Python and, after some archaeology on the Internet, octave as well. We install the MariaDB database from the standard repositories with the following command:

```
sudo apt-get install mariadb-server
```

Despite installing the package mariadb-server, the name of the program to access the database is still mysql.

Once MariaDB is installed, we create a database named readA0 that will contain roughly the same information as the flatfile in the previous section, namely a timestamp and measurement values. Creation of a new database must be done by mysql-user root and executed with administrator privileges using sudo. Without being asked for a password we log into the database with the following command:

```
sudo mysql -u root
```

We are immediately greeted by the prompt MariaDB [(none)]> where we create the new database with the following command:

```
create database readA0;
```

Here we specifically need to point out the necessity to complete every command by a semicolon. We verify that the database `readA0` exists with the `show databases;` command at the MariaDB prompt. Since later we will not always want to work with the database `readA0` as mysql-administrator, we create a database user. To do so, we declare that we want to work with `readA0` by writing

```
use readA0;
```

at the prompt, which causes the name of the database to be displayed as part of the prompt. Now we create the user `me` and then grant privileges to work with the database, with the following sequence of commands:

```
create user 'me'@'localhost' identified by 'pwpw';
grant all privileges on readA0.* to 'me'@'localhost';
```

where `pwpw` is the (too simple, invent a better one) password. Luckily the syntax is fairly self-explanatory. Note that the granting of privileges can be tailored in a rather detailed fashion into reading, writing, and other privileges, but for our simple example we stick to the simple version. Once the general administration of creating a database and assigning a mysql user is done, we exit from the mysql prompt by typing `quit;`.

But now we log onto MariaDB as user `me` by typing

```
mysql -u me -p
```

at the command prompt and entering the password. Alternatively, the command `mysql -u me -ppwpw` with the password `pwpw` immediately appended after the `-p` will directly log onto the mysql-prompt, but now as the user `me` who only has privileges to work with the `readA0` database. We verify this by executing the `show databases;` command, which shows `readA0` and some other administrative database that do not concern us here. In order to create a data structure, called `table` in our database, we select it with `use readA0;` and then enter

```
create table fdata (ts timestamp, A0 float, A1 float);
```

which creates a table `fdata` that stores a time stamp and two values in every row. Other options for things to store are `integer` and BLOB, the latter being a *binary large object*. The MariaDB manual, available at `https://mariadb.org/documentation/`, holds a wealth of detailed information.

The administrative task to create the table is done at this point, and we could `quit;`, but it is instructive to insert values into the newly created table by executing the following command at the prompt

```
insert into fdata (A0,A1) values (3.14,2.71);
```

and verify the contents of the database with the command

```
select * from fdata;
```

which prints the contents of the table. We observe that the timestamp variable was automatically filled with the time the values are inserted. The syntax to insert values is, again, rather self-explanatory. We `insert into` a table called `fdata` the variables (A0,A1) with values 3.14 and 2.71. Reading all values from the table is achieved by the `select` command. Specifying the asterisk causes the entire table `fdata` to be displayed. If we only want to print the timestamp `ts` and value `A1`, we can issue `select ts,A1 from fdata;`. These commands to create, insert, and select data from the table are standardized and are called

Structured Query Language or SQL. Please inspect tutorials and further information about SQL on the Internet. After this exercise we exit the mysql program by executing the quit; command.

Now we know the basic SQL commands to insert and retrieve data from tables. But rather than typing them from the mysql prompt, our next task is to use Python to execute the SQL commands. We add MySQL functionality, which also works with MariaDB, to Python by installing the python3-mysqldb package with the command

```
sudo apt-get install python3-mysqldb
```

at the command prompt, and are ready to access MariaDB databases from within Python. As a first task, we read the table dat from the database readA0 we had created earlier. The following Python script achieves just that.

```
# mysql_show.py, V. Ziemann, 221003
import MySQLdb
db=MySQLdb.connect("localhost","me","pwpw","readA0")
cur=db.cursor()
cur.execute("select * from fdata;")
reply=cur.fetchall()
print(reply)
# for r in reply:                          # loop over entries
#    print(r)                              # print each entry
#    print(str(r[0]), str(r[1]), str(r[2]))  # format nicely
db.close()
```

First the library with MySQL support is imported, and then we connect to the database readA0 on computer localhost, as user me with password pwpw. The MySQLdb.connect() function returns a handle db to the database, and executing it corresponds to logging onto the database and executing the use readA0; command at the mysql prompt. The next line creates a cursor cur, which is equivalent to the MariaDB prompt and allows us to enter database insertion or retrieval commands. In the next line we execute the select command to display the entire table. We retrieve the output from the command in the variable reply with the fetchall() call and print the reply before closing the database. Note the structure of first executing a MariaDB command and then retrieving the reply with the fetchall() function. The reply, however, is a list of entries in an unfamiliar form. It is possible to display each measurement entry on a separate line by using the lines commented out by a hash (#). The code loops over each entry in the list, and in the first commented example, the print command displays one entry at a time. In the second example, the print command uses the str() function to convert each entry to a string which makes it human readable. Finally, we close the database with the call to the db.close() function.

Our next task is to insert new entries with measurements into the database, and this is accomplished by the following Python script.

```
# serial2msyql.py, V Ziemann, 221003
import serial,time,MySQLdb
ser=serial.Serial("/dev/ttyUSB0",9600,timeout=1)
time.sleep(3)      # wait for serial to be ready
ser.write("A0?\n".encode('utf-8'))
reply=ser.readline().decode('utf-8')
val0=float(reply[3:].strip())
ser.write("A1?\n".encode('utf-8'))
reply=ser.readline().decode('utf-8')
```

```
val1=float(reply[3:].strip())
ser.close()
db=MySQLdb.connect("localhost","me","pwpw","readA0")
cur=db.cursor()
sql="insert into fdata (A0,A1) values (%f,%f);" % (val0,val1)
print(sql)
cur.execute(sql)
db.commit()
db.close()
```

The first part of the script is very similar to earlier scripts. It just queries the Arduino UNO on the serial line and converts the measurement values from analog pin A0 and A1 to float values, which are stored in the variables val0 and val1. In the second part we connect to the database readA0 and obtain a cursor. We then build the sql string with the MySQL insert command and execute it in the next command, before committing the changes to the database and closing it. Note that we could perform several transactions in a row that might leave the database in a bad intermediate state. To prevent this from happening, all insert commands are buffered and all changes are committed to the database simultaneously with the commit() function call. In order to automatically record data, we add the line

```
* * * * * /usr/bin/python3 /home/pi/python/serial2mysql.py
```

to the crontab file using the command crontab -e that we already discussed in the previous section on flatfile databases.

Now that we have a process that continuously fills the database, we want to use octave for postprocessing and to prepare plots for presentations. We therefore need a function to access the database from within octave, which turns out to be difficult, because neither MySQL nor MariaDB are supported by either the Raspberry Pi or octave repositories. Instead, we use code from the github site https://github.com/markuman/mex-mariadb that we must compile ourselves, which requires the installation of two development packages with

```
sudo apt-get install liboctave-dev libmariadb-dev
```

Once that is accomplished we use git clone to download the source code and then compile it with a call to make

```
git clone https://github.com/markuman/mex-mariadb.git
cd /home/pi/mex-mariadb
make
```

This creates the mex file file mariadb_.mex, which is the compiled binary file that is called from a wrapper function mariadb.m. In order to enable all octave scripts to use the mariadb() function, we add the line

```
addpath /home/pi/mex-mariadb
```

to the octave startup file /home/pi/.octaverc.

In a separate directory we prepare an octave script mariadb_read.m, which reads the values from the database and displays them.

```
% mariadb_read.m, V. Ziemann, 221003
clear all;
sql=mariadb('hostname','localhost','username','me','password','pwpw');
sql.database='readA0'; % select database
```

```
request='select * from fdata;';
d=sql.query(request);    % retrieve request
sql.command='quit;';     % close database
d(1,:)=[];               % remove column labels
tt=datenum(d(:,1),'yyyy-mm-dd HH:MM:SS');
d(:,1)=[];               % remove the cells with dates
a=str2double(d);         % convert rest to an array
plot(tt,a(:,1),'k',tt,a(:,2),'r');
datetick('x','ddd/HH:MM');
legend('A0','A1');
```

After adding the path to the functions to access MariaDB from the previous paragraph the `mariadb_read.m` script opens the database before selecting the `readA0` database. In the next line we specify the `request` with the SQL `select` command. The following `sql.query` then requests all data from table `fdata` and stores the retrieved data in the variable d before closing the database by executing the `quit;` command. The first row of the cell array d contains the column names `ts`, `A0`, and `A1`, which we already know and just delete. The first column `d(:,1)` contains the timestamps as strings that we convert with the `datenum()` function to a decimal format, given in days since January 1, 0000. We supply the format string as a second argument in order to aid the conversion. At this point we no longer need the information about the dates in the cell array d and therefore remove the first column. After this step, the reduced cell array d only contains strings of numerical values, which we convert to type `double` with the built-in function `str2double()`. We then plot the data and specify the tick marks on the horizontal axis to include the day of the week, hours, and minutes. Finally, we add a legend.

In the previous examples we always select all available data from the database. In many circumstances we prefer to restrict the displayed data to a smaller range. Instead of filtering in octave, we instruct the database to only return data from the restricted time window. To achieve this we use the following SQL query string:

```
request='select * from fdata where ts > "2022-10-03 17:30"
                    and ts < "2022-10-03 17:55";'
```

instead of the command `select * from fdata;` used earlier. Note that the command needs to be written onto a single line. This query string explains the restricted time window in clear text. We require the timestamp `ts` to be larger than some date and less than some other date. The values returned from the command `d=sql.query(request);` thus only contain data from within the requested time window.

Hopefully this short introduction to MySQL and its open-source pendent MariaDB helps you to access them from Python and octave in case a SQL database is needed in a project. But now we will turn to a second database, one that only stores values over a finite time-horizon and also thins them out the further back in time the data originate. This database is the round-robin database called `rrdtool` that we discuss in the next section.

5.6.3 RRDtool

The `rrdtool` [28] program was initially conceived as a tool for allowing computer-network administrators to present network traffic over different time horizons (last hour, day, week, month, year) in a convenient and flexible way. It generates graphical representations of data that can be shown in a web browser by automatically generating consolidated data, such as average, minimum, or maximum over some period of time. Rrdtool is particularly useful to generate plots of the measured data on the fly, and we later use it to display our measurement data, such as temperatures, on a web page. The round-robin database is implemented as a circular buffer that is filled up to the end and then wraps around and starts to overwrite the oldest values at the beginning of the buffer. Using `rrdtool` comprises three steps: creating the database, filling it with data, and extracting the data. But before delving into examples, we need to install the software with the following command:

```
sudo apt-get install rrdtool
```

after having updated the repositories with `sudo apt-get update` (just a friendly reminder not to forget the update step). Now we are ready to use the software.

In what follows, we assume that all files reside in the subdirectory `/home/pi/rrdtool`. The first step – creating the database – is achieved by the `rrdtool create` command that we enter at the command prompt of the Raspi. In this step we define the frequency of storing data, the type and valid range of data, and the way the stored data should be preprocessed; more on the last point later. The simplest example, namely to create a database `db1.rrd`, is shown in the following example:

```
rrdtool create db1.rrd --step 60 \
   DS:temp:GAUGE:180:-20:100 \
   RRA:AVERAGE:0.5:1:2880
```

where we can also omit the backslash and write the entire command on a single line. The first part of the command creates `db1.rrd` and the database stores values every 60 seconds. The second line defines the data source (`DS:`), a variable called `temp` that is of type `GAUGE`, which is rrdtool-speak for "measurement value." We require a valid data point to be uploaded to the database at least every 180 seconds before a value is marked as invalid. The expected range of values for the data points is between -20 and 100, which is reasonable for a temperature reading. The line starting with `RRA` defines the round-robin archive that contains averaged values; the number 0.5 is used internally and should not be changed; the number of data points to be averaged, here 1; and the total number of (averaged) data points that the database should hold. In the example we use 2880, which is the number of minutes in two days. Since we chose to average only one data point, we store all values, rather than actually averaging. The above `rrdtool` command creates the file `db1.rrd` in the directory where the command is executed. Note that we can create several data sources; for example, for temperature, humidity, and barometric pressure, by adding `DS:` statements. Moreover, note that the `RRA:` line creates one table in the database `db1.rrd`. We can define several more tables with additional `RRA:` statements. For example, adding `RRA:AVERAGE:0.5:30:336` will create a table with data averaged over 30 readings, thus one point every 30 minutes, and will store 336 values, which corresponds to one week, because there are 336 half-hour periods in a week. Other options, instead of `AVERAGE`, are `MIN` and `MAX`, which will store the minimum or maximum in the specified time period, respectively. Please consult `man rrdtool` for more options. Now that we have a database, we can start to fill it with data.

We fill the database with the `rrdtool update` command. The data we store are temperature measurements from an LM35 temperature sensor attached to an Arduino UNO, similar to the way we used it before. The following Python script sends T? to the UNO and receives the temperature data as a string that is similar to T 22.5.

```
# readtemp.py, V Ziemann, 221004
import serial, time
ser=serial.Serial("/dev/ttyUSB0",9600,timeout=1)
time.sleep(3)          # wait for serial to be ready
ser.write(b"T?\n")     # shorthand notation
```

```
reply=ser.readline().decode('utf-8')
print(reply[2:].strip())
ser.close()
```

The difference from earlier versions is that we print out the numerical value only. We call this Python program from a shell script file with the name `filldb1.sh`. It contains the following lines:

```
#!/bin/bash
DB=/home/pi/rrdtool/db1.rrd
TEMP=$(/usr/bin/python3 /home/pi/rrdtool/readtemp.py)
/usr/bin/rrdtool update $DB N:$TEMP
```

and we make it executable by executing `chmod +x filldb1.sh`. In the script, we first define a variable `DB` that contains the absolute path to the database file. It resides in `/home/pi/rrdtool/` as already mentioned above. On the next line we fill the variable `TEMP` with the output of the command between `$(` and `)`, but this is the temperature value that the command `/usr/bin/python3 /home/pi/rrdtool/readtemp.py` returns. Note that the absolute path is used for both the Python interpreter and the `readtemp.py` script. In the last line we execute the `rrdtool update` command with the database name stored in the variable `DB`, and fill it with the current time stamp, as indicated by `N:` and the temperature value stored in the variable `TEMP`. Executing `filldb1.py` from the command line sends a data point with the current time to the database, but that is rather inconvenient. A better solution is to run `filldb1.py` once a minute, automatically. We therefore update the crontab file by executing `crontab -e` and add the line

```
* * * * * /home/pi/rrdtool/filldb1.sh
```

to the end of the file. This will send a new data point to the database once every minute.

Finally, in the third step, we retrieve the data and generate a graph with the `rrdtool graph` command. Here is an example:

```
rrdtool graph db1.png -s -4h \
   -t "Temperature in my office" -v "T [C]" \
   DEF:t0=db1.rrd:temp:AVERAGE \
   LINE1:t0#FF0000:"Temperature";
```

Note that we also can write the entire command on a single line. It creates a graphics file `db1.png` with start of the time axis 4 hours back, `-s -4h` in time, a title specified after the `-t` option, and the vertical axis label after the `-v` option. In the next line the handle `t0` is defined to refer to the data coming from the `db1.rrd` database and as the `AVERAGE` of the variable `temp`. Compare this to the `rrdtool create` command, where `temp` was defined in the `DS:` part of the command and `AVERAGE` in the `RRA:` part. Finally, we define a displayed line to use the handle `t0`, specify the color by hexadecimal RGB values (here red: `#FF0000`), and give it the label `Temperature`. In Figure 5.9 we show the resulting graphics produced by the above `rrdtool graph` command. Note that only the last four hours are plotted, but the first part is missing, because the cron job was not running at that time. The data are therefore noted *invalid* in the database, and consequently not printed. The scale is adjusted automatically in the previous examples, but can also be specified explicitly. The command `man rrdtool` and the pointers therein provide a wealth of information on how to fine tune the output.

After being able to measure, store, and retrieve measurement values, we want to present them on a web page for online observation and continuous checking.

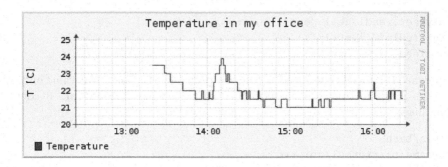

Figure 5.9 Graphics from `rrdtool`.

5.7 ONLINE PRESENTATION

In this section we describe how to present the measured data on a web page that is published by a web server. It turns out that the Raspi is quite capable of running a web server on the side while doing all the other things such as acquiring and storing data. The basic software needed is a web server, and we chose `apache2` [29], which is installed by executing

```
sudo apt-get install apache2
```

on the command line of the Raspi. After the installation is complete, we use a browser on any computer connected to the network the Raspi is connected to and enter the IP number of the Raspi in the address line. This also works on the browser installed on the Raspi; enter `localhost` as the web address, and you should be greeted by the web page shown in Figure 5.10. This page instructs us to replace the file named `index.html` in the directory

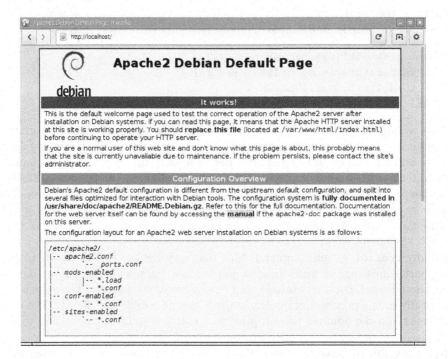

Figure 5.10 Web page from the Raspi after installation of the Apache2 web server.

/var/www/html/ with our own copy. Note that the **apache2** web server is configured to present a file named index.html by default, if only the address of the folder is given.

So, we need to prepare a simple web page, which is a specially formatted text file in a format called HTML – an acronym for *Hyper-Text Markup Language*. There are plenty of books on the subject, and tutorials can be found on the web in abundance. Here we only use the most basic features. We navigate to the directory /var/www/html/ and save index.html as a backup with sudo mv index.html index.html.bak. Then we start our favorite editor as superuser by prepending sudo, enter the following text, and save the contents as index.html.

```
<!DOCTYPE HTML>
<HTML>
  <HEAD>
    <TITLE>Raspi Web Server</TITLE>
  </HEAD>
  <BODY>
    <H1 ALIGN=CENTER>Raspi Web Server Main Page</H1>
    <HR SIZE=2 WIDTH=80%>
    <H3>Available Goodies:</H3>
    <UL>
      <LI> <A HREF="temp/">Temperature Graph</A> </LI>
      <LI> <A HREF="https://www.w3schools.com/html">
                  HTML Tutorial</A> </LI>
      <LI> <A HREF="https://www.w3resource.com/">
                  W3resource tutorials</A> </LI>
      <LI> <A HREF="https://www.raspberrypi.org">
                  Raspberry Pi web site</A> </LI>
      <LI> <A HREF="https://www.arduino.cc">Arduino web site</A> </LI>
    </UL>
  </BODY>
</HTML>
```

If no typos crept in, by entering the address http://localhost on a browser running locally on the Raspi we should see the page shown in Figure 5.11. This is how the web browser renders the contents of the file we just entered and now briefly discuss. At the top of the file we have the declaration of the document type; it is an HTML file. Then we have the opening tag <HTML> and a matching closing tag </HTML> at the end of the file. Anything between them describes the contents of the web page. Note that HTML tags (almost) always come in pairs: the tag name in angle brackets and a matching closing tag with the same name, but prepended with a slash. The next tags we encounter in the file are **HEAD** and **TITLE**, which describe things that do not show up on the page but appear in the title bar of the web browser. Finally, we reach the **BODY** tags, where we find the description of the web page proper. The <H1> tag declares a large headline, and the **ALIGN** directive specifies it to be centered on the web page. There are different levels of header tags, from H1 to H6. The next line defines a horizontal rule with the <HR> tag that covers 80 % of the width of the page. Then we add a smaller header with <H3> and an unnumbered list between the <UL> tags, with each list item described by <LI> tags. The <A> tag is called an anchor. It points to other websites specified in the **HREF** directive. The first list item points to a local directory temp/ under the directory where the file index.html resides. Since this does not yet exist, we need to create it.

Figure 5.11   The first self-made web page.

In the directory /var/www/html/, we create the temp/ subdirectory and copy a file, also named index.html, with the following contents into the newly created subdirectory /var/www/html/temp/:

```
<!DOCTYPE HTML>
<HTML>
  <HEAD>
    <TITLE>Raspi Web server</TITLE>
  </HEAD>
  <BODY>
    <H1 ALIGN=CENTER>Temperature</H1>
    <IMG SRC="db1.png" ALT="Temperature in my office">
  </BODY>
</HTML>
```

The file contents follows the same general layout as before, with the DOCTYPE declared first, followed by <HTML> tags. Next come the <HEAD> tags and then the <BODY> tags bracketing the displayed contents of the page. Here we also find a header and a new tag <IMG> to direct the web browser to display the image specified in the SRC directive. To make this work, we copy the file db1.png, the one we created with rrdtool graph in the previous section, to the directory /var/www/html/temp/. We always need to use sudo to edit or copy files to the system areas to which the files under /var/www/ belong. This we can avoid by enabling private web pages.

The private web pages commonly reside in a subdirectory called public_html under the user's home directory. As user pi we therefore create it by typing mkdir /home/pi/public_html. In order to use it we have to enable the userdir module by executing sudo a2enmod userdir at the command prompt, and restarting apache2 with the command sudo systemctl restart apache2, such that the newly enabled module is loaded. Then we copy the files index.html and db1.png from /var/www/html/temp/ to /home/pi/public_html and are ready to access the same web page as before, but now under the new address http://localhost/~pi. Any file we copy to the subdirectory /home/pi/public_html is then accessible from a browser at the address http: //localhost/~pi/ with the filename appended. For example, http://localhost/~pi/ db1.png only displays the graphics file in the browser window, but nothing else.

The contents on the web page in the previous example is static. We copy files to the public_html directory and then they are presented as is. If we want to update, for example, the temperature graph, we have to run **rrdtool graph** again and copy the new copy of the db1.png to public_html. But this is a task that we easily automatize with the help of a cron job. We prepare a file with the following contents:

```
#!/bin/bash
DB=/home/pi/rrdtool/db1.rrd
/usr/bin/rrdtool graph /home/pi/public_html/db1.png -s -4h \
     -t "Temperature in my office" -v "T [C]" \
     DEF:t0=$DB:temp:AVERAGE \
     LINE1:t0#FF0000:"Temperature";
/usr/bin/rrdtool graph /home/pi/public_html/db2.png -s -2d \
     -t "Temperature in my office" -v "T [C]" \
     DEF:t0=$DB:temp:AVERAGE \
     LINE1:t0#FF0000:"Temperature";
```

and name it makegraph.sh. Then we place the file in the /home/pi/rrdtool/ subdirectory and make it executable with chmod +x makegraph.sh. In the file, we first define the database to use and then have two almost identical copies of the **rrdtool graph** command from the previous section. But in the first case, we create db1.png for data from the last 4 hours, and in the second case we create db2.png for data from the last two days (-s -2d). Note that the graphics files db1.png and db2.png are specified to reside in the public_html directory, where they are accessible to the web server. Finally, we make the makegraph.sh script execute every 10 minutes by adding the following line to our crontab file with the command crontab -e

```
*/10 * * * * /home/pi/rrdtool/makegraph.sh > /dev/null
```

where */10 in the first column means *every 10 minutes* and the > /dev/null at the end means to suppress any output from the makegraph.sh command. So now we can update the to-be-displayed content, but we still need to coax the web browser to actually re-read that content at some interval. For this purpose the <META http-equiv=..> tag exists, and we include it between the HEAD tags, as shown in the following updated version of the public_html/index.html file

```
<!DOCTYPE HTML>
<HTML>
  <HEAD>
    <TITLE>Raspi Web server</TITLE>
    <META http-equiv="refresh" content="300">
  </HEAD>
  <BODY>
    <H1 ALIGN=CENTER>Temperature</H1>
    <H3>The last 4 hours</H3>
    <IMG SRC="db1.png" ALT="Temperature over 4 hours">
    <H3>The last 2 days</H3>
    <IMG SRC="db2.png" ALT="Temperature over 2 days">
  </BODY>
</HTML>
```

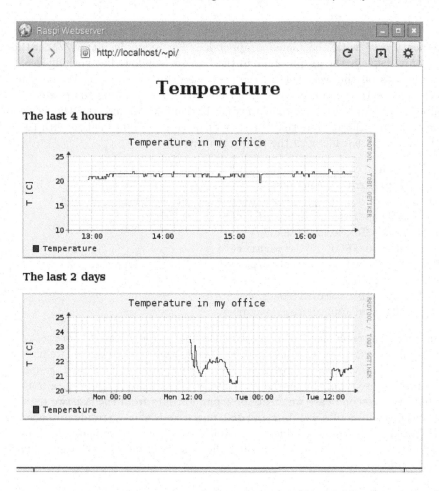

Figure 5.12 User `pi`'s home page that displays the continuously (but slowly) updated temperature.

which instructs the browser to reload the page every 300 seconds, as indicated by the META tag. At the same time, we added the second image that displays the temperature over the last two days. A screen shot of the web page is shown in Figure 5.12. We can change the update frequency in both the crontab file and in the META tab if we are impatient. They do not need to match.

At this point we can obtain, store, retrieve, and display measurements on an actively updated web page. The update mechanism that we implemented using cron jobs is very rudimentary, and there are better solutions; for example, `cgi-bin` or `php` server-side programs. They execute code on-demand at the press of a button on the web page, and update the displayed information on the fly, but that is beyond the scope of our presentation.

So far we have used the Raspi as the hub in our sensor network to query the sensor nodes, and store and present the data, but we can even turn the Raspi into a node of a larger control system such as EPICS. This is the topic of the next chapter.

QUESTIONS AND PROJECT IDEAS

1. Where do you find help about programs installed on the Raspi?

2. How do you copy files on the Raspi?

3. How do you rename files on the Raspi?

4. How do you copy files between the Raspi and your desktop computer?

5. Find out what the program `ping` does.

6. Find out what the program `touch` does.

7. Find out what the program `wireshark` does.

8. What is the `netmask`?

9. Start *Mathematica* from the *Programming* menu and play around with it. Find out what it can do.

10. Start the `synaptic` software installation program from Section 5.3, enter "games" as the search keyword and explore.

11. Start `python3`, execute `help()`, and follow the instructions.

12. Write a program that displays "Hello World" in Python.

13. Create a file with the name `look_at_me` in the home directory of user `pi`, and instruct the `cron` daemon to `touch` it every second Thursday of the month.

14. How do you get help about commands in `octave`?

15. Connect a USB web cam to the Raspi and make pictures with the `cheese` program. Use the `imagemagick` tools to cut a small portion from the image (crop) and convert it to another format, say gif.

16. Find out how to store integer values in a MariaDB database.

17. Find out how to store images in a MariaDB database.

18. Measure the brightness with an LDR attached to a NodeMCU and log it once per hour in a MariaDB database. If you are patient you can see see days getting longer and shorter as the seasons progress.

19. Log the steps that a stepper motor takes and visualize them on a web page with RRDtool.

20. Make a web page that shows your Arduino sketches.

21. Find out how to specify colors in HTML.

22. Explore the webpages mentioned on Figure 5.11.

23. Prepare a web page about yourself with picture and all other information you care to display.

## CHAPTER 6

Control System: EPICS

It is not too farfetched to visualized the *Experimental Physics and Instrumental Control System* (EPICS) as a home automation system for scientific laboratories. It is used in a number of laboratories to control large particle accelerators, such as the Advanced Photon Source (APS) in the United States, the Swiss light source (SLS) in Zürich, or the European Spallation Source (ESS) in Sweden. Other users of EPICS are the fusion reactor International Thermonuclear Experimental Reactor (ITER) in France and the W. M. Keck astronomical observatory on Hawaii.

EPICS is based on a number of independent computers called *input–output controllers* (IOC) that announce their capabilities on the network such that other computers can interact with them. Almost any type of computer can participate in an EPICS system, and support libraries for many programming languages, such as C, Python, and Octave, are available. So it is no surprise that a Raspi can also serve as an IOC and join an EPICS control system, no matter how big it is. We illustrate this by configuring the Raspi to communicate the measurements collected from locally attached microcontroller-based sensor nodes to EPICS. This makes the measurements accessible in a larger control system context.

The first task is to install the EPICS software on the Raspi, and we follow [30] in doing so. The steps are somewhat arcane and the instructions resemble a cookbook. Normally these steps are done by an experienced system administrator.

6.1 INSTALLATION

First we need to download the EPICS Base package from `epics.anl.gov`. At the time of writing, the current release is 7.0.7 and the downloaded package is called `base-7.0.7.tar.gz`. In order to avoid excessive use of `sudo` we create a subdirectory `/home/pi/epics` where all the software resides, but we also create a soft link of that directory to `/usr/local/epics` where it is available to all users on the Raspi. The detailed sequence of commands is the following:

```
cd /home/pi
mkdir epics
sudo ln -s /home/pi/epics /usr/local
cd epics
cp /home/pi/Downloads/base-7.0.7.tar.gz .
tar xzf base-7.0.7.tar.gz
```

where `tar` is an archiving program that unpacks (`x` option) compressed (`z` option) files (`f` option). Note the period '.' at the end of the `cp` command, which is the shorthand notation for the current directory, /home/pi/epics in this case. The last command creates a subdirectory under /home/pi/epics with the name base-7.0.7. In order to avoid writing the version number over and over again, we create a soft link with the ln -s command, which essentially creates an alias in the same directory by executing

```
ln -s base-7.0.7 base
```

and we can henceforth refer to the directory with the EPICS base package via /usr/local/epics/base by virtue of the soft link from /home/pi/epics to /usr/local/epics. Now we have the source code in place, but in order to use it we still need to compile it.

And for the compiler and later the executable programs to find the EPICS sources, we need to copy the following lines into the file /home/pi/.bash_aliases:

```
export EPICS_ROOT=/usr/local/epics
export EPICS_BASE=${EPICS_ROOT}/base
export EPICS_HOST_ARCH='${EPICS_BASE}/startup/EpicsHostArch'
export EPICS_BASE_BIN=${EPICS_BASE}/bin/${EPICS_HOST_ARCH}
export EPICS_BASE_LIB=${EPICS_BASE}/lib/${EPICS_HOST_ARCH}
if [ "" = "${LD_LIBRARY_PATH}" ]; then
    export LD_LIBRARY_PATH=${EPICS_BASE_LIB}
else
    export LD_LIBRARY_PATH=${EPICS_BASE_LIB}:${LD_LIBRARY_PATH}
fi
export PATH=${PATH}:${EPICS_BASE_BIN}
```

If the file does not exist, just create a new copy with the above contents. Note that the environment variable EPICS_ROOT points to the subdirectory /usr/local/epics, the one we created in an earlier step. The other variables are defined relative to that and point to the places where the libraries and executables reside. Once we complete this step, we need to open a new command window, because this rereads the .bash_alias file and we need these definitions before compiling the base system with

```
cd /home/pi/epics/base
make distclean
make -j 4
```

which takes about 20 minutes depending on the version of the Raspi. The option distclean makes sure that the epics sources are in a pristine state and the option -j 4 causes the compilation to start four jobs simultaneously, one for each CPU core. The terminal window is filled with the information about what part of the base system is currently compiled. Once the compilation completes without errors, we enter the command caget at the command prompt. It should respond with "no pv name specified..." and this indicates that the executables are available and located in the PATH. The "pv" is the name of a quantity that EPICS deals with, and is called a *process variable*. An example of such a name is the read-back current of power supply PSabc that might be called PSabc:current. Reading or changing the power supply current then refers to that process variable. Reading is done by entering caget PSabc:current on the command line. More on that topic comes in the next section. For the next test we start the softIoc program, and at the epics> prompt that appears, we type iocInit. This step should result in the response "iocrun: all initialization complete" and convinces us that we have successfully installed EPICS on the Raspi and can proceed to investigate it.

6.2 COMMUNICATING WITH EPICS

Before connecting the sensor nodes with the microcontrollers to EPICS, we explain the basic EPICS functionality by reading and controlling virtual parameters that only exist in the memory of the Raspi. Again, based on examples from [30], we prepare the following EPICS database file and name it `simple2.db`.

```
# simple2.db
record(bo,"raspi:trigger") {
  field(DESC,"trigger PV")
  field(ZNAM,"off")
  field(ONAM,"on")
}
record(stringout,"raspi:message") {
  field(DESC,"message on the RPi")
  field(VAL,"RPi default message")
}
record(calc, "raspi:random") {
  field(SCAN,"1 second")
  field(CALC,"RNDM")
}
record(ao,"raspi:A") {
  field(DESC,"variable A")
  field(VAL,"0")
}
record(ao,"raspi:B") {
  field(DESC,"variable B")
  field(VAL,"0")
}
record(calc,"raspi:C") {
  field(DESC,"sum of A and B")
  field(SCAN,"1 second")
  field(INPA,"raspi:A")
  field(INPB,"raspi:B")
  field(CALC,"A+B")
}
```

The file contains definitions of six process variables: `raspi:trigger`, `raspi:message`, `raspi:random`, `raspi:A`, `raspi:B`, and `raspi:C` in the `record` definitions. Each record has a type, which might be `bo`, `bi`, `ao`, or `ai`, standing for binary or analog input or output. Other types are `stringout` or `calc`. Let's describe the records one at a time. The first record is a binary output, `bo`, called `raspi:trigger`, which may be either 1 or 0 and contains three defining fields, a description, a text string for state zero (`ZNAM`), and one for state one (`ONAM`). The second record describes a string that is initialized in the `VAL` field. The third record is of type `calc` and performs some calculation, in this case a rather trivial one: It generates a new random number once per second as specified in the `SCAN` field. The next two records define process variables `raspi:A` and `raspi:B` that are initialized to zero. The sixth record is of type `calc` and calculates the sum of `raspi:A` and `raspi:B` once per second. Once the file `simple2.db` is saved, we start it with the following command:

```
softIoc -d simple2.db
```

which will start an EPICS server and publishes our six process variables. The latter we can

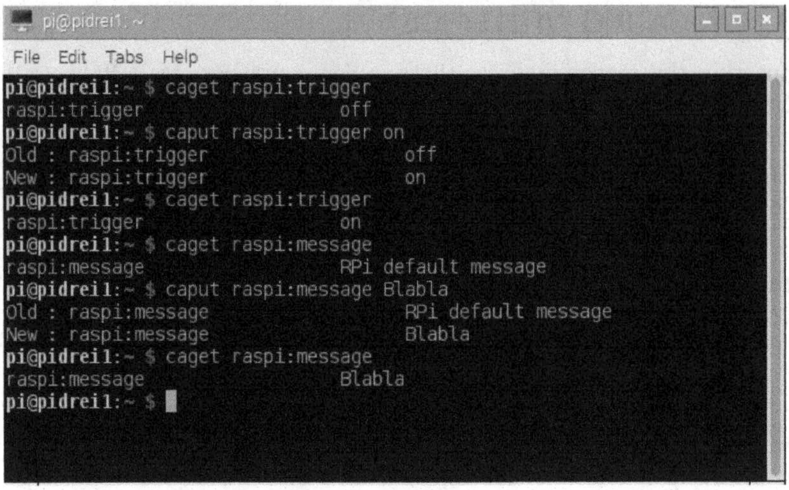

Figure 6.1 Using the EPICS commands `caget` and `caput`.

verify by typing `dbl` (database list) at the `epics>` prompt, which lists the process variables served from the running `softIoc` process.

We interact with the EPICS server from a second terminal window and enter `caget raspi:trigger` in it. The response is the name of the process variable and the current state, which initially is `off`. We change the state by entering `caput raspi:trigger on` and subsequently verify with `caget` that the state has indeed changed. Note that we change process variables with `caput` and retrieve their value with `caget`. We immediately try this with the `raspi:message` process variable by entering `caget raspi:message`, which displays the message from the `simple2.db` file. We change it with `caput raspi:message Blabla`, and the next `caget raspi:message` should display `Blabla` instead. Figure 6.1 shows these transactions. When reading a process variable with `caget`, we can suppress the echo of the variable name by using the `-t` (for terse) option and use `caget -t` instead. A companion program to `caget` is `camonitor`, which monitors one or several process variables and reports whenever one of them changes its value. We use it immediately to verify that the `raspi:random` process variable indeed produces a new random number once every second, by entering `camonitor raspi:random`. The last three records define process variables that are linked in such a way that the third `raspi:C` calculates the sum of the other two process variables. We test this functionality by using `caput` to enter the values 17 and 4 to `raspi:A` and `raspi:B`, respectively. Subsequently reading `raspi:C` will report 21. In passing we mention that a second computer on the same network, either a second Raspi or a desktop computer that has the EPICS base package installed, can access all process variables served by the first Raspi.

We already mentioned the command-line programs `caget` and `caput` to read and write EPICS variables. The same functionality is available from python after installing the `pyepics` package with

```
sudo pip install pyepics
```

from the command line. The following script

```
# read_write_epics.py, V. Ziemann, 221103
import epics
msg=epics.caget("raspi:message")
```

```
print("The message is: ", msg)
caput("raspi:message","Krusiduller")
```

first reads the epics variable `raspi:message` and displays its value nicely formatted before replacing it with the new message Krusiduller.

This dry run of EPICS worked without interfacing hardware. In order to communicate with our sensor nodes, we need two additional libraries that link EPICS to the hardware.

6.3 ASYN AND STREAM LIBRARIES

The external sensor nodes communicate with the Raspi via serial RS-232 lines, Bluetooth, Ethernet, or WLAN. EPICS requires interface libraries to be able to take over these communication channels. For this purpose we install two additional packages, `asyn` and `StreamDevice` into the newly created subdirectory `/home/pi/epics/modules/` with the following sequence of commands.

```
mkdir /home/pi/epics/modules
cd /home/pi/epics/modules
git clone https://github.com/epics-modules/asyn.git
```

Here we use `git` to create the subdirectory `/usr/local/epics/modules/asyn` and copy all files from the repository for the asyn package to it. At the time of writing release 4-43 was the most recent one. Before compiling the package, we need to edit the `./asyn/configure/RELEASE` file, change the `EPICS_BASE` variable to `/usr/local/epics/base`, and make sure that the lines starting with `IPAC` and `SNCSEQ` are commented out with a hash character in the first column. Once this is done we compile by entering `make -j 4` in the directory `/usr/local/epics/modules/asyn` and wait a few minutes for completion.

The second module we need to install is the `StreamDevice` package. Inside the subdirectory `/home/pi/epics/modules/` we directly download the source code with `git` by executing

```
git clone https://github.com/paulscherrerinstitute/StreamDevice.git
ln -s StreamDevice stream
cd stream
```

This creates a subdirectory `StreamDevice` and makes it known under the generic name `stream` before we use `cd` to enter it. To make the build process compatible with the standard make system, we need to remove the `GNUMakefile` – if it is present – from that subdirectory before editing `./configure/RELEASE` to make sure that `EPICS_BASE` points to `/usr/local/epics/base` and that the three line starting with `ASYN=` read

```
ASYN=/usr/local/epics/modules/asyn
#CALC=$(SUPPORT)/calc-3-7
#PCRE=$(SUPPORT)/pcre-7-2
```

Here, we added a hash to prevent including support for the `CALC` and `PCRE` modules. Finally, we initiate the compilation by running `make -j 4` in the `/home/pi/epics/modules/stream` subdirectory. This completes the preparation of the basic libraries, and we are ready to write IOCs that talk to our hardware, the microcontrollers with the sensors and actuators attached.

6.4   WRITING AN IOC

As examples, we link the temperature measurement on the UNO, connected by serial line, and on the NodeMCU, connected by WLAN, to EPICS, and publish the measurements as process variables. For this purpose, we create a subdirectory `/home/pi/epics/ioc` and a subdirectory `temp` in it. We change to the `temp` directory with the `cd` command and execute the following commands:

```
cd /home/pi/epics/ioc/temp
makeBaseApp.pl -t ioc temp
makeBaseApp.pl -i -t ioc temp
```

which creates the basic file infrastructure unde `/home/pi/epics/ioc/temp`. Then we add the two lines

```
ASYN=/usr/local/epics/modules/asyn
STREAM=/usr/local/epics/modules/stream
```

to the end of the file `./configure/RELEASE` and make sure that `EPICS_BASE` points to `/usr/local/epics/base`.

Next, we prepare the protocol file that describes the communication protocol we used earlier: send 'T?' and receive 'T 21.2'. The file resides in the subdirectory `./tempApp/Db` under the `temp` base directory for this IOC. Inside it we create the following file named `temperature.proto`

```
# ./tempApp/Db/temperature.proto
Terminator = CR LF;
get_temp {
    out "T?";
    in  "T %f";
    ExtraInput = Ignore;
}
```

The content is rather straightforward to understand. First we define the terminating characters that denote the end of a line, and a function `get_temp` that sends the string `T?` to the device and expects `T` and a float (`%f`) number in return. Moreover, any extra characters should be ignored. It is possible to have more than one function defined in the same protocol file. The protocol file is the lowest level to define the communication; the next higher level is the database file, which we already encountered in Section 6.2. Here we use the following file

```
# ./tempApp/Db/temperature.db
record(ai, "$(USER):temp") {
    field(DESC, "Temperature")
    field(SCAN, "10 second")
    field(DTYP, "stream")
    field(INP, "@temperature.proto get_temp $(PORT)")
}
```

This file defines an analog input `ai` record for a process variable with the name `$(USER):temp`. Here `$(USER)` will be defined in the calling program. The description and update rate of 10 seconds are defined in the first two fields. The third field declares the record to be of type `stream`, and as input function (INP) we use the function `get_temp` from the protocol file `temperature.proto`. We use the communication interface with the name supplied in the variable `$(PORT)`. We then have to edit `./tempApp/Db/Makefile` and add the line

```
DB += temperature.db
```

to it. Next, inside the subdirectory ./tempApp/src, we create the file xxxSupport.dbd with the following contents

```
# ./tempApp/src/xxxSupport.dbd
include "stream.dbd"
include "asyn.dbd"
registrar(drvAsynIPPortRegisterCommands)
registrar(drvAsynSerialPortRegisterCommands)
```

and add the line

```
temp_DBD += xxxSupport.dbd
```

after the line with temp_DBD += base.dbd in ./tempApp/src/Makefile. This instructs the build process to include the asyn and stream modules as well as support for serial communication and Internet protocol ports such as network sockets. Then, near the bottom of the file, following temp_LIBS += $(EPICS_BASE_IOC_LIBS), we add the lines

```
temp_LIBS += asyn
temp_LIBS += stream
```

which are needed to link against the two libraries.

In a last step, we define the startup program st.cmd for the IOC that is located in the ./iocBoot/ioctemp/ directory. In this file we add the following line

```
epicsEnvSet(STREAM_PROTOCOL_PATH,"../../tempApp/Db")
```

after the line with < envPaths, which describes the location of the protocol files. Following the line with temp_register RecordDeviceDriver pdbbase we add the definition of the serial port we intend to use. For the serial line to the UNO, this looks like

```
drvAsynSerialPortConfigure("SERIALPORT","/dev/ttyACM0",0,0,0)
asynSetOption("SERIALPORT",-1,"baud","9600")
asynSetOption("SERIALPORT",-1,"bits","8")
asynSetOption("SERIALPORT",-1,"parity","none")
asynSetOption("SERIALPORT",-1,"stop","1")
asynSetOption("SERIALPORT",-1,"clocal","Y")
asynSetOption("SERIALPORT",-1,"crtscts","N")
```

and can be referred to by its symbolic name SERIALPORT. Finally, we need to load the database records using the definition of the variables $(PORT) and $(USER)

```
dbLoadRecords("db/temperature.db","PORT='SERIALPORT',USER='raspi'")
```

where SERIALPORT replaces the place holder $(PORT) in the database record file temperature.db. Moreover, the name of the process variable is prepended by raspi, such that we can later access the temperature with the command caget raspi:temp. Now the software setup for the IOC is complete, and we compile it by running make in the directory /home/epics/ioc/temp/. Once the compilation successfully completes, we make the file st.cmd that is located in ./iocBoot/ioctemp/ executable by executing chmod +x st.cmd and run it with

```
./iocBoot/ioctemp/st.cmd
```

This starts the EPICS server and the process variable, here only `raspi:temp`, is published on the local network so that any computer with the EPICS base system installed and a working `caget` program can read the temperature from our Raspi.

It remains to connect the NodeMCU microcontroller to EPICS. Since the NodeMCU server from Section 4.6.3 listens on port 1137 at IP number `192.168.20.135` and uses the same protocol (send 'T?', receive 'T 22.1'), we just add the following two lines to the `st.cmd` file:

```
drvAsynIPPortConfigure("SOCKET1","192.168.20.135:1137",0,0,0)
dbLoadRecords("db/temperature.db","PORT='SOCKET1',USER='node'")
```

The first line defines a symbol `SOCKET1` that points to the port on the NodeMCU, and the second line instructs EPICS to use the same database file `temperature.db` as before, and link the communication to the `PORT` corresponding to the one defined in the previous line. Once we add the two lines, we need to compile the project again. This we do by issuing `make` in the directory `/home/epics/ioc/temp/` and restarting `st.cmd` by executing `./iocBoot/ioctemp/st.cmd` from the same directory. The result of this exercise is that EPICS now publishes two process variables, `raspi:temp` from before and `node:temp` from the NodeMCU connected by WLAN, which we can verify by entering the command `dbl` at the `epics>` prompt that `st.cmd` provides.

Most of the above actions we only need to do once, and adding additional sensors to the EPICS IOC is only moderately complex. It only requires writing the appropriate protocol file, say `new.proto` and database file `new.db`, and copying both to the `./tempApp/Db` directory. Then we need to add the database file to the project by adding `DB += new.db` to the Makefile in the same directory. Finally, we need to add `drvAsynXConfigure` to the `st.cmd` file and execute `dbLoadRecords` to link the database file to the appropriate `PORT`. Finally, we compile once again and and run the `st.cmd` command.

6.5 STARTING THE IOC AT BOOT TIME

In the description above we need to start the IOC `st.cmd` by hand and keep a terminal window with the running program open all the time. To remedy this inconvenience, we create a startup script named `epicsioc` that launches the IOC when the system boots. The file contains the following lines:

```
#!/bin/sh
#/etc/init.d/epicsioc
### BEGIN INIT INFO
# Provides:            empicsioc
# Required-Start:      $remote_fs $syslog
# Required-Stop:       $remote_fs $syslog
# Default-Start:       2 3 4 5
# Default-Stop:        0 1 6
# Short-Description: Simple script to start a program at boot
# Description:         Start and stop epicsioc server
### END INIT INFO
case "$1" in
  start)
    echo "Starting epicsioc"
    . /home/pi/.bash_aliases
    cd /home/pi/epics/ioc/temp/iocBoot/ioctemp
    /usr/bin/procServ -n "IOC" -L /tmp/epics.log -i ^D^C 20000 ./st.cmd
```

```
    ;;
  stop)
    echo "Stopping epicsioc"
    kill $(pidof procServ)
    ;;
  restart)
      $0 stop
      sleep 10
      $0 start
      ;;
  *)
    echo "Usage: /etc/init.d/epicsioc {start|stop|restart}"
    exit 1
    ;;
esac
exit 0
```

The file contains some INIT INFO that is required by the init program, which orchestrates the startup of the Raspios operating system. In the subsequent case structure, three actions are specified: start, stop, and restart. The start option reads the default EPICS environment variables and then changes to the subdirectory with the st.cmd program. There it executes the procServ program that in turn starts the st.cmd program. The purpose of procServ is to redirect the input and output of st.cmd to the network socket 20000 to which we can connect via telnet, and interact with the program in the same way we did earlier when we started it in a terminal window. The advantage of using procServ is that we can close the telnet program while st.cmd keeps running independently. The other options give the running process the name IOC, write a log file, and ignore the control sequences to close the process. The stop case determines the process id of procServ with pidof and terminates the process with the kill command. The restart case executes stop first, waits 10 seconds, and then executes start. The default case, denoted by *), displays a brief usage note.

Before actually running the program, we need to install the procserv and telnet packages with the normal installation procedure:

```
sudo apt-get install procserv telnet
```

In case one of the programs is already present in the current system, no new software is installed. Once all required programs are installed and the script is written, we copy it, using sudo, to the system directory /etc/init.d, where all startup scripts for the operating system reside. We make it executable using

```
sudo chmod 755 /etc/init.d/epicsioc
```

and start it by hand in order to verify that the script works as intended with the command

```
sudo systemctl start epicsioc
```

and if everything works we can install it permanently by executing

```
sudo systemctl enable epicsioc
```

which will automatically start the epicsioc service at boot time. Note that we can also stop and disable the service. If we want to connect to the now-running-autonomously process st.cmd, we use telnet when logged onto the Raspi

```
telnet localhost 20000
```

and see the normal output of the `st.cmd` program, plus some administrative info from `procServ` prepended by `@@@`. We can enter commands such as `dbl` to list all process variables, and immediately receive the output back on the telnet window. The port opened by `procServ` is only accessible from the Raspi itself. If we want to log from another computer, we use `ssh` to first log onto the Raspi, and then execute `telnet`. This can be automated by executing

```
ssh -t pi@192.168.10.22 telnet localhost 20000
```

from a remote desktop computer and where `192.168.10.22` is the IP-number of the Raspi. In this way, the Raspi works unattended as an Epics IOC but we can connect to it at any time using `telnet` in order to follow the performance and interact with `st.cmd`.

QUESTIONS AND PROJECT IDEAS

1. Why is a standardized control system advantageous to use in comparison to a home-grown system?

2. Write protocol and database files to interface the sensors discussed in previous chapters.

3. Write protocol and database files to connect an Arduino UNO running the sketch in Section 4.5.2 to control a DC motor.

4. Write protocol and database files to connect an Arduino UNO running the sketch in Section 4.5.3 to control a model-servo.

5. Write protocol and database files to connect an Arduino UNO running the sketch in Section 4.5.4 to control a stepper motor.

6. Research other control systems used in industry or research institutions.

7. Research the documentation of the stream library on how to treat information that appears on the RS-232 line unsolicited, such as a continuously appearing stream of position data from a GPS receiver.

8. In this chapter we address the Raspi by IP number `192.168.10.22` and in Section 5.4 by `192.168.20.1`. Explain, why this not a typo.

## CHAPTER 7

Messaging System: MQTT

EPICS is not the only system to integrate a large number of sensors and actuators in a common framework. Another such system is the *message queue telemetry transport* (MQTT) [6] protocol that was originally created by, among others, IBM, in order to collect information from widely distributed infrastructures such as oil pipelines in a battery-saving and energy-efficient as well as robust and secure way. Today it is commonly used to pass messages between devices that constitute the Internet of Things (IoT) and has become ISO-standard ISO/IEC 20922. MQTT uses a *broker* that passes messages from clients publishing data to those that subscribe to the data. The name of parameters – the data – that are passed around are referred to as "topic." They are organized hierarchically, such that the name

```
weatherstations/stationA/node1/temperature
```

refers to the temperature sensor on a weather station, called station A and connected to node 1. When subscribing to topics it is possible to use wildcards such as the plus-sign "+", which is a single-level wild-card, or the hash "#," which is a multi-level wild-card. Robustness and reliability of transmission is guaranteed by specifying three levels of *quality of service* (QoS), where the simplest level 0 implies that the sensor only publishes data without any acknowledgment from the broker. Levels 1 and 2 implement increasingly advanced levels of handshake signals that ensure the arrival of data. Even a *last will* is available, which is transmitted to subscribing clients should a publishing client disconnect. Normally the broker immediately transmits any received data to the subscribing clients, but it is possible to specify that the broker retains the last good data point and transmits it to subscribers in case the publisher is offline, and to new subscribers upon their first connection. Since MQTT was intended to operate across public networks, encryption and security are part of the protocol.

The core functionality of MQTT resides on the broker, such that clients can be very simple and connect and disconnect at will. This makes it possible to use clients that only wake up from a battery-saving deep sleep mode, perform a measurement, send it to the broker, and go back to sleep again. This attractive feature makes MQTT very popular for IoT applications and we therefore also discuss it as a complement to EPICS. In the following sections we operate a broker on the Raspberry Pi and use NodeMCUs as clients that publish and subscribe. In a final section we discuss a simple gateway that links MQTT to EPICS in order to benefit from the best of both worlds. In this way we can access lightweight clients connected over a public network across the gateway from our EPICS control system.

After this overview of MQTT, let us follow the theme of the book and describe a working system that provides the basic functionality.

DOI: 10.1201/9781003341703-7

7.1   BROKER

The `mosquitto` broker is widely supported on many platforms, and also on the Raspberry Pi. We install it, after bringing the system up to date with `sudo apt-get update` and `sudo apt-get upgrade`, using the command

```
sudo apt-get install mosquitto mosquitto-clients
```

where the `mosquitto` package contains the broker and the `mosquitto-clients` package contains the command line executables `mosquitto_pub` and `mosquitto_sub`. We use them to test the base functionality on the Raspi alone, without external clients.

Immediately after the installation, `mosquitto` is already running and is registered as a service that starts after every reboot. During our initial tests we inspect the logging output with the following command, which helps to debug the problem in case something unexpected happens:

```
sudo tail -f /var/log/mosquitto/mosquitto.log
```

where `tail -f` takes the filename as argument and shows the last lines of the file, but, instead of stopping the display at the end of the file, keeps appending newly generated messages. Now we open a second terminal window and start publishing MQTT data using the `mosquitto_pub` command-line client with

```
mosquitto_pub -d -h localhost -i Pub1 -t dummy/value -m 42 -r
```

where `mosquitto_pub` is the executable, `-d` enables debugging output, and `-h` specifies the IP address of the broker; here it is `localhost`, because the broker runs on the same Raspi where we run the `mosquitto_pub` client. The parameter specified after `-i` identifies the publisher, `-t` the topic, and `-m` specifies the message, here the value 42. Appending `-r` instructs the broker to retain the message and send it in case the publisher is offline. The debugging output can be disabled by omitting the `-d` option. Running `mosquitto_pub -h` gives a short overview of available commands, and the manual page, accessible with `man mosquitto_pub`, provides more information.

The `mosquitto_sub` executable provides the receiving end of MQTT message-passing. We run it in another terminal window where we execute

```
mosquitto_sub -d -h localhost -i Sub1 -t dummy/value
```

to subscribe to the topic `dummy/value` on the broker running on `localhost`. We identify the subscribing client by `Sub1` and enable debugging with `-d`. After starting the executable, we are greeted with either an empty line or with the last published message that has the retain flag `-r` enabled. The `mosquitto_sub` keeps running, and if we publish new messages in the second terminal with `mosquitto_pub`, they immediately appear in the subscriber window.

In the default configuration, this only works if both publisher and subscriber are on the same computer. In order to grant access to the broker from the network, we need to edit the configuration file `/etc/mosquitto/mosquitto.conf` and add the lines

```
listener 1833
allow_anonymous true
```

to the end of the file. We can also restrict access to the wireless interface by adding the IP number of that interface to the first line, which then reads `listener 1833 192.168.20.1`, but then we always need to refer to this address when specifying the broker. The command line arguments then contain `-h 192.168.20.1`, rather than `-h localhost`. The second line enables access to the broker without a password.

At this point we have a working broker whose functionality we verified with the command-line programs `mosquitto_pub` and `mosquitto_sub` running on the same Raspi as the broker. In the next section we replace these command-line clients by external clients running on NodeMCUs that publish temperature data and at the same time subscribe to another topic to turn a cooling fan on or off.

7.2   NODEMCU CLIENTS

The hardware connected to the NodeMCU is very simple in this case. We connect ground and supply pins of an LM35 temperature-sensor to the respective pins on the NodeMCU. The signal pin of the LM35 is connected to the single analog input A0. Second, we connect a transistor to an output pin as shown in previous chapters to turn the fan on and off. In order to use MQTT functionality in the Arduino IDE, we need to install one of the many available libraries. We select the "PubSubClient" library and install it from within the IDE by going to the library manager located at *Tools→Manage Libraries*, enter "PubSubClient" in the search field, find the one by author Nick O'Leary, and install it. The following sketches are based on the `mqtt_esp8266` example that is included in the installation.

It turns out that adapting the example code to write a sketch that serves as both publisher and subscriber is straightforward, and the following code fulfills that purpose.

```
// MQTT client, V. Ziemann, 170816
const char* ssid = "messnetz";
const char* password = "zxcvZXCV";
const char* broker = "192.168.20.1";
const int fan_pin=D4;
#include <Ticker.h>
volatile uint8_t do_something=0;
Ticker tick;
void tick_action() {do_something=1;}  // executed regularly
#include <ESP8266WiFi.h>
#include <PubSubClient.h>
WiFiClient espClient;
PubSubClient client(espClient);
void on_message(char* topic, byte* msg, unsigned int length) {
  char ch[30]; memcpy(ch,msg,length); ch[length]='\0';
  if (strstr(topic,"node1/fan")==topic) {
    int val=(int)atof(ch);
    Serial.print(" Fan="); Serial.println(val);
    if (val==0) {
      digitalWrite(fan_pin,HIGH);
    } else {
      digitalWrite(fan_pin,LOW);
    }
  } else if (strstr(topic,"node2/temp")==topic) {
    Serial.print("Temperature on node2 = "); Serial.println(ch);
  }
}
void setup() { //...............................setup
  pinMode(fan_pin,OUTPUT);
  Serial.begin(115200);
  WiFi.begin(ssid,password);
```

```
    while (WiFi.status() != WL_CONNECTED) {
      delay(500); Serial.print(".");
    }
    Serial.print("\nWifi connected to "); Serial.println(ssid);
    Serial.print("with IP address: "); Serial.println(WiFi.localIP());
    client.setServer(broker, 1883);  // 1883 = default MQTT port
    client.setCallback(on_message);  // execute when a message arrives
    tick.attach(5,tick_action); // execute tick_action every 5 seconds
  }
  void loop() {  //...............................loop
    while (!client.connected()) {
      if (client.connect("PubSub1")) {  // identification
        client.subscribe("node1/fan");  // external fan control
        client.subscribe("node2/temp"); // temp on other nodemcu
      } else {
        delay(5000);
      }
    }
    client.loop();
    if (do_something) {
      do_something=0;
      char message[30];
      int temperature=(int)(3.3*100*analogRead(A0)/1023.0);
      sprintf(message,"%d",temperature);
      client.publish("node1/temp",message);
    }
  }
```

In the first few lines of the sketch we define the usual network credentials and the IP number of the broker as well as the pin to which the fan is connected. Then we include the header for the Ticker library, which provides a timer that is executed at regular intervals. After declaring the variable do_something, we declare the tick object and define the function tick_action() to be executed in regular intervals. All this function does is to set the variable do_something to one. We declare do_something volatile, because it changes asynchronously from the main loop. Next we include the WiFi header, the MQTT PubSubClient functionality, and declare both WiFiClient and a PubSubClient named client. The following function named on_message() is executed every time a MQTT message arrives. Its arguments are the topic, the message, and the length of the message. Within the function we first convert the received message to a character string, because we want to handle it using the same mechanism we used in previous chapters. Then we enter the usual construction where we check which topic has arrived; if it is node1/fan we convert the message to an integer value val and turn a pin on or off, depending on whether val is zero or not. If the received topic is node2/temp we only display it on the serial line, but could easily add a test to turn the fan on or off, depending on the temperature received.

Once variables and auxiliary functions are declared, we define the setup() function and configure the mode of the used pins, the serial line, and WiFi. The function client.setServer() connects to the broker on the default MQTT port 1883, and the call to the client.setCallback() function registers the function on_message() to be executed when a new message arrives. Finally, we start the ticker process to execute the function tick_action() once every 5 seconds. In the loop() function we first ensure that the connection to the broker is up and running. If it is not running we connect to the broker with the call to the `client.connect()` function and provide the identification of the NodeMCU as `PubSub1`. In the above example using `mosquitto_sub`, this corresponds to the parameter following the `-i` command line switch. The subsequent calls to `client.subscribe()` register the strings used as argument with the broker, which subsequently sends any updated values. If the connection with the broker fails, a new attempt is made after 5 seconds. Once the connection to the broker is established, we call the `client.loop()` function to service any background activities related to MQTT, and finally check whether the variable `do_something` was set, which happens every time the ticker fires. If `do_something` was set, we reset it to zero and publish the parameter `node1/temp`.

In a terminal window on the Raspi, we can start `mosquitto_sub` to subscribe to the topic `node1/temp` and should see the updated temperature every 5 seconds. Furthermore, in order to test multiple NodeMCU clients to communicate, we can program a second NodeMCU with the same sketch, but swap the reference to `node1` and `node2`. Moreover, if we only want to publish values from a NodeMCU client, all code related to receiving messages can be removed from the sketch, such as the `on_message()` function and the two calls to `client.subscribe()`. If we want to use a client with subscription-only functionality, we can remove everything related to the tick function and the variable `do_something`. In any case, using the above sketch as a base should make it possible to serve almost any need to connect NodeMCUs to a MQTT network.

So far, the MQTT system is unrelated to other control systems such as EPICS. So, if we require interoperability, we need to provide a gateway that translates the message formats, and that is what we describe in the next section.

7.3   GATEWAY TO EPICS

We want the gateway to seamlessly translate between the systems and to achieve the following functionality: The gateway will listen on a network socket on port 51883 to communicate with EPICS, and uses default ports for anything else. On EPICS we intend to use stream-based protocol files such that all strings that EPICS sends to our gateway will have the format `topic value`, and we configure the gateway to publish this as topic `topic` with the message contents `value`. This is all we need to do to publish MQTT topics from EPICS. In order to receive MQTT messages on EPICS, we need to configure the gateway to subscribe to the messages and pass any incoming messages on to EPICS. For this we decide that any message from EPICS to the gateway having the format `SUBSCRIBE topic` will instruct the gateway to subscribe to `topic`. We also implement a command to `UNSUBSCRIBE` as a matter of order.

We chose to implement the gateway using the Python language, because it has powerful support for both conventional network sockets and for MQTT. The former is included in standard installations, and for MQTT we use the `paho.mqtt` library that we install with

```
sudo apt-get install python3-paho-mqtt
```

and are ready to write our gateway. The powerful libraries make the following Python-program rather compact.

```
# Epics to MQTT gateway, V. Ziemann, 221027
import socket,atexit,paho.mqtt.client
def cleanup():
  sock.close()
atexit.register(cleanup)

def on_message(c,u,msg):
```

```
        print("msg = ",msg.topic," ",msg.payload," ",msg.qos)
        outmsg=msg.topic + " " + msg.payload.decode('utf-8') + "\r\n"
        epics.send(outmsg.encode('utf-8'))
mqttc=paho.mqtt.client.Client()
mqttc.connect("localhost",1883)
mqttc.on_message=on_message
mqttc.loop_start()

sock=socket.socket(socket.AF_INET,socket.SOCK_STREAM)
sock.bind(('',51883))
sock.listen(1)
while 1:
    epics,address = sock.accept()
    print("Connected from ",address)
    while 1:
        msg=epics.recv(1024).decode('utf-8')
        if not msg: break
        words=msg.split()
        if len(words)<2: break;
        if words[0].upper()=="SUBSCRIBE":
            mqttc.subscribe(words[1],1)
            print("Subscribing ", words[0], words[1])
        elif words[0].upper()=="UNSUBSCRIBE":
            mqttc.unsubscribe(words[1])
            print("unsubscribing ", words[0], words[1])
        else:
            print("Publish:",words[0],words[1])
            mqttc.publish(words[0],words[1])
        pass
    epics.close()
    print("Disconnect from ",address)
```

At the start of the program, we import support for sockets, MQTT and `atexit`. The latter is used to execute code asynchronously when terminating the program, by registering the `cleanup()` function to close the network socket. Next we define the function `on_message`, which is called when a subscribed MQTT message arrives. After printing the message, which is useful for debugging, we compose the `outmsg`, which is of type string. Note that we append both a carriage return `\r` and a line feed `\n` in order to match the convention for line terminators used in EPICS. Since the payload is transmitted in binary form, we first convert it to a string with the `.decode()` method. Conversely, the message to send via the network socket named `epics` must be binary, which we accomplish with the `.encode()` method. Now we are ready to define the MQTT client `mqttc`, and connect it to the broker that runs on the same computer as the gateway, namely on `localhost`, and to the standard MQTT port 1883. On the following line we register the `on_message` function to be executed at every `on_message` event of `mqttc`, and start the MQTT event-loop with the `mqttc.loop_start()` function. Note that the sequence of connecting to the broker, registering a call-back function for arriving messages, and starting the event loop mimics the code in the sketch that runs on the NodeMCU. After configuring the MQTT connections, we define a socket `sock` that uses the IPv4 (`AF_INET`) protocol and TCP (`SOCK_STREAM`). We bind the socket to port 51883 where it listens for requests from EPICS. We instruct the socket to only accept one connection at a time and enter into an infinite loop in which the socket waits for an incoming request from EPICS in the function `sock.accept()`. This function blocks until a request arrives, when it returns a handle `epics` to the new connection and the IP number `address` of the connecting computer. Once the connection is established, the gateway enters a second loop and waits for a command on the `epics` socket that it converts to the string `msg` with the `.decode()` method. If `msg` has an invalid format, the connection closes. Otherwise, the received message `msg` is split into words, and branches depending on the first word. If, after conversion to uppercase, it is SUBSCRIBE, the gateway calls the `mqttc.subscribe()` function with the topic as argument. If it is UNSUBSCRIBE it calls the `mqttc.unsubscribe()` function. In all other cases the two words are interpreted as topic and value, and published using the `mqttc.publish()` function. If messages with incorrect format are received or the calling EPICS computer disconnects, the `epics` socket closes and prints a message. At this point the outer `while 1:` loop is still active and the gateway reverts to the `sock.accept()` function and waits for a new connection.

We start the gateway by executing

```
python3 epics2mqtt.py
```

from the command line, and writing a boot script similar to the one for EPICS from Section 6.5 is left as an exercise. The basic functionality of the gateway can be easily tested using `netcat` to emulate EPICS and send strings to the gateway. If the terminal window with the `mosquitto_sub` command from the end of Section 7.1 is still running, we can send messages to it with the command

```
echo "dummy/value 57" | netcat -C localhost 51883
```

which sends the string following `echo` to socket 51883 on the local computer, but this is where the gateway listens, and publishes this message on MQTT on our behalf.

If we connect to the gateway with the command `netcat -C localhost 51883` executed from the command line, we can issue the string SUBSCRIBE `node1/temp` to subscribe to the topic `node1/temp`. Executing the following command in another terminal window

```
mosquitto_pub -d -h localhost -i Pub1 -t node1/temp -m 23
```

will cause the string `node1/temp` 23 to appear in the window with `netcat` running. Issuing the command UNSUBSCRIBE `node1/temp` in `netcat` stops the subscription, and messages will no longer appear in the `netcat` window.

Finally, we prepare EPICS protocol and database files that implement the following behavior. Executing `caput node1/fan 1` publishes the topic `node1/fan` with message 1, and `caput SUBSCRIBE node1/temp` subscribes to the topic `node1/temp` such that we asynchronously receive the messages in EPICS. The first turns the fan on and the second reports the temperature measured by the LM35 on the NodeMCU. The protocol file to implement this behavior is stored in the following file, named `mqtt.proto`:

```
# ./tempApp/Db/mqtt.proto
Terminator = CR LF;
set_fan {out "node1/fan %i";}
subscribe {out "SUBSCRIBE %s";}
unsubscribe {out "UNSUBSCRIBE %s";}
get_temp {in "node1/temp %f";}
```

First we define the Terminator that matches the \r\n used in the gateway code, and then define functions that either input or output values in the same way we used in the previous chapter. Note that the functions have the MQTT names hard coded as character strings and that the functions to subscribe and unsubscribe have a string as argument. The corresponding database file, called `mqtt.db`,

```
# ./tempApp/Db/mqtt.db
record(ao, "node1/fan") {
    field(DESC, "Fan on node1")
    field(DTYP, "stream")
    field(OUT, "@mqtt.proto set_fan $(PORT)")
}
record(stringout, "SUBSCRIBE") {
    field(DESC, "subscribe to topic")
    field(DTYP, "stream")
    field(OUT, "@mqtt.proto subscribe $(PORT)")
}
record(stringout, "UNSUBSCRIBE") {
    field(DESC, "unsubscribe from topic")
    field(DTYP, "stream")
    field(OUT, "@mqtt.proto unsubscribe $(PORT)")
}
record(ai, "node1/temp") {
    field(DESC, "Temperature on node1")
    field(DTYP, "stream")
    field(INP, "@mqtt.proto get_temp $(PORT)")
    field(SCAN,"I/O Intr")
}
```

links the functions defined in the protocol file to process variables where the MQTT variables have the same name in EPICS, and we add two process variables handling subscription. Since they register a name of a variable, their EPICS record type is stringout. All variables that we receive from MQTT via the gateway arrive asynchronously, at the rate they are published, and we therefore need to use the line field(SCAN,"I/O Intr") in the database file, which handles data that arrive without explicitly being requested. Note that we only need to subscribe to values we want to read with caget. MQTT variables we want to set, such as turning on the fan, need no subscription.

After having defined the protocol and database file, we need to edit ./temp/tempApp/ Db/Makefile and add the line DB += mqtt.db. Finally, we add the following definition for the PORT in st.cmd in ./iocBoot/ioctemp/.

```
drvAsynIPPortConfigure("SOCKET","192.168.20.1:51883",0,0,0)
dbLoadRecords("db/mqtt.db","PORT='SOCKET',USER='mqtt'")
```

The IP address points to the IP on which both broker and gateway run, and the port on which the gateway listens. Once this is operational, we have access to MQTT topics from EPICS.

Whereas EPICS and MQTT provide frameworks to connect multiple sensors and actuators as well as presentation and controlling computers, enable *websockets* point-to-point communication between two partners with one of them running inside a web browser.

QUESTIONS AND PROJECT IDEAS

1. What is the purpose of the broker?

2. Find out about public brokers on the Internet.

3. Connect the humidity and barometric pressure sensor to a NodeMCU and have the data published via MQTT.

4. Control the brightness of a LED connected to a NodeMCU via MQTT.

5. Connect a DC motor with H bridge to a NodeMCU and control it via MQTT.

6. Connect a stepper motor with H bridge to a NodeMCU and control it via MQTT.

7. Connect a model-servo to a NodeMCU and control it via MQTT.

8. Prepare a boot script for the gateway following the example in Section 6.5.

9. Write gateway between MQTT and a MySQL database such that the database is automatically filled with newly arriving data samples.

10. Write gateway that fills an RRDtool database with data from MQTT.

11. Write a gateway to interface `octave` to MQTT such that a plot is continuously updated with new data.

12. Find an MQTT client for your smartphone and use it to read the temperature from the NodeMCU.

13. Discuss the similarities and differences between EPICS and MQTT. Under what circumstances do you prefer one or the other?

14. Write an MQTT client for the NodeMCU that mimics the query-response behavior: It receives a query, does something, and publishes a response.

## CHAPTER 8

Websockets

Occasionally, it is desirable to control the data acquisition hardware directly from an internet browser where the user interface is realized as an interactive webpage using *JavaScript*. In this way, even smartphones connected to the network can control the acquisition and display data in real time. Unfortunately, due to security concerns can normal sockets, such as those used the second example in Section 4.6.3, not be accessed from a browser. Instead, however, we can use a simpler interface called *websockets*, which allows us to exchange simple text-based messages between microcontrollers and browsers.

The text-based messages are formatted using the *JavaScript object notation* (JSON) where all data are specified by name-value pairs, enclosed in curly brackets. For example, we send a single temperature value from the microcontroller formatted as {"TEMP":"27.2"}. We can even add several pairs in the same message, such as {"X":"17", "Y":"4"}, or send arrays via {"DATA":"[17","4"]}, where the array values are enclosed in square brackets. We access the first value via DATA[0], which evaluates to 17.

In order to illustrate the communication, we will use the hardware from Section 4.6.3 with the LM35 temperature sensor connected to the analog input of a NodeMCU microcontroller. We program the NodeMCU to listen to JSON messages coming from a controlling browser that instructs it to start or stop measuring the temperature at regular intervals, pack the results in a JSON message, and send it back to the browser as soon as new data is available. Moreover, we start a regular webserver on the NodeMCU that serves the controlling webpage, shown in Figure 8.1, that any browser, here chromium, can access. On the top, buttons to start and stop the acquisition are displayed together with a menu to select the update rate of the measurements. A third button clears the logging display shown below the buttons. Continuously arriving data points are added to it until the screen is filled when the trace starts moving to the left as new samples arrive. Below the logging display status information is shown, normally the value of the most recent data.

Now, let's see how to bring this system to live; first we look at the code on the NodeMCU.

8.1 ON THE NODEMCU

The NodeMCU has to handle several tasks simultaneously. It must serve the HTML file of webpage from Figure 8.1, it has to listen to requests coming from the same page once it is running in the browser, and it must measure the temperature at regular intervals. We therefore add functionality, provided by additional libraries, for these tasks and open the *Library manager* in the *Tools* menu of the Arduino IDE to install the WebSockets and the ArduinoJSON libraries. Since the program is rather long, we'll discuss it in smaller chunks.

DOI: 10.1201/9781003341703-8

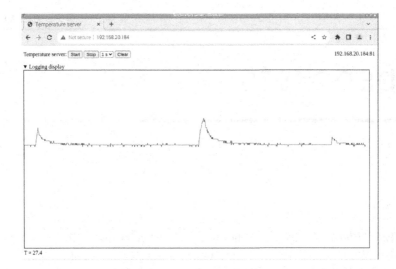

Figure 8.1 The webpage served by the NodeMCU from which the data acquisition is controlled and the data are displayed in real time illustrated by the bumps that are caused by placing a finger on the temperature sensor.

After defining the network credentials, we add support for WiFi in general, for the webserver, and for the server responsible for the websocket. Furthermore, we add a library to decode JSON messages and another one to enable support the SPIFFS filesystem that we use to store the webpage in the flash memory of the NodeMCU. With these declarations, we are ready to declare the webserver **server2** to listen on port 80, which is commonly used for html. Likewise we define **webSocket** to handle the websocket communication on port 81.

```
// Minimal websocket temperature server, V. Ziemann, 221017
const char* ssid     = "messnetz";
const char* password = "zxcvZXCV";
#include <ESP8266WiFi.h>
#include <ESP8266WebServer.h>
#include <WebSocketsServer.h>
#include <ArduinoJson.h>
#include "FS.h"                    // SPIFFS filesystem
ESP8266WebServer server2(80);                           // port 80
WebSocketsServer webSocket = WebSocketsServer(81);  // port 81
```

In the following lines we add the support for an automatically repeating process that we call **SampleSlow**, define the period at which to execute the process and two flags that signal availability of things to send to the browser. The function **sampleslow_action()** simply sets one of these flags, such that the temperature measurement is handled in the main loop. We also allocate a variable to hold information that we send to the browser and a variable to remember the handle to the websockets that we need in several parts of the program.

```
#include <Ticker.h>
Ticker SampleSlow;
int sample_period=1,data_available=0,info_available=0;
void sampleslow_action() {data_available=1;}
char info_string[70];
volatile uint8_t websock_num=0;
```

Now we are ready to specify how we want the websocket to react to requests coming from the other end of the communication channel – the browser. Any activity on the websocket, which is handled in the background, triggers the execution of the function webSocketEvent(). As input this function receives the handle num, which identifies the websocket, the type of the interaction, its payload, and its size. Depending on the type we use a switch-case construction to define the appropriate action. If the browser disconnects or connects, a short message is printed to the serial line. Any JSON-formatted string from the browser has type WStype_TEXT and the payload contains the command coming from the browser. Using the function deserializeJson(), provided by the ArduinoJSON library, we parse the payload and place the name-value pairs into the variable root which is of type DynamicJsonDocument. Now we can access the value of cmd, which specifies the command, as root["cmd"] and access the value of val as root["val"]. For example, the string {"cmd":"START","val":1} contains the command START and the value 1. In the subsequent part of the function, the reactions to the commands are coded. If a START command is received, the value is interpreted as the sample_period and used to register the sampleslow_action that is executed with the specified periodicity. Furthermore, the info_string is filled with a text that we send back to the browser. A STOP command just stops the periodic acquisition and any unknown command is displayed on the serial line for debugging purposes.

```
void webSocketEvent(uint8_t num, WStype_t type, uint8_t * payload,
    size_t length) {
  Serial.printf("webSocketEvent(%d, %d, ...)\r\n", num, type);
  websock_num=num;
  switch(type) {
    case WStype_DISCONNECTED:
      Serial.printf("[%u] Disconnected!\r\n", num);
      break;
    case WStype_CONNECTED:
      {
        IPAddress ip = webSocket.remoteIP(num);
        Serial.printf("[%u] Connected from %d.%d.%d.%d url: %s\r\n",
          num, ip[0], ip[1], ip[2], ip[3],payload);
        strcpy(info_string,"Websocket opened on ESP8266");
        info_available=1;
      }
      break;
    case WStype_TEXT:
    {
      Serial.printf("[%u] get Text: %s\r\n", num, payload);
      DynamicJsonDocument root(300);        //.........parse JSON
      deserializeJson(root,payload);
      const char *cmd = root["cmd"];
      const int val = root["val"];
      if (strstr(cmd,"START")) {
        sample_period=val;
        SampleSlow.attach(sample_period,sampleslow_action);
        strcpy(info_string,"Acquisition started"); info_available=1;
      } else if (strstr(cmd,"STOP")) {
        SampleSlow.detach();
        strcpy(info_string,"Acquisition stopped"); info_available=1;
```

```
      } else {
        Serial.println("Unknown command");
      }
      break;
    }
  }
}
```

At this point all support functions are defined, and we configure the NodeMCU in the setup() function. There we first initialize the serial line and then Wifi. A successful connection is then reported via the serial line. Next we call webSocket.begin() to start the websocket server and register the function webSocketEvent() to be executed when a websocket event occurs. We then test whether the SPIFFS filesystem is operational and, if that is the case, start server2, instruct it to serve the file webpage.html, and report the successful start to the serial line. We return later to the point of how to copy the file to the filesystem.

```
void setup() {
  Serial.begin(115200); delay(1000);
  WiFi.begin(ssid, password);
  while (WiFi.status() != WL_CONNECTED) {delay(500); Serial.print(".");}
  Serial.print("\nConnected to ");  Serial.print(ssid);
  Serial.print(" with IP address: "); Serial.println(WiFi.localIP());
  webSocket.begin();
  webSocket.onEvent(webSocketEvent);
  if (!SPIFFS.begin()) {
    Serial.println("ERROR: no SPIFFS filesystem found");
    return;
  } else {
    server2.begin();
    server2.serveStatic("/", SPIFFS, "/webpage.html");
    Serial.println("SPIFFS file system found and server started");
  }
}
```

After the NodeMCU is configured, the loop() function executes continuously by first giving control to server2 to take care of any requests and then do the same for the webSocket. If data_available is set to unity, which can only happen inside the sampleslow_action() function, we allocate some buffers, measure the temperature, convert the numerical value to a string with dtostrf() and assemble the JSON string in the variable out that we subsequently send back to the browser by executing the function webSocket.sendTXT(). Likewise, if info_available is set to unity, the info_string is packaged into a JSON package with name INFO and sent back to the browser.

```
void loop() {
  server2.handleClient();  // handle webserver on port 80
  webSocket.loop();        // handle websocket server on port 81
  if (data_available==1) {
    data_available=0;
    char buf[20],out[100];
    float temperature=analogRead(0)*100*3.3/1023.0;
    dtostrf(temperature,7,1,buf);
    (void) sprintf(out,"{\"%s\":\"%s\"}","TEMP",buf);
```

```
      webSocket.sendTXT(websock_num,out,strlen(out));
    }
    if (info_available==1) {
      info_available=0;
      char out[100];
      (void) sprintf(out,"{\"%s\":\"%s\"}","INFO",info_string);
      webSocket.sendTXT(websock_num,out,strlen(out));
    }
  }
```

This sketch thus takes care of handling the webserver, the websockets and the measurements.

After configuring the NodeMCU, we move on to programing the other end of the communication channel, which is the html file `webpage.html` that runs in a browser window.

8.2 IN THE BROWSER

Below we reproduce and comment the html file `webpage.html` responsible for Figure 8.1. It contains both html and JavaScript commands to orchestrate the communication between the Browser and the NodeMCU. The file contains the usual `!DOCTYPE` and `HTML` tags, here with specifying the language as English. Inside `HEAD` tags we specify some generic information, such as the title, before providing style information for several items that populate our webpage. The `#displayarea` will have a black border around the space that we use to display the measurement data as a red `#trace0`. The `#ip` entity will show the IP number of the connected NodeMCU on the right-hand side of the displayed webpage.

```
<!DOCTYPE HTML>
<HTML lang="en">
<HEAD><TITLE>Temperature server</TITLE>
  <META charset="UTF-8">
  <STYLE>
    #displayarea {border: 1px solid black; }
    #trace0 {fill: none; stroke: red; stroke-width: 1px;}
    #ip {float: right;}
  </STYLE>
</HEAD>
```

The following section defines the buttons and the menu near the top of webpage. The buttons are named with `id=` and have a so-called callback function assigned with `onclick=`, which defines the action caused by pressing the button. The menu to select the sample period is defined with `SELECT` tags, where the different selections are defined with `OPTION` tags. The item with `id='ip'` automatically floats to the right-hand margin, because we had specified that behavior in the `STYLE` declaration. The area with the graph that we see on Figure 8.1 is sandwiched between `DETAILS` tags, which allows us to minimize it. The item with `id='displayarea'` is a *Scalable Vector Graphic*. Inside the `SVG` tag we define the width and height of the area reserved for the graph. The `PATH` tag has the name `trace0` and will be filled with the data points to display. We initialize it with the svg command `d="M0 256"` which initializes `trace0` by moving (M) the starting point of the displayed line to the coordinates (0,256). Finally, we place some text into the line reserved to display status information.

```
<BODY>
<P> Temperature server:
```

```
<BUTTON id="start" type="button" onclick="start();">Start</BUTTON>
<BUTTON id="stop" type="button" onclick="stop();">Stop</BUTTON>
<SELECT onchange="setSamplePeriod(this.value);">
  <OPTION selected="selected" value="1">1 s</OPTION>
  <OPTION value="2">2 s</OPTION>
  <OPTION value="5">5 s</OPTION>
</SELECT>
<BUTTON id="clear" type="button" onclick="cleardisplay();">
        Clear</BUTTON>
<A id='ip'>IP address</A>
</P>
<DETAILS open><SUMMARY>Logging display</SUMMARY>
  <SVG id="displayarea" width="1024px" height="512px">
    <PATH id="trace0" d="M0 256" />
  </SVG>
</DETAILS>
<DIV id="status">Status window</DIV>
```

Up to this point we only defined the static webpage. In the following we make it dynamic with the help of JavaScript which is enclosed between SCRIPT tags. In the first line we determine the ip number of the computer that served the webpage, normally the NodeMCU, and append the port number :81. After initializing a number of variables needed to control and display the measurements, we display the address. Note that the entity in the current browser window that is named with id='ip' is accessible by document.getElementById('ip') and the method .innerHTML gives access to the content of that entity. Here we use it to change the displayed text to ipaddr. With the IP number known, we are ready to open the websocket and give it the name websock. We immediately specify its standard behavior by defining the methods for error handling as well as opening and closing the websocket; we simply write a descriptive message to the JavaScript console – we discuss access to it below – and to the status line at the bottom of the displayed webpage. The onmessage function breathes life into our webpage; it determines the response of the browser window to messages arriving from the NodeMCU. After writing the arriving event to the JavaScript console we parse the event.data with the built-in JSON.parse function, which extracts the JSON name-value pairs into the variable stuff. We then, for example, access the received information with stuff["INFO"] and, provided it is not undefined, write it to the status line on the browser window. If stuff["TEMP"] is not undefined, we have received a new temperature, whose value is stored in val. In the following line, we extract the attribute d, which contains the displayed data points, from trace0 and store it in dd to which we add the new temperature value, properly scaled by appending L xpix ypix to dd. Here xpix is the horizontal coordinate in pixels and ypix the corresponding vertical coordinate inside displayarea. Then we copy the updated dd back to d of trace0, which makes the new data point visible in the browser window. Finally we increment the current_position and check for overflow in which case we remove the first data point with the replace() function. Its first argument is the search string that starts with an M, ends with an L, and has any number of characters in between that are not an L ([^L]*). Then we move the entire trace0 to the left and add new data points on the right edge.

```
<SCRIPT>
  var ipaddr=location.hostname + ":81";
  var sample_period=1, current_position=0, Tmin=10, Tmax=40;
  document.getElementById('ip').innerHTML=ipaddr;
  var websock = new WebSocket('ws://' + ipaddr);
```

```
websock.onerror = function(evt) { console.log(evt); toStatus(evt) };
websock.onopen = function(evt) {console.log('websock open'); };
websock.onclose = function(evt) {
  console.log('websock close'); toStatus('websock close');
};
websock.onmessage=function(event) {
  console.log(event);
  var stuff=JSON.parse(event.data);
  var val=stuff["INFO"]; //............................info
  if ( val != undefined ) {toStatus(val);}
  val=stuff["TEMP"];       //....................temperature
  if ( val != undefined ) {
    dd=document.getElementById('trace0').getAttribute('d');
    dd += ' L' + current_position + ' '
        + (512-512*(val-Tmin)/(Tmax-Tmin));
    document.getElementById('trace0').setAttribute('d',dd);
    current_position += 1;
    toStatus('T = ' + val);
    if (current_position > 1024) {
      dd=document.getElementById('trace0').getAttribute('d')
        .replace(/M[^L]*L/, "M");
      document.getElementById('trace0').setAttribute('d',dd);
      document.getElementById('trace0').setAttribute
      ('transform','translate(' + (1024-current_position) + ',0)');
    }
  }
}
```

The SCRIPT continues with the definition of the function toStatus(), which just writes the text given as its argument to the status line and the definitions of the callback functions for the buttons. The function start() forms a JSON message with the two name-value pairs, one for cmd and one for val. It sends the start command to the NodeMCU with the sample_period given as val. Likewise stop() stops the acquisition and cleardisplay() resets the attribute d of trace0, which clears the displayed trace. The callback function for the selection menu in the SELECT tags setSamplePeriod() receives the value specified in the chosen OPTION tag as argument and assigns it to sample_period. Finally, the open tags for SCRIPT, BODY, and HTML are closed.

```
function toStatus(txt){document.getElementById('status').innerHTML=txt;}
function start() {
  websock.send(JSON.stringify({"cmd" : "START", "val" : sample_period}));
}
function stop() {
  websock.send(JSON.stringify({"cmd" : "STOP", "val" : "-1" }));
}
function setSamplePeriod(v) {
  toStatus("Setting sample period to "+ v + " ms");
  sample_period=v;
}
function cleardisplay() {
  dd = "M0 256";
  document.getElementById('trace0').setAttribute('d',dd);
```

```
    document.getElementById('trace0').setAttribute
            ('transform','translate(0)');
    current_position=0; toStatus("Display cleared");
  }
</SCRIPT></BODY></HTML>
```

We copy this webpage to a subdirectory named `data` under the subdirectory where `websocket_tempserver.ino` file resides. Then we install the program that copies the contents of the `data` subdirectory to the flash memory on the NodeMCU by downloading `ESP8266FS-0.5.0.zip` from the *releases page* on `https://github.com/esp8266/arduino-esp8266fs-plugin` and unzip it in the `/Arduino/tools/` subdirectory. After restarting the Arduino IDE, we find an entry *ESP8266 Sketch Data Upload* in the *Tools* menu. It uploads the webpage to the NodeMCU – but only if the *Serial monitor* is closed – such that the server on port 80 finds it and delivers it to connecting browsers.

Using this system is rather simple. On any browser that has access to the network with the NodeMCU we enter the IP number of the NodeMCU in the address bar, which requests the webpage shown in Figure 8.1 from the NodeMCU. Once visible in the browser, the JavaScript embeddded on that webpage connects to the websocket port 81 on the NodeMCU and establishes the two-way communication channel – the websocket. It is henceforth used to send JSON-formatted messages back and forth.

On the NodeMCU we can observe the communication via the serial monitor, because the messages are echoed to the serial line. In a chromium browser, we open the JavaScript console by pressing `Ctrl-Shift-I` and selecting the `Console` tab. It can also be found behind the three dots at the top-right of the browser under *More tools* and *Development tools*. Every argument of `console.log()` in the JavaScript code embedded in the webpage shows up there. Now we can observe how the communication between NodeMCU and browser is established when the page is reloaded by pressing F5 in the browser (Mozilla or Chrome).

Most examples we encountered so far were deliberately chosen to be moderately simple, in order to illustrate the mechanisms, but in the coming chapters we advance to more complex projects, and start with a weather station with distributed sensors.

QUESTIONS AND PROJECT IDEAS

1. Who initiates the communication: the NodeMCU or the browser?

2. Add a 10 s sampling time to the `SELECT` menu.

3. Add a 100 ms sampling time to the `SELECT` menu. Note that you need to use `SampleSlow.attach_ms()` on the NodeMCU to handle times shorter than a second.

4. Add a button to the webpage that turns a pin on the NodeMCU on and off.

5. Display the state of an input pin on the NodeMCU in a separate status line.

## CHAPTER 9

Example: Weather Station with Distributed Sensors

Our first project is a weather station that measures barometric pressure, humidity, temperature, and air quality at a number of locations inside and outside a building. We use **rrdtool** to prepare plots of the measurements over periods of 4 hours, 2 days, a week, and a month. Moreover, we prepare database files to enable integration into an EPICS control system. We select the NodeMCUs as microcontrollers for the sensor nodes because they are very easy to program, and flexible that they may be deployed all over the place without having to pull wires. Later in this chapter, we briefly show how to replace the NodeMCU by its smaller brethren, the ESP-01, which requires a little extra attention. Note that in either case we only show how to connect a single sensor node, but multiple copies only differ by their IP number. We can duplicate any interfacing software by simply changing the IP numbers appropriately.

We choose the BME680, already discussed in Section 4.4.3, to measure the barometric pressure, humidity, temperature, and air quality. Figure 9.1 shows the project assembled on a breadboard with NodeMCU to the right of the BME680. The only connections are the power rails for ground and 3.3 V, as well as two I2C lines for clock and data. The former connects pin D1 on the NodeMCU to SCK on the sensor where as the latter connects pin D2 to SDI. We also added a 100 nF decoupling capacitor to the power rails.

The program running on the microcontroller follows the template we used earlier. After defining the network credentials and including libraries for WiFi and the sensor, we define the WiFiserver **server** listening on port 1137 and the **bme** object, which allows us to communicate with the sensor. In the **setup()** function we first open the Serial line and configure the WiFi to use a static IP address. This is advantageous if we always want to find the weathernode at the same address rather than depending on the DHCP server on the Raspi to always assign the same. Then we establish the WLAN connection and then start the server, just as in Section 4.6.3. At the end of the **setup()** function, we ensure that the sensor is actually present before initializing it.

```
// Weathernode-BME680, V. Ziemann, 221028
const char* ssid     = "messnetz";
const char* password = "zxcvZXCV";
const int port=1137;
#include <ESP8266WiFi.h>
WiFiServer server(port);
#include <Adafruit_BME680.h>
Adafruit_BME680 bme;
```

DOI: 10.1201/9781003341703-9

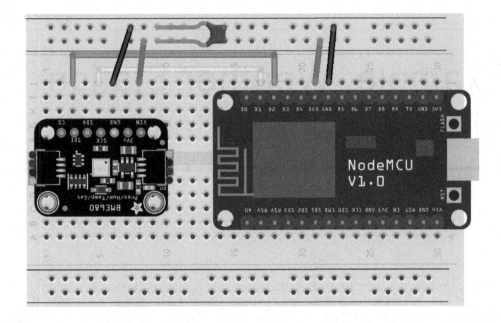

Figure 9.1 The weathernode circuit with the NodeMCU on the right and the BME680 sensor on the left[†].

```
char status[30] = "OK";
void setup() {   //........................................setup
  Serial.begin(115200); delay(1000);
  IPAddress ip(192, 168, 20, 56);          // define static IP
  IPAddress gw(192, 168, 20, 1);           // gateway
  IPAddress subnet(255, 255, 255, 0);      // netmask
  WiFi.config(ip,gw,subnet);               // configure static IP
  WiFi.begin(ssid,password);
  while (WiFi.status() != WL_CONNECTED) {delay(500); Serial.print(".");}
  Serial.print("\nConnected to ");  Serial.print(ssid);
  Serial.print(" with IP address: "); Serial.println(WiFi.localIP());
  server.begin();
  // Wire.begin(0,2);  // for ESP-01 only
  if (!bme.begin()) {  // default address
    strcpy(status,"Error: BME680 not found");
    Serial.println(status);
  }
  bme.setTemperatureOversampling(BME680_OS_8X); // configure BME680
  bme.setHumidityOversampling(BME680_OS_2X);
  bme.setPressureOversampling(BME680_OS_4X);
  bme.setIIRFilterSize(BME680_FILTER_SIZE_3);
  bme.setGasHeater(320, 150); // 320*C for 150 ms
}
void loop() { //........................................loop
  char line[30];
  WiFiClient client = server.available();
```

```
    while (client) {
      while(!client.available()) {
        delay(5);
        if (!client.connected()) {client.stop(); break;}
      }
      if (!client.connected()) {client.stop(); break;}
      client.readStringUntil('\n').toCharArray(line,30);
      bme.performReading();
      float T=bme.temperature;              // Celsius
      float P=bme.pressure/100.0;           // hPa=mbar
      float H=bme.humidity;                 // percent
      float VOC=bme.gas_resistance/1000.0;  // kOhm
      if (strstr(line,"V?")==line) {
        client.print("WeatherNode-BME680, Status ");
        client.println(status);
      } else if (strstr(line,"T?")==line) {
        client.print("T "); client.println(T,2);
      } else if (strstr(line,"P?")==line) {
        client.print("P "); client.println(P,2);
      } else if (strstr(line,"H?")==line) {
        client.print("H "); client.println(H,2);
      } else if (strstr(line,"VOC?")==line) {
        client.print("VOC "); client.println(VOC,2);
      } else if (strstr(line,"ALL?")==line) {
        sprintf(line,"%.2f:%.2f:%.1f:%.2f",T,P,H,VOC);
        client.print(line);  % no println!
        client.flush(); client.stop();
      } else {
        client.println("Unknown command, disconnecting!");
        client.flush(); client.stop();
      }
      yield();
    }
  }
```

In the `loop()` function, we first wait for a client to connect and then for a request from the client, before reading out the sensor and storing the measurement values in variables. The previously-used query-response construction follows and is used to decode the request for T?, P?, H?, and VOC?. We added an additional request for ALL? that returns a all four values suitably formatted for `rrdtool`, sends the string without newline character appended, and closes the socket unconditionally. Likewise, any unknown command causes the network connection to be closed. Finally, the `yield()` function permits the microcontroller to do internal bookkeeping.

On the Raspi, we use a `cron` job to query the microcontroller once every minute to read the measurement values and place them into a `rrdtool` data base. Since most of the formatting is already done on the NodeMCU, the following command

```
echo "ALL?" | netcat -C 192.168.20.56 1137
```

returns a colon-separated string, such as `26.55:1003.02:45.1:24.51` with the four measurement values for temperature, pressure, humidity, and air quality. Note that we use the built-in `echo` command to pipe the string ALL? to the `netcat` program that we had earlier used to directly communicate with network sockets.

The following shell script uses this construction with `echo` and `netcat` to store the colon-separated string into the shell variable `TEMP` that is used to feed the measurement values into the `rrdtool` database.

```
#!/bin/bash
# fill_weather_db.sh
DB=/home/pi/rrdtool/weather.rrd
TEMP=$(echo "ALL?" | netcat -C 192.168.20.56 1137)
/usr/bin/rrdtool update $DB N:$TEMP
```

After saving the script we need to make it executable with the command `chmod +x fill_weather_db.sh`. But before we start to fill the database with the `rrdtool update` command, we must create it, where we assume the following command is executed in the `/home/pi/rrdtool` directory.

```
rrdtool create weather.rrd --step 60 \
    DS:T:GAUGE:180:-20:100 \
    DS:P:GAUGE:180:900:1100 \
    DS:H:GAUGE:180:0:100 \
    DS:VOC:GAUGE:180:0:1000 \
    RRA:AVERAGE:0.5:1:2880 \
    RRA:AVERAGE:0.5:10:2880 \
    RRA:AVERAGE:0.5:60:2880
```

It creates the `weather.rrd` database file that expects values once every 60 seconds, and contains four measurement columns (`DS`) for the temperature, pressure, humidity, and air quality. Then it defines the archive (`RRA`), which is filled by single averages for 2880 samples, which amounts to 2 days. The next two lines define averages over 10 samples or 10 minutes, for a total of 2880 samples or 20 days, and finally, hourly averages over 60 samples for 120 days. In order to automatically fill the database, we start the crontab editor with `crontab -e` and add the line

```
* * * * * /home/pi/rrdtool/fill_weather_db.sh
```

to the end of the file, which causes the script to be executed once a minute and store the values in the `weather.rrd` database.

Now we just have to wait until some measurements make it into the database, and then create graphs in the same way we did in Section 5.6.3 using the `rrdtool graph` command. The command to create one set of graphs for temperature, pressure, and humidity called `weathergraph.sh`, is the following:

```
#!/bin/bash
S=$1
DB=/home/pi/rrdtool/weather.rrd
PICDIR=/home/pi/public_html/weather
/usr/bin/rrdtool graph $PICDIR/temperature${S}.png -s $S \
    -w 800 -h 200 \
    -t "Temperature" -v "T [C]" -l 16 -u 32 -r \
    DEF:t0=$DB:T:AVERAGE LINE1:t0#FF0000:"Temperature";
/usr/bin/rrdtool graph  $PICDIR/pressure${S}.png -s $S \
    -w 800 -h 200 \
    -t "Barometric Pressure" -v "P [mbar]" -A -l 975 -u 1025 -r \
    DEF:t0=$DB:P:AVERAGE LINE1:t0#FF0000:"Pressure, range=[975,1025]";
/usr/bin/rrdtool graph $PICDIR/humidity${S}.png -s $S \
```

```
        -w 800 -h 200 \
        -t "Humidity" -v "Humidity [%]" -l 0 -u 100 \
        DEF:t0=$DB:H:AVERAGE LINE1:t0#FF0000:"Humidity";
/usr/bin/rrdtool graph $PICDIR/VOC${S}.png -s $S \
        -w 800 -h 200 \
        -t "Air quality" -v "Air  quality [kOhm]" \
        DEF:t0=$DB:VOC:AVERAGE LINE1:t0#FF0000:"Air quality";
```

It starts by copying the first command line argument, which should contain the start time of the display to the variable S, declares the database file to use, and the location where the graphs are stored. For this we choose a directory under /home/pi/public_html, which is accessible to the apache2 web server. The first rrdtool graph produces the temperature plot. Note that the graph file carries the time period appended to its name. The second rrdtool command produces the pressure, the third the humidity graph, and the last one the air quality, all in the way discussed in Section 5.6.3. In order to produce graphs with time spans of 8 hours, 1 day, 1 week, 1 month, and 3 months, we call weathergraph.sh with different starting dates as command line arguments, and place the five commands in a separate script, called allweathergraph.sh. It is reproduced here.

```
#!/bin/bash
# /home/pi/rrdtool/allweathergraph.sh
/home/pi/rrdtool/weathergraph.sh -8h > /dev/null
/home/pi/rrdtool/weathergraph.sh -1d > /dev/null
/home/pi/rrdtool/weathergraph.sh -1w > /dev/null
/home/pi/rrdtool/weathergraph.sh -1m > /dev/null
/home/pi/rrdtool/weathergraph.sh -3m > /dev/null
```

The call to weathergraph.sh uses the absolute path because we want to run it from a cron job to create new graphs every 10 minutes. We also redirect the output into the big bit-bucket /dev/null. We execute this script regularly as a cron job with the command crontab -e in order to add the line

```
*/10 * * * * /home/pi/rrdtool/allweathergraph.sh
```

in the crontab file.

It remains to prepare the index.html file in the directory with the graphs. The following HTML file presents the data as the simple web page shown in Figure 9.2.

```
<!DOCTYPE HTML>
<HTML>
  <HEAD>
    <TITLE>Raspi Weather Station</TITLE>
    <META http-equiv="refresh" content="120">
  </HEAD>
  <BODY>
    <H1 ALIGN=CENTER>Raspi Weather Station (8 hours)</H1>
    Display data for:
    <A HREF=index.html>8 hours</A>
    <A HREF=index-1d.html>1 day</A>
    <A HREF=index-1w.html>1 week</A>
    <A HREF=index-1m.html>1 month</A>
    <A HREF=index-3m.html>3 months</A>
    <HR>
```

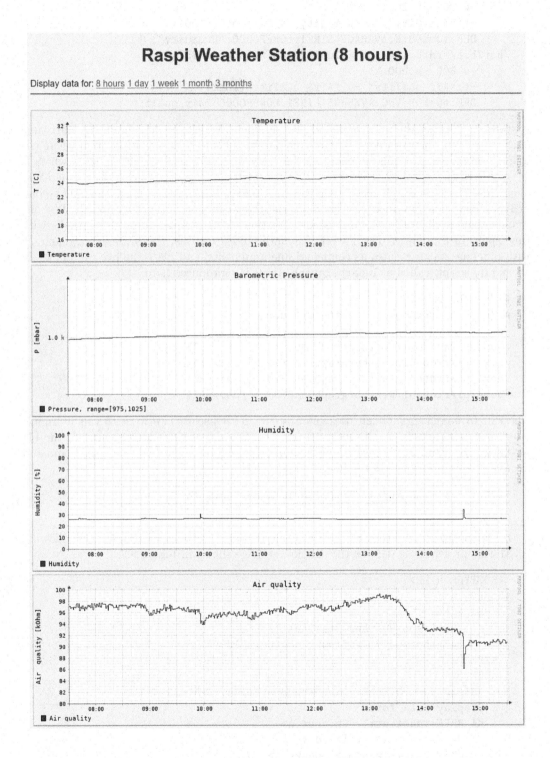

Figure 9.2 The weather station web page served by the Raspi.

```
    <BR>
    <IMG SRC="temperature-8h.png" ALT="Temperature">
    <IMG SRC="pressure-8h.png" ALT="Pressure">
    <IMG SRC="humidity-8h.png" ALT="Humidity">
    <IMG SRC="VOC-8h.png" ALT="Humidity">
  </BODY>
</HTML>
```

The file contains the already-known header information with the title `Raspi Weather Station`, and the automatic refresh rate of 120 seconds. Then a line with options of different time ranges follows. Each entry points to a different web page residing in the same directory, but serving the graphs for the other ranges. Remember, the graphs are automatically updated by the cron job, but the different `index-XX.html` pages are constant. We need to point out that the reported temperature is affected by the little heater for the MQ-x sensor inside the BME680 and the reported value is therefore too high by about four degrees. Moreover, I caused the small spikes in the humidity and corresponding dips in air quality graphs by exhaling on the sensor.

Having the weather data displayed for different time ranges is certainly adequate for home use, but in a laboratory we want to integrate the data from the sensor nodes into the control system. If we already have the EPICS system running, all we have to do is to prepare the protocol and database files for our weather data. The former, written in a more compact way, is the following.

```
# ./tempApp/Db/weather.proto
Terminator = CR LF;
get_T {out "T?"; in "T %f"; ExtraInput = Ignore; }
get_P {out "P?"; in "P %f"; ExtraInput = Ignore; }
get_H {out "H?"; in "H %f"; ExtraInput = Ignore; }
get_VOC {out "VOC?"; in "VOC %f"; ExtraInput = Ignore; }
```

It follows the already-known pattern of sending a query with a question mark appended, and then receives the query, followed by the measurement value. To accompany this file, which orchestrates the low-level communication, we need a data base file, named `weather.db`, with the following contents.

```
# ./tempApp/Db/weather.db
record(ai, "$(USER):T") {
   field(DESC, "Temperature")
   field(SCAN, "10 second")
   field(DTYP, "stream")
   field(INP, "@weather.proto get_T $(PORT)")
}
record(ai, "$(USER):P") {
   field(DESC, "Pressure")
   field(SCAN, "10 second")
   field(DTYP, "stream")
   field(INP, "@weather.proto get_P $(PORT)")
}
record(ai, "$(USER):H") {
   field(DESC, "Humidity")
   field(SCAN, "10 second")
   field(DTYP, "stream")
   field(INP, "@weather.proto get_H $(PORT)")
```

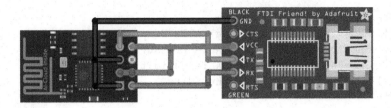

Figure 9.3 Connections to program an ESP-01 with a USB-to-serial converter[†].

```
}
record(ai, "$(USER):VOC") {
    field(DESC, "Air quality")
    field(SCAN, "10 second")
    field(DTYP, "stream")
    field(INP, "@weather.proto get_VOC $(PORT)")
}
```

Here we find analog input `ai` records that define the process variables `$(USER):XX`, and link them to the respective protocol functions. Moreover, we need to include the new database file in the `Makefile` by adding the line

```
DB += weather.db
```

and add the following two lines to the `./temp/iocBoot/ioctemp/st.cmd` EPICS command file

```
drvAsynIPPortConfigure("SOCKET","192.168.20.56:1137",0,0,0)
dbLoadRecords("db/weather.db","PORT='SOCKET',USER='weather'")
```

and finally, call `make` in the base directory of the `temp` hierarchy. Then we execute `st.cmd` as before, either from the command line or automatically with an init script as discussed at the end of Section 6.5. The process variables are now accessible from any computer on the network. To access the barometric pressure, we issue `caget weather:P`, and similarly for the other variables. In this way, they can be seamlessly included in logging, monitoring, or other display programs that EPICS provides. Adding more sensors is as simple as adding protocol and database files as well as including them in the `Makefile` in the `Db` directory, and adding two lines in the `st.cmd` program.

With its on-board USB connector and many input-output pins, the NodeMCU is a very convenient development platform, but in this example, we only use two pins for I2C and, once the NodeMCU is up and running, we no longer need the USB interface or the serial line. All communication can be done via the WLAN interface. It is therefore prudent to migrate the project to the ESP-01, which is also less expensive, an attractive feature if we plan to deploy multiple weathernodes. On the other hand, the absence of a USB interface makes programming the ESP-01 a bit more tricky. We can, however, directly connect the ESP-01 with its eight exposed pins to an USB-to-serial converter, often they are referred to as FTDI converters, named after the company that initially offered them. We need to ensure, however, that VCC from the converter is 3.3 V rather than the more common 5 V. If that is not the case, a voltage regulator, as discussed in Section 2.2.5, will help.

Figure 9.3 shows the ESP-01 on the left-hand side. Its eight pins come in two columns of four each. The pins on the left are labeled RXD, GP0, GP2, and GND from top to bottom. The pins on the right are labeled VCC, RST, CH_PP, and TXD. In order to program the ESP-01, we

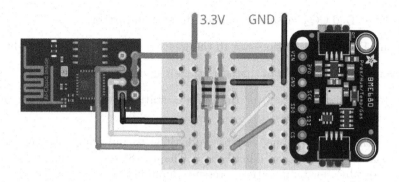

Figure 9.4   Weathernode with ESP-01 connected to a BME680[†].

connect its GND and VCC to the corresponding pins of the USB-to-serail converter. We also connect CH_PD and GP2 to VCC and GP0 to GND, which puts the ESP-01 into programming mode. Finally, we connect the serial-receive pin RXD to the serial-transmit pin TX on the converter and the TXD pin to RX. With the wiring in place we attach the USB-to-serial converter to our desktop computer, select the port from the *Tools→Port* menu, and select *Generic ESP8266 Module* in the *Tools→Board* menu in the Arduino IDE.

We can essentially use the same program as for the NodeMCU, shown earlier in this chapter. All we need to do is to remove all references to Serial, because we will not use the Serial line. Moreover, we uncomment the line with Wire.begin(0,2); in the function setup(), which is necessary, because the default I2C pins that are used on the NodeMCU are not routed to exposed pins on the ESP-01. Instead, we use pins GP0 and GP2 for the I2C lines SDA and SCL, respectively. Finally, we compile and download the sketch the our ESP-01.

Now we disconnect the programmed ESP-01 from the USB-to-serial converter and, following the wiring from Figure 9.4, connect it to the BME680 instead. Ground and 3.3 V supply rails are connected to VCC and GND pins on both devices. Note that also the pins labeled RST and CH_PD are connected to VCC. The I2C lines connect GP0 on the ESP-01 to SCK in the BME680. Likewise, GP2 is connected to SDI. Note that we must provide external pull-up resistors to the I2C lines; we use $10\,k\Omega$ resistors, but their value is not critical. For good measure, we should also provide $100\,nF$ bypass capacitors (not shown) between GND and VCC. Once we connect the supply voltage, this circuit will behave the same way as the one shown in Figure 9.1, albeit with a smaller footprint.

QUESTIONS AND PROJECT IDEAS

1. What is the average barometric pressure at sea level, on a rainy day, in a heavy storm, in a hurricane, or in La Paz?

2. What is the typical humidity where you live, in a rain forest, in Death Valley, or at the La Silla Observatory in Chile?

3. Add a separate temperature sensor, for example an LM35 or DS18b20, to crosscheck the value reported by the BME680.

4. Add a separate humidity sensor such as the HYT221 or DHT11 to crosscheck the value reported by the BME680.

5. Add a dust sensor to turn the weather station into an environmental logging station.

6. Add the SCD30 CO_2 sensor to enhance the system further.

7. Add an LDR or a phototransistor to the weather station to log the brightness.

8. Add a remote-controlled fan that stirs the air upon request.

9. Rewrite the client program for the weather station running on the NodeMCU and give it an MQTT interface.

10. Rewrite the sketch running on the NodeMCU to publish a webpage with graphs of measurement values, as described in Chapter 8, that continuously updates the displayed values for temperature, pressure, humidity, and air quality via websockets.

## CHAPTER 10

Example: Geophones

In the second example, we use the SM-24 geophone from Figure 2.12 to record ground vibration spectra. Our plan is to attach the SM-24 to a battery-powered microcontroller that samples 1024 values at a high rate and transmits the measurements via WLAN to the host computer, our Raspi. There we use octave to postprocess and Fourier-transform the samples, as well as present the results. Besides using octave, we also make the data available to EPICS, such that the standard EPICS programs for display and post-processing can be used.

For the sensor node, we use the NodeMCU microcontroller and the `Ticker.h` library that allows sampling with a rate of 1000 times per second. Since the frequency range of the SM-24 sensor is from 10 to 240 Hz, the sampling rate of 1 kHz is four times the maximum frequency and should be adequate. The SM-24 sensor only produces a very small and bipolar output voltage in the mV range. We therefore need to amplify the voltage in order to match the input voltage range of 0 to 3.3 V of the ADC on the NodeMCU.

We base the amplifier on the circuit shown in Figure 2.21, but increase the amplification to $\times 100$ and add a few components to adapt it to our sensor and arrive at the circuit shown in Figure 10.1. The amplification is mostly given by the ratio of R_3 and R_7 to R_1 and R_2. We also add a 1.5 kΩ resistor R_9 across the input terminals in order to damp the resonant peak at 10 Hz, which the SM-24 sensor exhibits according to its datasheet. If we consider the internal coil resistance (375 Ω) of the sensor, mounting an additional 2.2 μF capacitor across the input terminals creates a low-pass filter with a cutoff frequency of around 200 Hz. As discussed in Section 2.2.4, this avoids aliasing of higher frequencies into the digitizing bandwidth from 0 Hz to the Nyquist frequency of 500 Hz. The other components have the same functionality as discussed in Section 2.2.2.

After the signal from the geophone is amplified, we use the ADC on the NodeMCU microcontroller and collect 1024 samples: one sample every millisecond, and then pass the digitized samples to the host computer via WLAN.

The code for this project first defines the credentials and port number for the WLAN, and then import support to create the `server()` in the next line. The constants `npts` and `sample_period` specify the number of samples to acquire and the number of milliseconds to wait between acquisitions, respectively. The `Ticker.h` library adds support for timed and interrupt-driven functions. The variables declared next are related to filling the `sample_buffer`. The function `samplefast_action()` is called automatically by the timer and is executed once a millisecond. In this function we first read the built-in ADC and place the value into the `sample_buffer` before incrementing the variable `isamp`, such that the next sample ends up in the following slot in the array. Once the desired number `npts` of samples are acquired, we disable the interrupt with the call to `SampleFast.detach()`, set the sample pointer `isamp` to zero, and set the `sample_buffer_ready` flag to signal the main

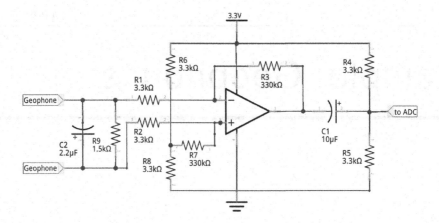

Figure 10.1 The amplifier to interface an SM-24 geophone to the NodeMCU[†].

program that the acquisition of the **npts** samples is complete. In the **setup()** function, we first set the mode of the pin with the built-in LED and turn it on before enabling the serial line for debugging, and connect to the WLAN. Then we start the server process with the call to **server.begin()** and report everything to the serial line, before turning the LED off.

```
// Minimal time-series-server, V. Ziemann, 170324
const char* ssid = "messnetz";
const char* password = "zxcvZXCV";
const int port = 1137;
#include <ESP8266WiFi.h>
WiFiServer server(port);
const uint16_t npts=1024;  // number of samples
const int sample_period=1; // ms
#include <Ticker.h>
Ticker SampleFast;
uint16_t sample_buffer[npts];
volatile uint16_t isamp=0,sample_buffer_ready=0;
char line[30];
void samplefast_action() { //............samplefast_action
  sample_buffer[isamp]=analogRead(0);
  isamp++;
  if (npts == isamp) {
    SampleFast.detach();
    isamp=0;
    sample_buffer_ready=1;
  }
}
void setup() { //....................................setup
  pinMode(LED_BUILTIN,OUTPUT);
  digitalWrite(LED_BUILTIN,LOW);
  Serial.begin(115200);
  WiFi.begin(ssid,password);
  while (WiFi.status() != WL_CONNECTED) {
```

```
      delay(500); Serial.print(".");
    }
    Serial.println("");
    Serial.print("Wifi connected to "); Serial.println(ssid);
    Serial.print("Server IP address: "); Serial.println(WiFi.localIP());
    server.begin();
    Serial.print("Server started on port "); Serial.println(port);
    digitalWrite(LED_BUILTIN,HIGH);
}
void loop() { //.....................................loop
    WiFiClient client = server.available();
    while (client) {
      while (!client.available()) {
        delay(5);
        if (!client.connected()) break;
      }
      client.readStringUntil('\n').toCharArray(line,30);
      Serial.print("Received: "); Serial.println(line)
      if (strstr(line,"WF?")==line) {
        digitalWrite(LED_BUILTIN,LOW);
        sample_buffer_ready=0;
        SampleFast.attach_ms(sample_period,samplefast_action);
        while (!sample_buffer_ready) {delay(2);} // wait until done
        for (int i=0;i<npts-1;i++) {
          client.print(sample_buffer[i]); client.print(", ");
        }
        client.println(sample_buffer[npts-1]);
        digitalWrite(LED_BUILTIN,HIGH);
      } else {
        Serial.println("unknown command, disconnecting");
        client.stop();
      }
      client.flush();
    }
    yield();
}
```

In the loop() function we wait for a client to connect, and then parse the request. If the query-string is WF?, we turn on the LED and ensure the variable sample_buffer_ready is zero before we start the acquisition with the call to SampleFast.attach_ms(). The arguments of the function are the periodicity in ms to call the function specified as the second argument. This launches the automatic acquisition that continues in the background. All we need to do in the loop() function is to monitor whether sample_buffer_ready is still zero, in which case we wait a bit longer. But once it becomes nonzero, which happens in the samplefast_action() function after the desired number of samples is collected, we break the waiting loop. Finally, we are ready to send the values, separated by commas, with client.print() function calls, back to the client and turn the LED off.

We program this sketch into the NodeMCU, whence it waits for a client to connect and to request samples. Here we use octave to request a time series of samples from the geophone and display both the received raw time-series data and its Fourier transform, the spectrum. The code that achieves this is the following:

```
% getTimeSeries.m, V. Ziemann, 221202
s=tcpclient("192.168.20.144",1137);
npts=1024;
write(s,"WF?\n");
pause(1);
data=zeros(1,npts);
for i=1:npts
  data(i)=str2double(tcp_getvalue(s));
end
clear s;
subplot(2,1,1); plot(data); xlim([0,npts])
xlabel('Time [ms]'); ylabel('Amplitude [ADC bits]')
data=data-mean(data);
fftdata=2*abs(fft(data))/npts;
frequency=(0:(npts/2-1))*500/(npts/2);
subplot(2,1,2); plot(frequency,fftdata(1:npts/2))
xlabel('Frequency [Hz]'); ylabel('Spectral density [ADC bits]')
print('spectrum.png','-S1000,700');
```

Also, the octave script follows earlier examples. We first define the socket s in the call to the tcpclient() function and then specify the number of points npts to acquire. This has to match the number of samples declared on the NodeMCU. Next, we send the query string WF? to the socket, wait a while, and then collect npts samples, which we store in the array data(). For this we use the function tcp_getvalue() that extracts one comma-separated value from the socket. We discuss that function in more detail below. The received value is encoded as a character string, and we therefore need to convert it to a float value with the str2double() function. Once all samples are received and stored in data(), we close the socket.

In the following lines, we create two subplots of which the upper contains the raw samples. Since the sample period is a millisecond, we state that as the horizontal axis label. The vertical axis is just the raw ADC conversion, which comes from the 10-bit ADC on the NodeMCU, and lies between 0 and 1023. For the spectrum that we show in the lower subplot, we first subtract the mean of the values in order to avoid a huge spectral peak at zero. It is caused by placing the signal in mid-range of the ADC with the preamplifier. Then we Fourier transform the samples and create an array frequency with the frequency values from zero to the Nyquist frequency before plotting the spectrum and labeling the axes. Finally, we use the print() function to create an image file that contains the displayed plot with the specified size in pixels, 1000×700 in this case.

In the octave script, we use the function tcp_getvalue to read a single sample from the input stream. It is similar to the queryResponse.m function we used in Section 5.5.4. Here is the octave code.

```
% tcp_getvalue.m, V. Ziemann, 221202
function out=tcp_getvalue(dev)
i=1;
int_array=uint8(1);
while true
  val=read(dev,1);
  if ((val==',') || (val==0xA)) break; end
  int_array(i)=val;
  i=i+1;
```

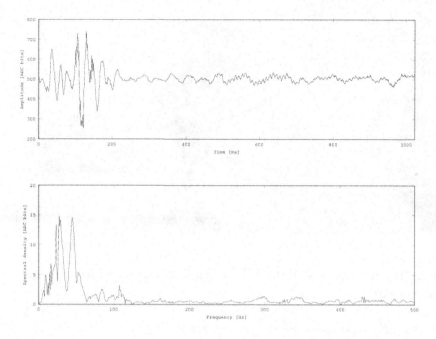

Figure 10.2   The raw time series from the geophone and the corresponding spectrum.

```
end
out=char(int_array);
```

We need to provide the handle `dev` to the socket as input parameter, and the function returns a character string that contains all characters until either a comma or the next end-of-line character. In the function, we read one character at a time from the socket and stop reading once a comma or end-of-line character `0xA` appears. Then we return all received data after conversion to characters.

We show the sensor node with SM-24 geophone, amplifier, and NodeMCU in Figure 10.3. Running the octave script with the NodeMCU connected to the same WLAN as the Raspi produces a file `spectrum.png`, shown in Figure 10.2. Running the octave script again will overwrite the file, and tapping with a finger on the table while recording samples results in a much more noisy spectrum.

Recording waveforms in octave and processing them further is convenient to produce plots of the spectra, but if we want to include the geophones in our EPICS control system, we need to provide the database and prototype files. We start with the database record for the time series or waveform. It is reproduced below,

```
# .../Db/geophone.db
record(waveform, "$(USER):wf") {
  field(DESC,"Geophone waveform")
  field(DTYP,"stream")
  field(SCAN,"10 second")
  field(NELM,"1024")
  field(FTVL,"FLOAT")
```

Figure 10.3 The prototype with the SM-24 geophone connected to the amplifier board and the NodeMCU.

```
    field(INP,"@geophone.proto get_wf $(PORT)")
}
```

where we have to declare a `waveform` record, because we want to acquire the entire stream of 1024 samples in one step. The first three fields provide the description, the declaration as a `stream` device, and the rate at which the waveforms are collected. The next two fields declare that we record 1024 float values, before linking to the protocol file `geophone.proto` and the function `get_wf` in the last field of type INP. After adding the line DB += geophone.db to .../Db/Makefile we prepare the referenced protocol file `geophone.proto` that is reproduced below.

```
# .../Db/geophone.proto
Terminator = CR LF;
get_wf {
  ExtraInput = Ignore;
  replyTimeout=2000;
  out "WF?";
  separator=",";
  in "%f";
}
```

Here we first define the termination string of one command as CR LF, and then define the `get_wf` function. For robustness, we require it to ignore any input that does not make sense, and wait 2000 ms for the reply from the NodeMCU. The request for a new waveform consists of sending WF? and receiving floats ("%f"), separated by commas. The number of values to receive from the NodeMCU, 1024, is already specified in the `geophone.db` file.

In a final step, we need to tell EPICS where to find the NodeMCU and what socket to connect to. This is done, as discussed in Chapter 6 on EPICS, in the `st.cmd` command file, which contains the following two lines:

```
drvAsynIPPortConfigure("SOCKET2","192.168.20.144:1137",0,0,0)
dbLoadRecords("db/geophone.db","PORT='SOCKET2',USER='geophone'")
```

The first defines the IP and port number of the connected NodeMCU device, and the second line specifies the database file geophone.db that describes the protocol that is used when communicating with the NodeMCU. After adding DB += geophone.db to the Makefile, we recompile the EPICS IOC and then execute st.cmd from the command line, or, once we make sure that everything works as expected, we create an init file to start the IOC at boot time, following the procedure we discussed in Section 6.5. At this point, with a running IOC continuously retrieving waveforms from the NodeMCU, we can obtain the most recent waveform with caget geophone:wf, which first returns the number of samples, here 1024, followed by the 1024 samples. Other EPICS programs, on any computer connected to the same network, can also obtain the same waveforms, a new one every 10 seconds, as specified in the geophone.db record file.

In order to illustrate access to epics variables from python we now prepare python script that generates the plots shown in Figure 10.2. First we execute sudo pip install pyepics to install the latest version of the pyepics package, which provides functions epics.caget() and epics.caput() to read and set epics process variables that we use in the following script.

```
# show_epics_waveform.py, V. Ziemann, 221103
import matplotlib.pyplot as plt  # for plotting
import numpy as np               # for fft
import epics                     # epics access
data=epics.caget("geophone:wf");
plt.subplot(2,1,1)
plt.plot(data);
plt.xlabel("Samples")
plt.ylabel("Amplitude")
plt.subplot(2,1,2)
fftvals=np.fft.fft(data)
plt.plot(abs(fftvals))
plt.ylabel("FFT")
plt.show()
```

After importing support for plotting and numerical calculations, just as described in Section 5.6, this script imports support for epics and uses it to retrieve the waveform, which is stored in the array data. The upper subplot shows the raw data trace and the lower subplot shows its Fourier transform, calculated with the numpy function fft.fft(). Adding scaling and further numerical calculations is straightforward.

QUESTIONS AND PROJECT IDEAS

1. Discuss the advantage of using interrupt-driven acquisition over measuring 1000 samples in a loop with a delay(1) between acquisitions.

2. What is the purpose of the low-pass filter discussed in this chapter?

3. What is the maximum frequency we can determine uniquely when sampling at a rate of 1000 samples per second?

4. What is the resolution (smallest detectable difference of frequencies) of the spectrum shown in Figure 10.2? How can you increase it?

5. Discuss why the sample rate of the Ticker.h library is limited to 1000 samples per second.

6. Use a pin diode as sensor and sample it at a rate of 1000 samples per second. Explore different flickering light sources such as lamps, TV screens, or computer monitors.

7. Use the `tone()` function to generate an oscillating signal on an output pin. Then sample that pin with the built-in ADC and observe the signal. What is the highest frequency you can uniquely observe? What happens when you go beyond that frequency?

8. Instead of the geophone, use a microphone and "observe" your voice.

9. Add an MCP3304 ADC to the NodeMCU and connect two geophones, fill two `sample_buffer` arrays with data in the interrupt handler, and send both waveforms to octave. If the geophones are placed on a large table far apart, try out to determine the speed of sound by tapping on the table and determine the shift in time between the two waveforms.

10. Convert the octave script `getTimeSeries.m` to Python and use `numpy` for the Fourier transform and `matplotlib` to display the data.

11. Rewrite the sketch running on the NodeMCU to publish a webpage that receives the time-series data via websockets and shows the upper graph from Figure 10.2. Search the Internet to find a JavaScript implementation of a Fourier transform and display also the lower graph.

## CHAPTER 11

Example: Monitor for the Color of Water

In this example, we determine the color of water, or more accurately stated, the absorption of material dissolved in water. An example is algae, which often proliferate during the summer and turn the water green because algae predominantly absorb the complementary color, which is red. In our experiment, we investigate the absorption of light with different colors by turning a three-color LED on and off, and observe the resulting modulation of the signal recorded by a phototransistor. If red light is absorbed, the resulting modulation depth of red is reduced, likewise for the other colors. By comparing the situation with LED on versus off, the system removes some of the influence of ambient background light, at least to some extent. On the left in Figure 11.1, we illustrate the absorption measurement with the RGB-LED on the left and the phototransistor on the right, with the absorbing material in between. We note that we can also use such a system to determine variations in the reflection from a surface, as is shown on the right in Figure 11.1.

We construct the setup shown in Figure 11.1, using an Arduino UNO, and an RGB-LED, which houses a red, a green, and a blue LED in the same housing with a single connector for a common cathode or anode. In our case we use one with a common anode that is connected to the positive supply voltage. The respective color lights up if the controlling pin on the UNO is LOW. The phototransistor is an SFH3310, but any other model, sensitive in the visual spectral range, should work. The very simple setup is shown in Figure 11.2.

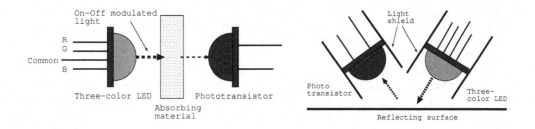

Figure 11.1 The setup to measure the color-dependent absorption (left) and the reflection from a surface (right).

DOI: 10.1201/9781003341703-11

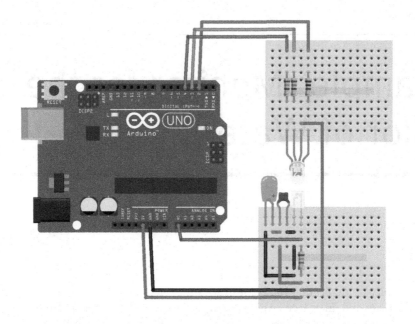

Figure 11.2 The setup to measure the water color with an Arduino UNO[†].

On the left we see the Arduino UNO and on the bottom right a small breadboard with the SFH3310 phototransistor. The emitter is connected to ground, and the collector is connected via a $68\,k\Omega$ resistor to the positive supply voltage. The resistor depends somewhat on the phototransistor and is determined by the requirement for a clearly visible modulation depth on the analog input pin, which is connected to the collector with the blue wire. A little experimenting will result in a reasonable value for the resistance. We add a $10\,\mu F$ and a $100\,nF$ capacitor to buffer the supply voltage. The RGD-LED is placed on the upper breadboard with the common anode connected to the positive supply voltage, and the three "color pins" via current-limiting resistors to pins D2, D3, and D4 on the UNO. We chose $220\,\Omega$ for the red LED, $180\,\Omega$ for the green, and $150\,\Omega$ for the blue in order to account for the different voltage drops of the different-color LEDs.

In the sketch that runs on the UNO, we need to toggle the LEDs one at a time and synchronously record the intensities recorded by the phototransistor. This is achieved by the following code.

```
// Color monitor, V. Ziemann,170823
int repeat=20;
void alloff() { //......................alloff
  digitalWrite(2,HIGH);  // red
  digitalWrite(3,HIGH);  // green
  digitalWrite(4,HIGH);  // blue
}
float measure(int pin, int repeat) { //.............measure
  int hi,lo;
  float sum=0;
  alloff();
  delay(10);
  for (int i=0;i<repeat;i++) {
```

```
      digitalWrite(pin,LOW);
      delay(5);
      lo=analogRead(0);
      digitalWrite(pin,HIGH);
      delay(5);
      hi=analogRead(0);
      sum+=(hi-lo);
    }
    return sum/repeat;
}
void setup() { //........................setup
    Serial.begin(9600); while (!Serial) {;}
    pinMode(2,OUTPUT);   // red
    pinMode(3,OUTPUT);   // green
    pinMode(4,OUTPUT);   // blue
    alloff();
    digitalWrite(2,LOW);
}
void loop() { //........................loop
    if (Serial.available()) {
      char line[30];
      Serial.readStringUntil('\n').toCharArray(line,30);
      if (strstr(line,"OFF")==line) {
        alloff();
      } else if (strstr(line,"RED")==line) {
        alloff();
        digitalWrite(2,LOW);
      } else if (strstr(line,"GREEN")==line) {
        alloff();
        digitalWrite(3,LOW);
      } else if (strstr(line,"BLUE")==line) {
        alloff();
        digitalWrite(4,LOW);
      } else if (strstr(line,"COLOR?")==line) {
        float red=measure(2,repeat);     // red
        float green=measure(3,repeat);   // green
        float blue=measure(4,repeat);    // blue
        Serial.print("COLOR "); Serial.print(red);
        Serial.print("\t");Serial.print(green);
        Serial.print("\t");Serial.println(blue);
      } else if (strstr(line,"REPEAT?")==line) {
        Serial.print("REPEAT "); Serial.println(repeat);
      } else if (strstr(line,"REPEAT ")==line) {
        repeat=(int)atof(&line[6]);
      }
    }
    delay(10);
}
```

First we declare a variable **repeat** and a function to turn all the LEDs off. The input
parameters of the function **measure**() are the pin to be toggled and the number of times

Figure 11.3 The color sensor with the RGB-LED on the left and the phototransistor on the right.

to repeat the measurement. We add a `delay` of 5 ms in order for the voltage to stabilize after the LED is switched on or off, before measuring the voltage at the collector of the phototransistor with the `analogRead()` function. The difference of the reading with LED on and off is accumulated in the variable `sum`, and finally the average of the accumulated differences is returned to the calling program. In the `setup()` function, we initialize the serial communication and the pins for the RGB-LED, turn all LEDs off, and then turn the red LED on in order to indicate that the system is up and running. The `loop()` function uses the same mechanism we use throughout this book to turn all LEDs `OFF` or turn one of the colors `RED`, `GREEN`, or `BLUE` on. If the query is `COLOR?` the system measures the variation on the phototransistor when toggling the LEDs on and off, and returns three numbers for the three color-variations. Finally, the query `REPEAT nnn` sets the variable `repeat` to `nnn` and `REPEAT?` returns the current value.

This system measures the absorption of any material placed between the LED and the phototransistor, which are pointing at each other as shown in Figure 11.1, and in Figure 11.3 where we show a simple prototype. Moreover, in order to measure the absorption in water, we need to encapsulate the LED and transistor in a protective coating, for which we use two layers of heat-shrink insulation material to cover the solder connection of wires to the pins of the LED and phototransistor.

If we want to deploy the system at a somewhat remote location, it is rather straightforward to extend the MQTT client program from Chapter 7 to monitor the color and publish the results once every minute. For this purpose, we use a NodeMCU instead of a UNO, and connect the RGB-LED to pins `D2`, `D3`, `D4` on the NodeMCU, and the collector of the phototransistor to the analog input pin `A0`. The following sketch implements this functionality.

```
// MQTT client water color, V. Ziemann, 170906
const char* ssid = "messnetz";
const char* password = "zxcvZXCV";
const char* broker = "192.168.20.1";
#include <Ticker.h>
volatile uint8_t do_something=0;
Ticker tick;
void tick_action() {do_something=1;}  // executed regularly
#include <ESP8266WiFi.h>
#include <PubSubClient.h>
WiFiClient espClient;
PubSubClient client(espClient);
float measure(int pin, int repeat) {  //.............measure
```

```
    int hi,lo;
    float sum=0;
    digitalWrite(D2,HIGH);
    digitalWrite(D3,HIGH);
    digitalWrite(D4,HIGH);
    delay(10);
    for (int i=0;i<repeat;i++) {
      digitalWrite(pin,LOW);
      delay(5);
      lo=analogRead(A0);
      digitalWrite(pin,HIGH);
      delay(5);
      hi=analogRead(A0);
      sum+=(hi-lo);
    }
    return sum/repeat;
}
void setup() { //................................setup
  pinMode(D2,OUTPUT);
  pinMode(D3,OUTPUT);
  pinMode(D4,OUTPUT);
  Serial.begin(115200);
  WiFi.begin(ssid,password);
  while (WiFi.status() != WL_CONNECTED) {
    delay(500); Serial.print(".");
  }
  Serial.print("\nWifi connected to "); Serial.println(ssid);
  Serial.print("with IP address: "); Serial.println(WiFi.localIP());
  client.setServer(broker, 1883);  // 1883 = default MQTT port
  tick.attach(60,tick_action); // execute tick_action every 60 seconds
}
void loop() {  //................................loop
  while (!client.connected()) {  // try to connect to broker
    if (!client.connect("PubSub1")) {delay(5000);}
  }
  client.loop();
  if (do_something) {
    do_something=0;
    char message[30];
    float red=measure(D2,20);    // red
    float green=measure(D3,20);  // green
    float blue=measure(D4,20);   // blue
    sprintf(message,"%d %d %d",(int)red,(int)green,(int)blue);
    client.publish("node1/color",message);
  }
  yield();
}
```

We first declare the network credentials and the IP address of the broker before including the Ticker.h library. It orchestrates the repeated measurements by calling the tick_action() function at well-defined time-intervals. Inside the function we only set the variable do_something to one, because this variable is monitored in the loop() function and triggers some activity, if it is set. Next, we include the WiFi libraries, and declare a WiFiClient and the MQTT PubSubClient named client, before defining the measure() function, which is a straight copy from the previous example. The setup() function declares the mode of the pins, configures the serial line, connects to WLAN, configures the broker, and starts the ticker to call tick_action every 60 seconds, which triggers the measurements. In the loop() function, we first ensure that the NodeMCU is connected to the broker before calling client.loop() to service any background activities pertaining to MQTT. Then we test whether the variables do_something is set, and execute the measurement with the measure() function for each color, construct a string that contains the three values, and publish the measurement.

Using this system, it is possible to submerge a sensor similar to the one shown in Figure 11.3 in a garden pond, place the NodeMCU in an enclosure next to the pond, possibly battery powered, and make sure it is within reach of the WLAN messnetz. That is all that is needed to monitor algae in the pond.

QUESTIONS AND PROJECT IDEAS

1. How do the reported color-variations change if you remove the capacitors from the small breadboard? Explain why!

2. How do the reported color-variations change if the waiting time of 5 ms in the measure() function is reduced to 1 ms? What happens if it is increased?

3. Connect the RGB-LED to output pins with pulse-width modulation capability and write a sketch that allows you to set a number of standard colors, such as orange or magenta.

4. Write a sketch to set the color of the LEDs via MQTT.

5. Build a reflectivity sensor similar to the one shown on the right in Figure 11.1, and connect it to the Arduino UNO.

## CHAPTER 12

Example: Impedance Measurements

In a later part of this chapter, we use the AD5933 impedance analyzer circuit to determine impedances in the frequency domain. We start, however, by measuring the unknown capacitance of a capacitor by first charging it, then turning off the charging supply and discharging it through a parallel resistor, while repeatedly measuring the voltage drop across the capacitor. The voltage drops exponentially $U \propto e^{-t/\tau}$ with time constant $\tau = RC$, where R is the discharging resistor and C is the unknown capacitance. Thus, from a linear least-squares fit to the logarithm of the voltage, we can determine the time constant, and from the known resistor value, also the unknown capacitance.

Figure 12.1 shows the experimental setup. The Arduino UNO is on the right and the small breadboard showing the capacitor with the $R = 33\,\text{k}\Omega$ resistor to discharge the capacitor connected in parallel; both are connected to ground and analog pin A0 on the Arduino. The capacitor can be charged by pulling digital pin D2 high; it is connected by a wire and a $220\,\Omega$ resistor. The latter resistor limits the initially flowing current, when the capacitor is fully discharged.

The task of the UNO is to first charge the capacitor, by configuring the digital output pin D2 as OUTPUT and setting its value to HIGH. Once a measurement is requested, the D2 is reconfigured as input with pinMode(2,INPUT), and setting the output nevertheless to LOW, to ensure that the internal pull-up resistors are *not used*. Once the charging voltage is disconnected, we repeatedly read analog pin A0 from an interrupt service routine until the required number of samples is collected in memory. Once the raw data are collected, we make a linear least-squares fit to the data and determine the capacitance from the slope. This plan of action is realized in the following Arduino sketch.

```
// Capacitance measurement, V. Ziemann, 170629
const int npts=100;
volatile int isamp=0,sample_buffer_ready=0;
uint16_t sample_buffer[npts],nsamp=npts,timestep=5;
float R=33e3;   // 33 kOhm
#include <MsTimer2.h>
void timer_action() {  //.................timer_action
  sample_buffer[isamp]=analogRead(A0);
  isamp++;
  if (nsamp == isamp) {
    MsTimer2::stop();
    isamp=0;
```

DOI: 10.1201/9781003341703-12

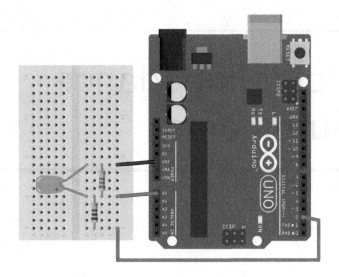

Figure 12.1 The setup to measure the capacitance[†].

```
      sample_buffer_ready=1;
    }
  }
  double linfit(int n, uint16_t y[]) {  //........linfit
    double ay0=0,ay1=0;
    double S0=n;
    double S1=0.5*n*(n+1);
    double S2=n*(n+1.0)*(2.0*n+1)/6.0;
    for (int k=0;k<n;k++) {
      ay0+=(k+1)*log(y[k]);
      ay1+=log(y[k]);
    }
    return (S0*ay0-S1*ay1)/(S2*S0-S1*S1);
  }
  void setup() {  //...............................setup
    Serial.begin(9600);
    while (!Serial) {delay(10);}
    pinMode(2,OUTPUT);
    digitalWrite(2,HIGH);
  }
  void loop() {  //.................................loop
    if (Serial.available()) {
      char line[30];
      Serial.readStringUntil('\n').toCharArray(line,30);
      if (strstr(line,"CAP?")) {
        nsamp=(int)atof(&line[5]);
        nsamp=min(nsamp,npts);
        if (nsamp==0) nsamp=npts;
        pinMode(2,INPUT);
        digitalWrite(2,LOW); // disables internal pullup
        MsTimer2::set(timestep, timer_action);
```

```
        MsTimer2::start();
        sample_buffer[isamp]=analogRead(A0);
        isamp++;
      } else if (strstr(line,"WF?")) {
        Serial.print("WF "); Serial.println(nsamp);
        for (int i=0; i<nsamp; i++) Serial.println(sample_buffer[i]);
      } else if (strstr(line,"TIMESTEP ")) {
        timestep=(int)atof(&line[9]);
        Serial.print("TIMESTEP "); Serial.println(timestep);
      } else if (strstr(line,"TIMESTEP?")) {
        Serial.print("TIMESTEP "); Serial.println(timestep);
      } else if (strstr(line,"RESISTOR ")) {
        R=atof(&line[9]);
        Serial.print("RESISTOR "); Serial.println(R);
      } else if (strstr(line,"RESISTOR?")) {
        Serial.print("RESISTOR "); Serial.println(R);
      }
    }
    if (sample_buffer_ready==1) {
      sample_buffer_ready=0;
      pinMode(2,OUTPUT);  // start charging capacitor
      digitalWrite(2,HIGH);
      delay(100);
      double slope=linfit(nsamp,(uint16_t)sample_buffer);
      double capacitance=-1e6*timestep*1e-3/(slope*R);  // in uF
      Serial.print("CAP "); Serial.println(capacitance,4);
      if (sample_buffer[0] < 250*sample_buffer[nsamp-1]) {
        Serial.println("***Time too short, double TIMESTEP");
      }
    }
    delay(1);
  }
```

The sketch follows the usual format, where we first declare a number of variables in which `npts` is the maximum number of points that can be acquired. The data samples are stored in the `sample_buffer` once every `timestep` milliseconds. We also declare the resistance `R` of the resistor used to discharge the capacitor. Since the `Ticker.h` library we use in other chapters for the NodeMCU is unavailable for the UNO, we include the `MsTimer2.h` header file and library instead. It provides the functionality to repeatedly generate an interrupt and call an interrupt-service routine called `timer_action()` in our sketch. This function is called every `timestep` milliseconds. It first reads analog pin A0, then stores the value in the `sample_buffer`, and increments `isamp`. If `isamp` reaches `nsamp`, the timer is stopped to prevent the acquisition of further data points. Then `isamp` is set to zero to prepare for the next acquisition and the variable `sample_buffer_ready` is set to 1 to indicate that an acquisition is complete and data is ready for further processing. Technically we can process the data within the `timer_action` function, but normally this routine should contain only time-critical actions, such as acquiring samples. It should be kept as compact as possible, because it is called asynchronously to all other processing and we should avoid excessive disturbances. The way to handle this problem is to flag that the acquisition is ready, and in the main program, check for this flag and perform the postprocessing. In this way, the time-critical and the slower processes are efficiently decoupled. The `linfit()` function takes the acquired data as well as the number of data points as input parameters, internally performs the least-squares fit to the logarithm of the data points, and returns the slope of the fit. The discussion of the algorithm is somewhat technical, and is deferred to Appendix B.

The `setup()` function initializes the serial line, declares digital pin D2 as output, and sets it to `HIGH` to start charging the capacitor. The `loop()` function uses the standard query-response protocol, and a capacitance measurement starts with the command `CAP? nnn`, where `nnn` is the desired number of measurement points; per default the maximum number `npts` is acquired. Then digital pin D2 is placed in high-impedance input mode. We also ensure that the internal pull-up resistor is disabled, and start the timer process with the `MsTimer::set()` function. It takes the time between samples and the function to call as argument, and then we start the timer. Before leaving this subroutine, we take the first data sample. The command `WF?` can be used to retrieve the waveform data of the last acquisition. This is useful for crosschecking the fitting in, for example, octave. The remaining commands are used to set and read the `TIMESTEP` or the `RESISTOR` values. Once the handling of the commands from the serial line is complete, we test whether the variable `sample_buffer_ready` is set, which indicates that an acquisition is complete and we can start postprocessing the raw data. First we reset the `sample_buffer_ready` to zero, to prevent repeated calls, and then start charging the capacitor again by configuring pin D2 as output, and wait a little while. Next we calculate the `slope` from the samples. Since the slope is inversely proportional to the time constant $\tau = RC$, we can solve this for the capacitance C with the equation defining the `capacitance`. The factor `1e6` causes the displayed value to be shown in μF. Finally, we check whether the exponential decay of the voltage actually extends over a sufficiently long time, and display a warning if that is not the case.

We can measure capacitances by sending commands from other programs that communicate over the serial line. From octave we use the following program to initiate a measurement, and plot the data as shown in Figure 12.2.

```
% capacitance_plot.m, V. Ziemann, 221103
% pkg load instrument-control
close all; clear all
s=serialport('/dev/ttyACM0',9600);
pause(3);
flush(s);
write(s,"CAP? 100");
pause(1);
reply=serialReadline(s);
capacitance=str2double(reply(4:end));
write(s,"TIMESTEP?");
reply=serialReadline(s);
if reply(1:8)=="TIMESTEP"
  timestep=str2double(reply(9:end));
else
  disp(reply)
  clear s;
  return
end
write(s,"WF?");
reply=serialReadline(s);
nsteps=str2num(reply(3:end));
xx=zeros(nsteps,1); yy=xx;
```

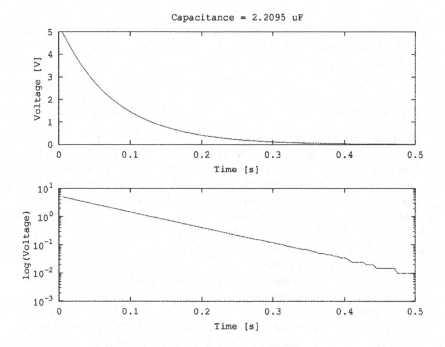

Figure 12.2 The waveform of the voltage on the capacitor as a function of time, while a 2.2 μF Tantal capacitor discharges through a 33 kΩ resistor on a linear scale (above) and on a logarithmic scale (below).

```
for k=1:nsteps
    xx(k)=k*timestep*1e-3;                    % in seconds
    yy(k)=str2double(serialReadline(s))*5/1023; % in Volt
end
clear s;
subplot(2,1,1); plot(xx,yy);
xlabel("Time [s]"); ylabel("Voltage [V]");
title(["Capacitance = " num2str(capacitance) " uF"]);
subplot(2,1,2); semilogy(xx,yy);
xlabel("Time [s]"); ylabel("log(Voltage)");
print('capacitance.png','-S1000,700')
```

We need to keep in mind to load the instrument-control toolbox in order to use the serial communication functionality and clear the workspace, before actually opening the serial line. First we remove any remaining characters from the input queue, and initiate a measurement with the CAP? command. After a short waiting period, we read the reply with the serialReadline() function, which is tailored after the queryResponse() function from Section 5.5.4. It is reproduced here.

```
% get response up to termination character
function out=serialReadline(dev,term_char)
    if (nargin==1) term_char=10; end  % defaults to LF=0x0A
    i=1;
    int_array=uint8(1);
```

```
while true                        % loop forever
    val=read(dev,1);              % and read one byte
    if (val==term_char) break; end  % until term_char appears
    int_array(i)=val;             % stuff byte in output
    i=i+1;
end
out=char(int_array);              % convert to characters
```

This function reads characters one at a time until an end-of-line character is found, and then returns the received characters to the calling program, where the numerical value is extracted. For the plot, we need to know the TIMESTEP used in the measurement, and request it from the UNO with the TIMESTEP? command. If the reply starts with TIMESTEP, all is well, and we extract the numerical value. If we retrieve something else from the serial input queue, the time window for the measurements was too short, and the measurement is invalid. In that case, the program closes the serial port and returns control to the octave command prompt. If, on the other hand, we obtain the numerical value of the time step, we request the waveform data with the WF? command. The first returned line contains the number of points acquired. The following lines contain the data points. Each new line is retrieved with a call to serialReadline() until all data points are copied to the array yy. Finally, we plot the curves with linear and logarithmic vertical scale. The latter shows a clear linear dependence, which indicates the exponential dependence of the voltage on time with a well-defined time constant. In the last line, we produce a png file shown in Figure 12.2 that can be used to document our work.

Instead of analyzing the behavior of a device in the time domain, we can also analyze it in the frequency domain by driving with a sinusoidal current whose frequency we vary through our *device under test* (DUT). Simultaneously, we measure the voltage drop over the device as a function of the frequency. This is the operating principle of a *network analyzer*, which allows us to deduce the magnitude and phase of the device's impedance from the ratio of voltage and current. For this task we use the AD5933 impedance network analyzer, which features a direct digital synthesizer, already discussed in Section 4.5.6, to generate frequencies f up to around 100 kHz that are passed through the DUT. At the same time it measures the voltage-drop across the DUT with a high-speed ADC. Internally it performs a digital Fourier transform at the excitation frequency – both with cosine and with sine – and reports back the two values, commonly referred to as IQ-data for *in-phase* (I) and *quadrature* (Q) values. From these two values we can deduce the amplitude of the impedance as $\sqrt{I^2 + Q^2}$ and the phase from $\arctan(I/Q)$.

The output current of the AD5933 is limited which in turn limits the smallest resistance we can measure. We therefore add an op-amp wired as a line buffer to its output (pin 6 on the AD5933) as shown in the schematics in the upper part of Figure 12.3. At the same time, the two 47 kΩ resistors and the 47 nF capacitor adjust the bias voltage on the DUT to half of the supply voltage, which matches the internally constrained offset voltage of the input (pin 5) to the AD5933. The 1 kΩ resistor between pin 4 and 5 is the feedback resistor of an op-amp inside the AD5933 that is wired as a transimpedance amplifier. It converts the input current to a voltage that is then digitized with an ADC inside the AD5933. This integrated circuit is controlled via I2C lines that require 10 kΩ pullup resistors.

The lower part of Figure 12.3 shows the realization of this circuit on a breadboard that is controlled by an UNO. We recognize the AD5933 in the upper half of the breadboard with the 10 kΩ pullup resistors to SDA on pin 15 and SCL on pin 16 on its right-hand side and the 1 kΩ feedback resistor on its left-hand side. Pins 9 to 11 are connected to the positive supply voltage and pins 12 to 14 are to ground. Here we need to point out that we soldered the AD5933, which comes in a breadboard-incompatible format (SSOP), onto a so-called

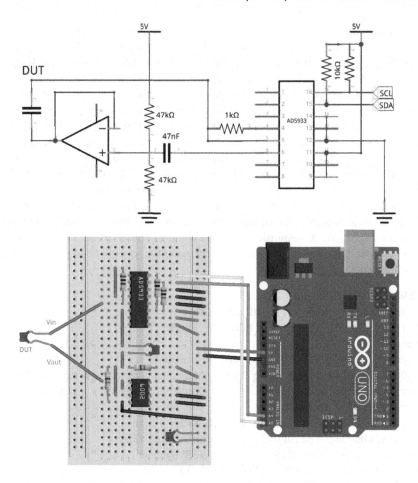

Figure 12.3 The schematics for measuring a DUT with an AD5933 and an op-amp to increase the measurement range (top) and the corresponding circuit on a breadboard, controlled by an UNO[†].

Winslow carrier board with a breadboard-compatible pin spacing (DIP). The power rails are connected to those on the UNO and the I2C lines SDA and SCL to pins A4 and A5 on the UNO, respectively. Pin 6 of the AD5933 is routed via a 47 nF capacitor to the positive input terminal of one of the op-amps inside the MCP6002 below; the two 47 kΩ resistor bias the input to half the supply voltage. The negative input terminal (pin 2) is connected to the output on pin 1 of the op-amp and further to the lower end of the DUT whose other end is directly connected to the input pin 5 of the AD5933. The second op-amp inside the MCP6002 is not used and we therefore tie its input pins to ground which is good practice to reduce noise in the circuit.

Bringing this circuit to life mostly involves controlling the AD5933 which requires carefully studying its datasheet. In order to make this effort reusable in other projects we collect this circuit-specific code in the file AD5933.h that is reproduced below. First, we declare the I2C address of the AD5933, initialize variables for the starting frequency and the frequency increment, as well as the frequency MCLK of the internal oscillator from which all oscillations are derived. The variable npts describes the number of frequency points, mode sets the output voltage range, here about 2 V peak-to-peak, and pga controls an internal ×5 amplifier.

The function `AD5933_config()` receives these parameters as input and programs the internal registers of the AD5933 according to the description in the datasheet. In particular, the control registers 0x80 and 0x81 set the mode of operation for `mode` and `pga`. Registers 0x8A and 0x8B set the time to wait after setting a new frequency and before starting to measure. Registers 0x82 to 0x84 receive an integer number, related to MCLK, that determines the output frequency according to an equation found in the datasheet. First the most-significant byte is written to 0x82, then the next byte to 0x83 and the least-significant byte to 0x84. In the next step the frequency increment `finc` is converted to an integer number that is likewise loaded to registers 0x85 to 0x87. Last, `npts` is written, one byte at a time, to 0x88 and 0x89. When leaving this function all registers of the AD5933 are loaded with the desired values and ready to enable the output, which requires issuing a sequence of commands that are specified in the datasheet. We provide a generic function `AD5933_command()` that receives cryptic byte values as input and sets configures the device. Immediately below, we define functions that give mnemonic names to some of the actions. The last two functions read status bits that indicate whether a measurement is complete and that the frequency sweep has finished.

```
// AD5933.h, V. Ziemann, 221118
const int AD5933=0x0D;  // I2C address
float freq=5000.0, finc=500, MCLK=16.776E6;
int npts=200;
byte mode=0,pga=1;  // ~2 Vpp, amp = times 1
void AD5933_config(float freq,float finc,int npts,byte mode,byte pga){
  I2Cwrite(AD5933,0x80,0x00);    // Control registers, reset
  I2Cwrite(AD5933,0x81,0x10);
  I2Cwrite(AD5933,0x80,((mode & 0x03)<<1) | pga);
  I2Cwrite(AD5933,0x8A,0x00);    // Settling time
  I2Cwrite(AD5933,0x8B,0xFF);    // 255 cycles
  long n=(long)(freq/(MCLK/4)*pow(2,27));
  I2Cwrite(AD5933,0x82,(n>>16) & 0xFF);      // Starting frequency
  I2Cwrite(AD5933,0x83,(n>>8) & 0xFF);
  I2Cwrite(AD5933,0x84,n & 0xFF);
  long m=(long)(finc/(MCLK/4)*pow(2,27));
  I2Cwrite(AD5933,0x85,(m>>16) & 0xFF);      // Frequency increment
  I2Cwrite(AD5933,0x86,(m>>8) & 0xFF);
  I2Cwrite(AD5933,0x87,m & 0xFF);
  I2Cwrite(AD5933,0x88,(npts>>8) & 0xFF);    // Number of increments
  I2Cwrite(AD5933,0x89,npts & 0xFF);
}
// cmd: 0x10=init, 0x20=start, 0x30=increment, 0x40=repeat
//      0x90=temp, 0xA0=Pwdn, 0xB0=Stdby
void AD5933_command(byte cmd) {
  uint8_t b=I2Cread(AD5933,0x80);
  I2Cwrite(AD5933,0x80,(b & 0x0F) | cmd);
}
void AD5933_standby() {AD5933_command(0xB0);}
void AD5933_init() {AD5933_command(0x10);}
void AD5933_increment() {AD5933_command(0x30);}
void AD5933_power_down() {AD5933_command(0xA0);}
int valid_data() {return ((I2Cread(AD5933,0x8F) & 0x02) == 0x02);}
int sweep_done() {return ((I2Cread(AD5933,0x8F) & 0x04) == 0x04);}
```

```
void AD5933_sweep(float freq, float finc, int npts, byte mode,
                  byte pga, int16_t *Ival, int16_t *Qval) {
  AD5933_config(freq,finc,npts,mode,pga);
  AD5933_standby();      // Standby
  AD5933_init();         // Init
  delay(200);            // Initial settle time
  AD5933_command(0x20);  // Start sweep
  int ic=0;
  while (!sweep_done()) {
    while (!valid_data()) {delay(10);}  // wait for data ready
    int16_t Itmp=(I2Cread(AD5933,0x94) << 8) | I2Cread(AD5933,0x95);
    int16_t Qtmp=(I2Cread(AD5933,0x96) << 8) | I2Cread(AD5933,0x97);
    if (Ival==NULL) {
      float mag=sqrt((float)Itmp*(float)Itmp+(float)Qtmp*(float)Qtmp);
      Serial.print(Itmp); Serial.print("\t"); Serial.print(Qtmp);
      Serial.print("\t"); Serial.println(mag);
    } else {
      Ival[ic]=Itmp; Qval[ic]=Qtmp; ic++;
    }
    AD5933_increment();   // increment frequency
  }
}
```

The function AD5933_sweep() performs the frequency sweep by following the sequence laid out in the datasheet and as long as the sweep is not completed. It checks the status bit whether a measurement is done, at which point the registers with the values for I and Q are read out and either written to the serial line or copied to the arrays IVAL[] and Qval[] before incrementing the frequency. Note that Itmp, Qtmp, Ival[] and Qval[] must be specified as int16_t in order to treat the sign bit correctly.

Inside AD5933.h we use the functions I2Cwrite() and I2Cread() from I2Crw.h on page 74 in Section 4.4.3. Both I2Crw.h and AD5933.h must reside in the same subdirectory as the sketch shown below. The setup() function only initializes serial and I2C communication. The loop() function follows the well-proven scheme of reacting to commands received via the serial line. The command SWEEP starts a sweep with the current parameters whereas OUTON produces a constant-frequency output and OUTOFF turns it off. All parameters can be changed with the commands FREQ, FINC, NPTS, MODE, and PGA, followed by a numerical value. Finally, STATE? returns the values of the current parameters.

```
// AD5933Serial, V. Ziemann, 221118
#include <Wire.h>
#include "I2Crw.h"
#include "AD5933.h"
void setup() {
  Serial.begin(115200); delay(1000);
  Wire.begin();
}
void loop() {
  char line[30];
  if (Serial.available()) {
    Serial.readStringUntil('\n').toCharArray(line,30);
    if (strstr(line,"SWEEP")) {
      AD5933_sweep(freq,finc,npts,mode,pga);
```

```
    } else if (strstr(line,"OUTON")) {
      AD5933_config(freq,finc,npts,mode,pga);
      AD5933_standby();
      AD5933_init();
    } else if (strstr(line,"OUTOFF")) {
      AD5933_power_down();
    } else if (strstr(line,"FREQ ")) {freq=atof(&line[5]);
    } else if (strstr(line,"FINC ")) {finc=atof(&line[5]);
    } else if (strstr(line,"NPTS ")) {npts=atoi(&line[5]);
    } else if (strstr(line,"MODE ")) {mode=atoi(&line[5]);
    } else if (strstr(line,"PGA ")) {pga=atoi(&line[4]);
    } else if (strstr(line,"STATE?")) {
      char msg[80];
      sprintf(msg,"STATE %ld %ld %d %d %d",\
                  (long)freq,(long)finc,npts,mode,pga);
      Serial.println(msg);
    }
  }
}
```

This sketch now allows us to control the AD5933 via the serial line for which we use octave. The following script opens the serial port s and uses write() to write the desired parameters to the AD5933 before verifying that they are properly set with STATE?. We then parse the reply to place the values into the array p. Then we start the SWEEP, wait a little while to ensure that some data has arrived, before checking its presence with f=fread(s) and concatenating any arriving data in the string reply. If no more data arrives, we close the serial line and reformat the text string reply into the array data that has three columns with I, Q, and the magnitude for each of the frequency points.

```
% AD5933nwa.m, V. Ziemann, 221202
clear all; close all;
waiting_factor=1;    % increase for very low frequencies
s=serialport("/dev/ttyUSB0",115200)  % open serial line
pause(3); flush(s)
write(s,"MODE 3\n");
write(s,"PGA 1\n");
write(s,"FREQ 10e3\n");
write(s,"FINC 300\n");
write(s,"NPTS 200\n");
write(s,"STATE?\n"); reply=serialReadline(s)
p=sscanf(reply,"STATE %d %d %d %d %d")
write(s,"SWEEP\n");
pause(2*waiting_factor);
reply="";
while length((r=fread(s))) != 0
  reply=strcat(reply,char(r));
  pause(0.5*waiting_factor);
endwhile
clear s   % close serial line
[a,count] = sscanf(reply,"%g");
data=reshape(a,[3,count/3])';    % I,Q,mag
```

Note that the raw measured data are returned in units of ADC counts. In order to obtain the impedance in Ohms and the phase in degrees, we need to calibrate the device for each chosen set of parameters with a known resistor $\hat{R}$ as DUT. For this purpose we typically chose $\hat{R}$ to be equal to the feedback resistor ($1\,\mathrm{k}\Omega$ in Figure 12.3) and fit a second-order polynomial `polyabs` to the magnitude and another polynomial `polyphase` to the phase. Note that we use the `atan2()` function, which takes the signs of its two arguments into account to find the angle in the correct quadrant. After plotting the magnitude, the phase and the raw I and Q signals in three plots, we specify the value `Rcal` of the calibration resistor. *Only* if the DUT is a calibration resistor, we uncomment the line with `dlmwrite()` to write the two polynomials to the file `calibpoly.dat`.

```
f=(p(1):p(2):p(1)+p(2)*p(3))/1000; % frequency in kHz
polyabs=polyfit(f',data(:,3),2)    % magnitude
phase=atan2(data(:,2),data(:,1))*180/pi;
polyphase=polyfit(f',phase,2)      % phase in degree
subplot(3,1,1); plot(f,data(:,3),'k',f,polyval(polyabs,f),'r-.');
xlabel("frequency [kHz]"); ylabel("abs(Z)")
legend("data","fit");
subplot(3,1,2); plot(f,phase,'k',f,polyval(polyphase,f),'r-.');
xlabel("frequency [kHz]"); ylabel("phase(Z) [deg]")
legend('phase','fit');
subplot(3,1,3); plot(f,data(:,1),'b',2,f,data(:,2),'r-.');
xlabel("frequency [kHz]"); ylabel("re(Z),im(Z)")
legend("real(Z)","imag(Z)");
Rcal=1;  // calibration resistor
%dlmwrite("calibpoly.dat",[polyabs',polyphase'],"\t")
```

Later, when measuring an unknown impedance, we read the previously saved polynomials and use them to scale the absolute value of the impedance and calculate the difference in phase to that of the resistor. This finally gives us the impedance in Ohms and the phase difference of the DUT relative to that of the calibration resistor.

```
figure;  %...............Impedance and phase
cal=dlmread("calibpoly.dat","\t");
absval=data(:,3);
Zabs=Rcal*polyval(cal(:,1),f)'./absval;
phase=atan2(data(:,2),data(:,1))*180/pi;
Zphase=phase-polyval(cal(:,2),f)';
subplot(2,1,1); plot(f,Zabs,'k')
xlabel("frequency [kHz]"); ylabel("abs(Z) [kOhm]")
subplot(2,1,2); plot(f,Zphase,'r')
xlabel("frequency [kHz]"); ylabel("phase(Z) [deg]")
%capacitance_nF=mean(1./(2e6*pi*Zabs.*f'))*1e9   % nF
```

The left-hand plots in Figure 12.4 show the raw values of a measurement where the DUT is a $C = 2.2\,\mathrm{nF}$ capacitor. The three plots show magnitude, the phase, as well as I and Q from top to bottom. The right-hand plots show the values after the device was calibrated with a $1\,\mathrm{k}\Omega$ resistor. We observe the inverse $1/2\pi f C$ dependence on the frequency on the top and a phase difference of close to -90 degree with respect to the calibration resistor. The commented-out line at the end of the script calculates the estimated capacitance from averaging $C = 1/2\pi f Z$, which reports 2.1 nF for our capacitor, which is within its specified 10 % tolerance.

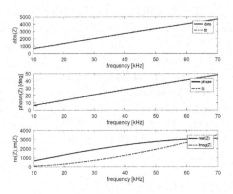 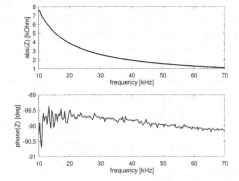

Figure 12.4 The raw output from measuring a 2.2 nF capacitor as DUT is shown on the left-hand side: magnitude of the impedance (top), the phase (middle), as well as raw I and Q values (bottom). The plot on the right-hand side shows the magnitude and phase of the impedance, converted to physical values relative to a calibration resistor.

QUESTIONS AND PROJECT IDEAS

1. Use the discharge method to verify the capacitance of capacitors found in your lab.

2. Use the AD5933 to measure the resistance of an unknown resistor.

3. Describe the plots, corresponding to those shown on the right-hand side in Figure 12.4 you expect when measuring an inductance.

4. Verify the inductance of an inductor found in your lab.

5. Calculate the resonance frequency of a 2.2 nF capacitor and a 22 mH inductor connected in parallel and then measure it. Repeat, after connecting them in series.

## CHAPTER 13

Example: Data Acquisition System

In this example we control both digital and analog signals produced and measured with an ESP32 microcontroller from a browser using websockets, a system similar to the temperature server in Chapter 8. We will control the digital output pin DO18, either to set its value or to automatically toggle it at a selectable repetition rate. We will also either set the DAC on pin 25 to a specified value or repeatedly ramp it between the minimum and maximum voltage to produce a simple sawtooth signal with selectable rate. In order to measure voltages, we read input pin DI21 and the two ADC channels on pins 32 and 33 at a selectable rate up to 200 samples per second. This provides the functionality of an, albeit slow, logic analyzer for the digital and oscilloscope for the analog channels.

Figure 13.1 illustrates the user interface running in a browser window. The top row shows buttons to start and stop the acquisition, to select the time between samples and to clear the display. The color of the text field below shows the state of input pin DI21 – red for off and green for on. The checkbox turns output pin DO18 on or off and the selection menu on the right toggles DO18 at a selectable rate. The slider labeled DAC25 sets the output voltage of the DAC while the selection menu to its right selects a sawtooth signal with different rates. The two bars labeled ADC32 and ADC33 show the input voltage on the two ADCs. The graph window displays the corresponding traces, ADC32 in red and ADC33 in blue. When saving the image shown in Figure 13.1, the sawtooth-generating DAC25 output was connected to ADC32. About halfway through the acquisition, pin DO18 was turned on to toggle once per second. It was directly connected to ADC33 such that the second trace shows a toggling voltage. The first status line below the display shows information generated on the browser, and the second line shows information received from the ESP32.

Now, let us turn to the code that brings the ESP32 to life. As in Chapter 8 we start with the credentials to connect to the WLAN and include support for WiFi, webserver, websockets, JSON, and the SPIFFS filesystem. Note that the latter has a different name from the one used earlier on the NodeMCU. Then we define the webserver to listen on port 80 and the websockets on port 81. The ticker is used to generate the repetitive actions for sampling the input pins and for controlling the output pins. After defining a number variables we define a several functions that are called by the tickers, here `sampleslow_action()` reads the input pins, stores their values in the array `samples`, and signals the availability of new data by setting the `output_ready` flag. The function `ramp25_action()` increments the 8-bit variable `dac25` and writes it to DAC25. We do not need to worry about values above 255, because the higher bits are automatically truncated. The function `pulse18_action()` simply toggles DO18.

DOI: 10.1201/9781003341703-13

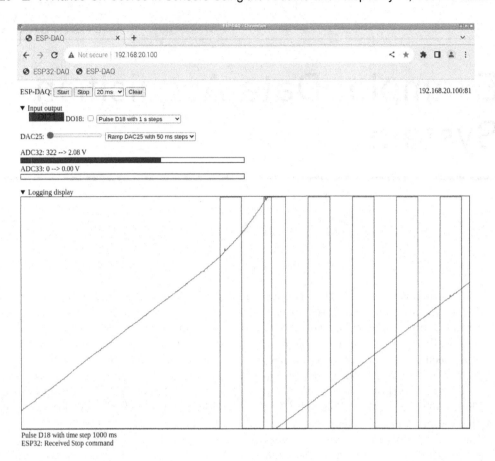

Figure 13.1 Web interface for the ESP32-based data acquisition system.

```
// ESP32 DAQ, V. Ziemann, 221018
const char* ssid     = "messnetz";
const char* password = "zxcvZXCV";
#include <WiFi.h>
#include <WebServer.h>
#include <WebSocketsServer.h>
#include <ArduinoJson.h>
#include <SPIFFS.h>
WebServer server2(80);                              // port 80
WebSocketsServer webSocket = WebSocketsServer(81);  // port 81
#include <Ticker.h>
Ticker SampleSlow,Ramp25Ticker,Pulse18Ticker;
volatile uint8_t websock_num=0,info_available=0,output_ready=0;
int sample_period=100,samples[3];
uint8_t dac25val=0,do18=0;
char info_buffer[80];
char out[300]; DynamicJsonDocument doc(300);
void sampleslow_action() { //.....................sampleslow_action
  samples[0]=analogRead(32)/8;  // fit into 512 pixels vertically
  samples[1]=analogRead(33)/8;
```

```
    samples[2]=digitalRead(21);
    output_ready=1;
}
void ramp25_action() {dac25val++; dacWrite(25,dac25val);}
void pulse18_action() {do18=!do18; digitalWrite(18,do18);}
```

The next code section defines functions that handle the communication with the browser. The function `handle_notfound()` simply returns a reminder about how to access the web interface. The following functions define the interaction with the websocket. The function `sendMSG()` formats a JSON package from name and value received as input and packs it into the `info_buffer`, whose availability is signalled to the main program by setting `info_available` to unity. The function `webSocketEvent` is the callback function that is executed every time something related to the websocket happens, such as a disconnecting or connecting client, which are simply reported to the serial line. Note that we use `sendMSG()` to also report back to the browser that the connection was established. Note that `sendMSG()` only formats the message, which is later sent outside the scope of the callback function. The next `case` handles JSON-formatted messages coming from the browser, from which we extract the name-value pairs for the name `cmd` and `val`. Depending on the `cmd` the `switch-case` construction starts or stops the data acquisition. It starts by registering `sampleslow_action()` for automatic execution at the `sample_period` specified by `val`. Other commands set the output pin DO18 and the DAC25 to the values specified in `val` or register "tickers" to ramp the DAC or toggle pin DO18.

```
void handle_notfound() {
  server2.send(404,"text/plain","not found, use http://ip-address/");
}
void sendMSG(char *nam, const char *msg) {
  (void) sprintf(info_buffer,"{\"%s\":\"%s\"}",nam,msg);
  info_available=1;
}
void webSocketEvent(uint8_t num, WStype_t type, uint8_t * payload,
                    size_t length) {
  Serial.printf("webSocketEvent(%d, %d, ...)\r\n", num, type);
  websock_num=num;
  switch(type) {
    case WStype_DISCONNECTED:
      Serial.printf("[%u] Disconnected!\r\n", num);
      break;
    case WStype_CONNECTED:
      {
        IPAddress ip = webSocket.remoteIP(num);
        Serial.printf("[%u] Connected from %d.%d.%d.%d url: %s\r\n",
          num, ip[0], ip[1], ip[2], ip[3], payload);
      }
      sendMSG("INFO","ESP32: Successfully connected");
      break;
    case WStype_TEXT:
      {
        Serial.printf("[%u] get Text: %s\r\n", num, payload);
        DynamicJsonDocument root(300);
        deserializeJson(root,payload);
        const char *cmd = root["cmd"];
```

```
      const long val = root["val"];
      if (strstr(cmd,"START")) {
        sendMSG("INFO","ESP32: Received Start command");
        sample_period=val;
        SampleSlow.attach_ms(sample_period,sampleslow_action);
        Serial.print("sample_period = "); Serial.println(sample_period);
      } else if (strstr(cmd,"STOP")) {
        sendMSG("INFO","ESP32: Received Stop command");
        SampleSlow.detach();
      } else if (strstr(cmd,"DAC25")) {
        dacWrite(25,val);
        sendMSG("INFO","ESP32: set DAC25 to requested value");
      } else if (strstr(cmd,"DO18")) { digitalWrite(18,val);
      } else if (strstr(cmd,"RAMP25")) {
        if (val > 0) {Ramp25Ticker.attach_ms(val,ramp25_action);
        } else {Ramp25Ticker.detach();}
       } else if (strstr(cmd,"PULSE18")) {
      if (val > 0) {Pulse18Ticker.attach_ms(val,pulse18_action);
        } else {Pulse18Ticker.detach();}
      } else {
        Serial.println("Unknown command");
        sendMSG("INFO","ESP32: Unknown command received");
      }
    }
  }
}
```

In the setup() function we first configure the digital pins, the serial line, and the WLAN interface, before starting the websocket and register the callback function webSocketEvent(). Then we start server2 to dish out the webpage, register the callback function to do the dishing, but only if the SPIFFS filesystem, which stores the html file, is present.

```
void setup() {
  pinMode(21,INPUT_PULLUP);
  pinMode(18,OUTPUT);
  dacWrite(25,0); // initialize DAC
  Serial.begin(115200); delay(1000);
  WiFi.begin(ssid, password);
  while (WiFi.status() != WL_CONNECTED) {delay(500); Serial.print(".");}
  Serial.print("\nConnected to ");  Serial.print(ssid);
  Serial.print(" with IP address: "); Serial.println(WiFi.localIP());
  webSocket.begin();
  webSocket.onEvent(webSocketEvent);
  if (!SPIFFS.begin(true)) {
    Serial.println("ERROR: no SPIFFS filesystem found"); return;
  } else {
    server2.begin();
    server2.serveStatic("/", SPIFFS, "/esp32-daq.html");
    server2.onNotFound(handle_notfound);
    Serial.println("SPIFFS file system found and server started");
  }
}
```

In the `loop()` function we first let the web server and the websocket interface handle their events and then check whether the `info_available` flag is set, in which case we send the JSON-formatted message that is stored in the `info_buffer` to the browser. Likewise, if `output_ready` is set, we first clear and then assemble a `DynamicJsonDocument` named `doc`. We call it `ADC` and add the measurement values stored in `samples` before converting this JSON-formatted message into the character array `out` that we send to the browser.

```
void loop() {
  server2.handleClient();
  webSocket.loop();
  if (info_available==1) {
    info_available=0;
    webSocket.sendTXT(websock_num,info_buffer,strlen(info_buffer));
  }
  if (output_ready==1) {
    output_ready=0;
    doc.to<JsonObject>();  // clear DynamicJsonDocument doc
    for (int k=0;k<3;k++) {doc["ADC"][k]=samples[k];}
    serializeJson(doc,out);
    webSocket.sendTXT(websock_num,out,strlen(out));
  }
  yield();
}
```

With the ESP32 programmed, we now turn to the webpage and the JavaScript code that constitutes the "other" end of the websocket communication channel. The header follows the description from Chapter 8, only here we specify two traces.

```
<!DOCTYPE HTML>
<HTML lang="en">
<HEAD>
  <TITLE>ESP-DAQ</TITLE>
  <META charset="UTF-8">
  <STYLE>
    #displayarea { border: 1px solid black; }
    #adc32 { border: 1px solid black; }
    #adc33 { border: 1px solid black; }
    #trace0 { fill: none; stroke: red; stroke-width: 1px;}
    #trace1 { fill: none; stroke: blue; stroke-width: 1px;}
    #ip {float: right;}
  </STYLE>
</HEAD>
```

The following section specifies the appearance of the web page shown in Figure 13.1. The buttons on the top, the selection menu to choose the sampling time, and the displayed IP number are defined analogously to Chapter 8.

```
<BODY>
<P> ESP-DAQ:
  <button id="start" type="button" onclick="start();">Start</button>
  <button id="stop" type="button" onclick="stop();">Stop</button>
  <SELECT onchange="setSamplePeriod(this.value);">
  <OPTGROUP label="Roll mode">
```

```
    <OPTION value="5">5 ms</OPTION>
    <OPTION value="10">10 ms</OPTION>
    <OPTION value="20">20 ms</OPTION>
    <OPTION value="50">50 ms</OPTION>
    <OPTION selected="selected" value="100">100 ms</OPTION>
    <OPTION value="200">200 ms</OPTION>
    <OPTION value="500">500 ms</OPTION>
    <OPTION value="1000">1 s</OPTION>
    <OPTION value="2000">2 s</OPTION>
    <OPTION value="5000">5 s</OPTION>
  </OPTGROUP>
  </SELECT>
  <button id="clear" type="button" onclick="cleardisplay();">
    Clear</button>
  <A id='ip'>IP address</A>
</P>
```

In the next section we open a DETAIL which later enables us to hide the section or make it visible by clicking on a little triangle next to the text specified in the SUMMARY. Inside the DETAIL, the CANVAS named Din displays the state of the input pin D21. It is followed by the definition of the checkbox that we use to change the state of output pin DO18. The selection menu lets us choose the period at which the pin is toggled by passing the selected value to the callback function setDO(), which assembles a JSON message and sends it to the ESP32. Then we define an INPUT slider that we use to set the output voltage of DAC25 followed by two elements of type CANVAS that show the current value reported by ADC32 and ADC33 before closing the DETAIL.

```
<P><DETAILS open> <SUMMARY>Input output</SUMMARY>
<CANVAS id="Din" width="100" height="20"> </CANVAS>
DO18:<INPUT type="checkbox" id="DO18" onchange="setDO(18,this.checked)"/>
<SELECT onchange="pulsepin18(this.value);">
    <OPTION value="0">Pulse D18 OFF</OPTION>
    <OPTION value="1">Pulse D18 with 1 ms steps</OPTION>
    <OPTION value="10">Pulse D18 with 10 ms steps</OPTION>
    <OPTION value="100">Pulse D18 with 100 ms steps</OPTION>
    <OPTION value="1000">Pulse D18 with 1 s steps</OPTION>
    <OPTION value="2000">Pulse D18 with 2 s steps</OPTION>
    <OPTION value="5000">Pulse D18 with 5 s steps</OPTION>
</SELECT> </P>
<P>DAC25: <INPUT type="range" min="0" max="255" step="1" value="0"
                onchange="updatedac25(this.value)" />
<SELECT onchange="rampdac25(this.value);">
    <OPTION value="0">Ramp DAC25 OFF</OPTION>
    <OPTION value="10">Ramp DAC25 with 10 ms steps</OPTION>
    <OPTION value="50">Ramp DAC25 with 50 ms steps</OPTION>
</SELECT></P>
<P><DIV id="adc32val">ADC32: unknown</DIV>
<CANVAS id="adc32" width="512" height="10"> </CANVAS>
<DIV id="adc33val">ADC33: unknown</DIV>
<CANVAS id="adc33" width="512" height="10"> </CANVAS>
</DETAILS></P>
```

In the following section we open a second DETAIL which contains the displayarea to show the voltages recorded by the ADCs. The displayarea is defined as a scalable vector graphic in between SVG tags. Each trace is defined as a PATH, just as in Chapter 8. After closing the DETAIL, we define the two status lines that are shown at the bottom of the webpage.

```
<DETAILS open>
  <SUMMARY>Logging display</SUMMARY>
  <SVG id="displayarea" width="1024px" height="512px">
    <PATH id="trace0" d="M0 256" />
    <PATH id="trace1" d="M0 256" />
  </SVG>
</DETAILS>
<DIV id="status">Status window</DIV>
<DIV id="reply">Reply from ESP32</DIV>
```

The following JavaScript makes the webpage interactive. We first declare some variables and determine the IP address of the ESP32. We store it, including the websocket port 81, in the variable ipaddr and display it in the webpage element ip. Then we open the websocket before defining the callback functions for error handling, opening, and closing the websocket. The functions toStatus() and toReply() copy the text given as argument to the respective lines at the bottom of the webpage. The functions start() and stop() are callback functions for the respective buttons on the webpage. They prepare JSON-formatted messages that instruct the ESP32 what to do.

```
<SCRIPT>
  var sample_period=100, current_position=0;
  var ipaddr=location.hostname + ":81";
  document.getElementById('ip').innerHTML=ipaddr;
  show_noconnect();
  var websock = new WebSocket('ws://' + ipaddr);
  websock.onerror = function(evt) { console.log(evt); toStatus(evt) };
  websock.onopen = function(evt) { console.log('websock open'); };
  websock.onclose = function(evt) {
    console.log('websock close');
    toStatus('websock close');
  };
  function toStatus(txt){
    document.getElementById('status').innerHTML=txt;}
  function toReply(txt) {
    document.getElementById('reply').innerHTML=txt;}
  function start() {
   websock.send(JSON.stringify({"cmd":"START", "val":sample_period}));
  }
  function stop() {
    websock.send(JSON.stringify({ "cmd" : "STOP", "val" : "-1" }));
    show_noconnect();
  }
```

The .onmessage method is executed whenever a new message from the ESP32 arrives. It first parses the messages and puts the message into the variable stuff, where we first probe whether it is a message of type INFO in which case we just pass it to the toReply() function which displays it below the graph. If the message is of type ADC, it contains the array with the measurement values. The first two elements, val[0] and val[1] contain the values from ADC32 and ADC33 on the ESP32. We convert the values to volts and display the value in the webpage before also draw a colored rectangle that illustrates the value as a horizontal bar; a red one for ADC32 and a blue one for ADC33. Finally we change the color of the field Din which displays DI21 according to the state of the pin. The remainder of the function first extracts attribute d of the PATH associated with each trace and appends the coordinates of the new sample, before copying it back to d. Finally, we increment current_position. If the new position exceeds the right-hand limits, we remove the left-most data point and translate the two traces towards the left.

```javascript
websock.onmessage=function(event) {
  console.log(event);
  var stuff=JSON.parse(event.data);
  var val=stuff["INFO"]; //............................info
  if ( val != undefined ) {toReply(val);}
  val=stuff["ADC"];        //.....................measurements
  if ( val != undefined ) {
    document.getElementById('adc32val').innerHTML=
    "ADC32: " + val[0] + " --> " + (val[0]*3.3/511).toFixed(2) + " V";
    c=document.getElementById('adc32').getContext("2d");    //..ADC32
    c.clearRect(0,0,512,10);
    c.fillStyle = "#FF0000"; c.fillRect(0,0,val[0],10);
    document.getElementById('adc33val').innerHTML=
    "ADC33: " + val[1] + " --> " + (val[1]*3.3/511).toFixed(2) + " V";
    c=document.getElementById('adc33').getContext("2d");    //..ADC33
    c.clearRect(0,0,512,10);
    c.fillStyle = "#0000FF"; c.fillRect(0,0,val[1],10);
    c=document.getElementById('Din').getContext("2d");      //...DI21
    if (val[2]==1) {c.fillStyle = "#00FF00";}
    else {c.fillStyle = "#FF0000";}
    c.fillRect(20,0,100,20);
    c.font = "20px Arial"; c.fillStyle = "#000000";
    c.fillText("DI21",50,17);
    dd=document.getElementById('trace0').getAttribute('d');
    dd += ' L' + current_position + ' ' + (512-val[0]);
    document.getElementById('trace0').setAttribute('d',dd);
    dd=document.getElementById('trace1').getAttribute('d');
    dd += ' L' + current_position + ' ' + (512-val[1]);
    document.getElementById('trace1').setAttribute('d',dd);
    current_position += 1;
    if (current_position > 1024) {
      dd=document.getElementById('trace0').getAttribute('d')
        .replace(/M[^L]*L/, "M");
      document.getElementById('trace0').setAttribute('d',dd);
      document.getElementById('trace0').setAttribute(
        'transform','translate(' + (1024-current_position) + ',0)');
      dd=document.getElementById('trace1').getAttribute('d')
        .replace(/M[^L]*L/, "M");
      document.getElementById('trace1').setAttribute('d',dd);
      document.getElementById('trace1').setAttribute(
        'transform','translate(' + (1024-current_position) + ',0)');
    }
  }
}
```

The next group of JavaScript functions first defines the callback functions to set the sample_period from the selection menu and to clear the display. The function show_noconnect() displays a grey button for the state of the input pin DI21, indicating that the data acquisition is not yet active. Then we define functions to send the messages that cause the ESP32 to either set DAC25 to a fixed value or to ramp it and create the sawtooth signal visible in Figure 13.1. Next come functions to set the state of an output pin and to toggle it continuously. Finally, we close the tags for SCRIPT, BODY, and HTML.

```
function setSamplePeriod(v) {
  toStatus("Setting sample period to "+ v + " ms");
  sample_period=v;
}
function cleardisplay() {
  dd = "M0 512";
  document.getElementById('trace0').setAttribute('d',dd);
  document.getElementById('trace0').setAttribute
      ('transform','translate(0)');
  document.getElementById('trace1').setAttribute('d',dd);
  document.getElementById('trace1').setAttribute
      ('transform','translate(0)');
  current_position=0;
}
function show_noconnect() {
  c=document.getElementById('Din').getContext("2d");
  c.fillStyle = "#DDDDDD"
  for (i=2;i<3;i++) {c.fillRect(20+110*(i-2),0,100,20);}
  c.font = "20px Arial"; c.fillStyle = "#000000";
  c.fillText("DI21",50,17);
}
function updatedac25(v) {
  volts=v*3.3/255;
  toStatus("Set DAC25 = " + v + " -> " + volts.toFixed(2) + " V");
  websock.send(JSON.stringify({ "cmd" : "DAC25", "val" : v }));
}
function rampdac25(v) {
  toStatus("Ramp DAC25 with time step " + v);
  websock.send(JSON.stringify({ "cmd" : "RAMP25", "val" : v }));
}
function setDO(v,s) {
  vv=s?1:0;
  toStatus("Status changed on DO"+ v + " is " + s + " or " + vv);
  websock.send(JSON.stringify({ "cmd" : "DO"+v, "val" : + vv }));
}
function pulsepin18(v) {
  toStatus("Pulse D18 with time step " + v + " ms");
  websock.send(JSON.stringify({ "cmd" : "PULSE18", "val" : v }));
}
</SCRIPT> </BODY> </HTML>
```

In order to transfer this webpage to the ESP32 we create a subdirectory named data under the subdirectory where ESP_minidaq.ino file resides and copy the html file to it. Then we install the program that copies the contents of the data subdirectory to the flash memory on the ESP32 by downloading `ESP32FS-1.0.zip` from the *releases page* on `https://github.com/me-no-dev/arduino-esp32fs-plugin` and unzip it in the `/Arduino/tools/` subdirectory. After restarting the Arduino IDE, we will find an entry *ESP32 Sketch Data Upload* in the *Tools* menu that uploads the webpage to the ESP32, such that the server on port 80 finds it and delivers it to connecting browsers. If the upload fails, ensure that the *Serial monitor* window is closed and try again!

At this point, we have established a two-way communication channel between the microcontroller and the browser that has many advantages, because the user interface is entirely specified by the HTML and JavaScript code running on the browser, which very flexible and easy to extend, so we immediately use it for our next project.

QUESTIONS AND PROJECT IDEAS

1. Produce a 500 Hz rectangular signal on some output pin of the ESP32.

2. Make the minimum and maximum amplitudes of the sawtooth signal adjustable, either use a selection menu or sliders on the web interface.

3. Use the second DAC on pin 26 to output a sine like signal by (slow) direct digital synthesis.

4. Do the same for a triangular output voltage that rises linearly to the maximum and linearly goes back to zero.

5. Add the AD9850 DDS board and make it controllable through the web interface.

6. Connect your favorite sensor to the ESP32 and make it controllable through the web interface.

7. Program four output pins to produce the on-off pattern that drives stepper motors.

8. Add one button to the web interface that closes the websocket and another one to reconnect if the connection was interrupted.

9. Save the measurement samples in an array on the browser using JavaScript, such that the numerical values are displayed at the press of a button.

## CHAPTER 14

Example: Fast Acquisition

Occasionally it is desirable to observe and generate signals at rates higher than the few hundered samples per second from Chapter 13, for example to analyze communication channels, such as RS-232, or to observe acoustic signals. Luckily, the ESP32 is capable of sampling at rates in excess of 200 000 samples per second by directly transferring samples from the ADC into memory without using the CPU using a mode called *direct memory access* (DMA). It uses the I2S (Inter-IC Sound) support on the ESP32 chip that is normally dedicated to digitally transmitting music. Moreover, the ESP has built-in frequency generators for sinusoidal and rectangular signals in the same frequency range.

It is straightforward to add these features to our data acquisition system by including the following code snippet – based on the ESP32 example HiFreq_ADC – to configure the high-speed mode in our sketch. It includes the header file `driver/i2s.h` for I2S support and then configures I2S to sample at the specified `i2s_sample_rate`. It reads the analog input from input pin 32, which is connected to `ADC1_CHANNEL_4`. Replace the _4 by _7 to connect it to pin 35. We set the attenuation of the input voltages such that it covers the range from zero to the supply voltage. Note that error handling is rather basic; the program just prints a short message to the serial line.

```
#include <driver/i2s.h>
bool i2s_adc_enabled=false;
void i2sInit(uint32_t i2s_sample_rate){
  i2s_config_t i2s_config = {
    .mode = (i2s_mode_t)(I2S_MODE_MASTER | I2S_MODE_RX
                                | I2S_MODE_ADC_BUILT_IN),
    .sample_rate =  i2s_sample_rate,
    .bits_per_sample = I2S_BITS_PER_SAMPLE_16BIT,
    .channel_format = I2S_CHANNEL_FMT_ONLY_LEFT,
    .communication_format = I2S_COMM_FORMAT_STAND_I2S,
    .intr_alloc_flags = ESP_INTR_FLAG_LEVEL1,
    .dma_buf_count = 8,
    .dma_buf_len = 1024,    // 1024 is maximum length
    .use_apll = true,       // needed for low frequencies
    .tx_desc_auto_clear = false,
    .fixed_mclk = 0
  };
  if(ESP_OK != i2s_driver_install(I2S_NUM_0, &i2s_config, 0, NULL)){
    Serial.print("Error installing I2S.");
  }
  if(ESP_OK != i2s_set_adc_mode(ADC_UNIT_1, ADC1_CHANNEL_4)){ // pin 32
```

DOI: 10.1201/9781003341703-14

```
    Serial.print("Error setting up ADC.");
  }
  if(ESP_OK != adc1_config_channel_atten(ADC1_CHANNEL_4, ADC_ATTEN_DB_11)){
    Serial.print("Error setting up ADC attenuation.");
  }
  if(ESP_OK != i2s_adc_enable(I2S_NUM_0)){
    Serial.print("Error enabling ADC.");
  }
  Serial.print("I2S ADC setup ok\n");
}
```

The sinusoidal frequency generator also uses the I2S hardware on the ESP32. After including three header files, two variables define the frequency and output scale of the signal, before the function cwDACinit(), which is based on the very detailed information from https://github.com/krzychb/dac-cosine, configures the generator by directly writing to registers on the ESP32. We point out that the fudge factor 0.95 that appears in the calculation of frequency_step is needed to compensate the _APPROX and to produce the specified output frequency. Note that setting the desired frequency to zero disables the output.

```
#include <driver/dac.h>
#include <soc/sens_reg.h>
#include "soc/rtc.h"
float dac25freq=0;
int dac25scale=0;    // output full scale
void cwDACinit(float freq, int scale, int offset) { //...cwDACinit
  if (abs(freq) < 1e-3) {dac_output_disable(DAC_CHANNEL_1); return;}
  int frequency_step=max(1.0,floor(0.5+0.95*65536.0*freq
                      /RTC_FAST_CLK_FREQ_APPROX));
  SET_PERI_REG_MASK(SENS_SAR_DAC_CTRL1_REG, SENS_SW_TONE_EN);
  SET_PERI_REG_MASK(SENS_SAR_DAC_CTRL2_REG, SENS_DAC_CW_EN1_M);
  SET_PERI_REG_BITS(SENS_SAR_DAC_CTRL2_REG, SENS_DAC_INV1, 2,
      SENS_DAC_INV1_S);
  SET_PERI_REG_BITS(SENS_SAR_DAC_CTRL1_REG, SENS_SW_FSTEP,
      frequency_step, SENS_SW_FSTEP_S);
  SET_PERI_REG_BITS(SENS_SAR_DAC_CTRL2_REG, SENS_DAC_SCALE1,
      scale, SENS_DAC_SCALE1_S);
  SET_PERI_REG_BITS(SENS_SAR_DAC_CTRL2_REG, SENS_DAC_DC1,
      offset, SENS_DAC_DC1_S);
  dac_output_enable(DAC_CHANNEL_1);
}
```

After the changes to the sketch we now address the webpage that the ESP32 publishes.

In the file with the webpage esp32-daq.html, we update the function webSocketEvent() where the response to JSON-encoded commands arrives. In particular, the START command now has to distinguish between starting the acquisition in slow or high-speed mode. We use the convention that positive values of sample_period still define the sampling speed in the slow acqusition mode. Negative values, however, initialize the high-speed acquisition by calling i2sInit(). Note that we need to disable and uninstall the i2s_adc driver before reconfiguring it with the new i2s_sample_rate. Likewise, the STOP section disables the slow acqusition if the sample_period is positive and the high-speed acqusition if it is negative. Note also that we reenable the frequency generator, because it also uses the I2S module and disabling the I2S driver stopped it. Finally, the new command SETDAC configures the frequency generator to output the frequency specified in `val`; negative values change the output voltage scale.

```
                 :
   if (strstr(cmd,"START")) {
       sendMSG("INFO","ESP32: Received Start command");
       sample_period=val;
       if (sample_period > 0) {
         SampleSlow.attach_ms(sample_period,sampleslow_action);
         Serial.print("sample_period = "); Serial.println(sample_period);
       } else {
         if (i2s_adc_enabled) {
           i2s_adc_disable(I2S_NUM_0); i2s_driver_uninstall(I2S_NUM_0);}
         uint32_t i2s_sample_rate=(uint32_t) abs(sample_period);
         i2sInit(i2s_sample_rate);
         i2s_adc_enabled=true;
       }
   } else if (strstr(cmd,"STOP")) {
       if (sample_period>0) {
         sendMSG("INFO","ESP32: Received Stop command");
         SampleSlow.detach();
       } else {
         if (ESP_OK != i2s_adc_disable(I2S_NUM_0)) {
           Serial.printf("Error disabling ADC.");
         }
         if (ESP_OK != i2s_driver_uninstall(I2S_NUM_0)) {
           Serial.printf("Error uninstalling I2S driver.");
         }
         i2s_adc_enabled=false;
         if (dac25freq>0) {cwDACinit(dac25freq,dac25scale,0);}
       }
   } else if (strstr(cmd,"SETDAC")) {
       Serial.printf("DAC25 frequency = %d\n",val);
       if (val < 0) {dac25scale=-val-1;} else {dac25freq=val;}
       cwDACinit(dac25freq,dac25scale,0);
   } else if (strstr(cmd,"DAC25")) {
                 :
```

We configure the rectangular signal generator to continuously output a 1 kHz signal on pin 27 by adding the following line

```
ledcSetup(0, 1000, 2); ledcAttachPin(27, 0); ledcWrite(0, 2);
```

to the end of the **setup()** function. It first configures PWM channel 0 to operate at 1 kHz with 2-bit resolution (four states), then attaches that channel to pn 27 and writes a value of 2, which gives the channel a duty cycle of 50 % (two out of four).

Finally, at the end of the **loop()** function, we add code to acquire data with **i2s_read()** and reorder the samples in the **buffer[]**, which is an artefact of the I2S system that usually sends stereo signals for left and right channels. Unfortunately, the convention of coding these data do not agree with the temporal order of the arriving samples, which makes the reordering necessary. Once the **buffer** is ready, we assemble the JSON array **WF0** and dispatch it to the browser, where it is displayed. We add a short delay at the end of the function to avoid flooding the wireless network and give a little time to other activities, such as the user interface.

```
if (i2s_adc_enabled) {
  size_t bytes_read;
  uint16_t buffer[1024], swap; //={0};
  i2s_read(I2S_NUM_0, &buffer, sizeof(buffer), &bytes_read, 15);
  for (int j=0; j<bytes_read/2; j+=2) {  // swap left and right channel
    swap=buffer[j+1]; buffer[j+1]=buffer[j]; buffer[j]=swap;
  }
  doc.to<JsonObject>();  // clear DynamicJsonDocument doc
  for (int j=0;j<bytes_read/2;j++){doc["WF0"][j]=(buffer[j]&0x0FFF)>>3;}
  serializeJson(doc,out);
  webSocket.sendTXT(websock_num,out,strlen(out));
  delay(2000);
}
```

Finally, we have to increase the size of out and doc, both are defined near the top of the sketch, from 300 to 20000 in order to account for the much longer dispatched waveform WF0.

We control the new functionality on the ESP32 from the browser by adding code to the file esp32-daq.html. To the SELECT menu for the sampling speed we add an OPTGROUP to send negative frequencies to the ESP32, which uses them to configure the high-speed ADC. with frequencies between $1\,\text{kS/s}$ and $1\,\text{MS/s}$.

```
<OPTGROUP label="Hi-speed I2S ADC">
  <OPTION value="-1000">1 kSamples/s</OPTION>
  <OPTION value="-10000">10 kSamples/s</OPTION>
  <OPTION value="-20000">20 kSamples/s</OPTION>
  <OPTION value="-50000">50 kSamples/s</OPTION>
  <OPTION value="-100000">100 kSamples/s</OPTION>
  <OPTION value="-200000">200 kSamples/s</OPTION>
  <OPTION value="-277777">277.777 kSamples/s</OPTION>
  <OPTION value="-500000">500 kSamples/s</OPTION>
  <OPTION value="-1000000">1 MSamples/s</OPTION>
</OPTGROUP>
</SELECT>
```

Likewise, we control the sinusoidal frequency generator with an additional SELECT menu. Note that sending positive values specify the frequency and negative values specify the output scale in steps that are hardcoded into the ESP32.

```
<SELECT onchange="setDacFrequency(this.value);">
  <OPTGROUP label="DAC frequency">
    <OPTION value="0">DAC25 OFF</OPTION>
    <OPTION value="2000">DAC25=2 kHz</OPTION>
    <OPTION value="5000">DAC25=5 kHz</OPTION>
    <OPTION value="10000">DAC25=10 kHz</OPTION>
    <OPTION value="20000">DAC25=20 kHz</OPTION>
    <OPTION value="50000">DAC25=50 kHz</OPTION>
    <OPTION value="100000">DAC25=100 kHz</OPTION>
    <OPTION value="200000">DAC25=200 kHz</OPTION>
  </OPTGROUP>
```

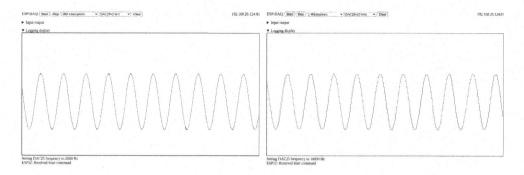

Figure 14.1 Sampling an internally generated 2 kHz signal with 200 kS/s (left) and a 10 kHz signal with 1 MS/s.

```
<OPTGROUP label="DAC output scale">
  <OPTION value="-1">1/1 scale</OPTION>
  <OPTION value="-2">~1/2 scale</OPTION>
  <OPTION value="-3">~1/4 scale</OPTION>
  <OPTION value="-4">~1/6 scale</OPTION>
</OPTGROUP>
```

Finally, the callback function setDacFrequency() is defined near the end of the JavaScript code. It receives the value f from the OPTION in the SELECT section and dispatches a JSON-formatted string to the ESP32.

```
function setDacFrequency(f) {
  toStatus("Setting DAC25 frequency to "+ f + " Hz");
  websock.send(JSON.stringify({ "cmd" : "SETDAC", "val" : f }));
}
document.getElementById("displayarea").addEventListener
        ('mousedown', showCoordinates, false);
function showCoordinates(event) {
  rect=document.getElementById("displayarea").getBoundingClientRect();
  toStatus("Mouse position: " + (event.clientX-rect.left) + " "
      + (rect.height-event.clientY+rect.top));
}
```

The showCoordinates() function writes the coordinates of the pixel where the mouse is clicked on the displayarea to the status line, which is convenient to determine the length of pulses.

We check out the system by generating a 2 kHz sinusoidal signal on pin 25, sample it with 200 kS/s and show the waveform on the left-hand side in Figure 14.1, which shows a smooth sinusoidal curve. The figure on the right-hand side shows a 10 kHz sine wave, sampled at 1 MS/s. We observe distinct steps in the displayed waveform indicating that the ADC reports the same value multiple times rather than sampling at a higher rate. The useful limit for the sampling frequency, taken from the HiFreq_ADC example, is 277 kS/s, which motivates including that value in the SELECT menu.

We can now use the system to analyze the signals from a TSOP2238 infra-red receiver that was already mentioned in Section 4.6.4. Its output pin is directly connected to the input for the ADC on pin 32 of the ESP32. The left-hand side in Figure 14.2 shows the signal caused by pressing the OFF button on a Samsung TV remote controller. We observe that the idle state is a high (3.3 V) signal level that is interrupted by a sequence of pulses where

Figure 14.2 The infra-red signal from a Samsung TV remote controller, recorded with a TSOP2238 receiver, whose output is sampled at 10 kS/s (left) and the start of the pulse sampled at 50 kS/s (right).

the signal level is low (0 V). The duration of the entire sequence is about 60 ms. Increasing the sampling rate to 50 kS/s and trying a few times to catch the start of the pulse, we obtain the figure on the right-hand side. By clicking with the mouse on the transitions to determine the pixel values, we can easily measure the duration of signals. We find that the long negative-going pulse at the start is 4.6 ms long and is followed by a 4.4 ms long period where the level is high. The following short pulses approximately are low for 0.6 ms and high for 0.4 ms. Other parts of the pulse can be worked out with patience, but the lack of a trigger system makes this somewhat inconvenient. Adding one is therefore first on the list of project ideas.

QUESTIONS AND PROJECT IDEAS

1. Add functionality to trigger the acquisition on a falling or rising edge.

2. Change the code to read the fast ADC from pin 33 instead.

3. Add a menu to select the frequency and pulse width of the generator for the rectangular signals on pin 27.

4. Double the size of the array `buffer[]` supplied to the `i2s_read()` function and record longer waveforms. What else do you need to change in order to display them on the webpage?

5. Add an amplifier to measure and then display the small voltages from a microphone.

6. Discuss adding an op-amp, wired as a line buffer, to the DAC output and a transimpedance amplifier to the ADC input in order to measure the impedance of components like we did in Chapter 12.

## CHAPTER 15

Example: Medical Sensing

In this chapter, we measure electrocardiograms (ECGs), the oxygen saturation level in our blood, and the bioimpedance with a NodeMCU, which controls the three sensors and displays their measurements using the web interface from Chapter 13, already shown in Figure 13.1. As an added bonus, we add circuitry to simulate the impedance of the human body and ECGs as well. Moreover, we provide the option to save measured values to the internal flash disk on the NodeMCU and later replay them, which opens the possibility to log data, for example, while exercising. Importantly, note that this project only illustrates the measurement processes and is not suitable for medical diagnosis!

The hardware with the NodeMCU on the right and the three sensors is shown in Figure 15.1. Next to the controller we see the breakout board [31] with the AD8232 to measure ECGs. It is accompanied by a cable that connects three electrodes to a 3.5 mm jack, which is shown just below the symbol of the heart. The leads to the electrodes are color-coded: the black electrode should be connected to the right lower arm (RA), the blue one to the left lower arm (LA), and the red one to the right leg (RL), though other placements are also possible. The breakout board senses, amplifies, and cleans up the small voltage measured between the black and the blue electrodes, uses the red one as reference, and provides an analog voltage on its output pin. This output signal we route directly to the analog input pin A0 of the NodeMCU. Furthermore, signals to indicate whether the electrodes are connected are routed to digital input pins D3 and D4.

The small breadboard shown on the right contains an ECG simulator, which allows us to test the AD8232 without connecting a patient. We store the voltages that describe

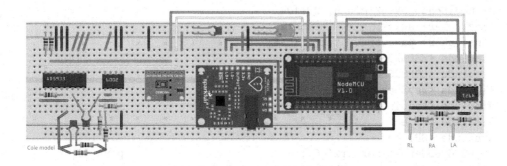

Figure 15.1 The project assembled on a solder-less breadboard. The circuit at the bottom left is a simple electronic model of the human body and the small breadboard on the right is part of the ECG simulator[†].

DOI: 10.1201/9781003341703-15

the ECG in tabular form on the NodeMCU and periodically pass values to the MCP4921 digital-to-analog converter that we already discussed in Section 4.5.5. Apart from the power rails, its SPI-control wires SCK, MOSI, and CS are connected to pins D5, D7, and D8 on the NodeMCU, respectively. The three resistors ($10\,k\Omega$, $10\,\Omega$, and $10\,k\Omega$), also placed on the small breadboard, reduce the signal between RL and RA to the level of mV, normally produced by the body. We directly connect them to the corresponding electrodes at the end of the cable that is attached to the jack on the AD8232 breakout board.

The MAX30102-based pulse oximeter, already shown in Figure 2.40, is also mounted on a small breakout board that is placed to the left of the AD8232. As already mentioned in Section 2.3.3, it measures the saturation level of oxygen in blood (SPO2) by comparing the reflection of red and infrared light measured with a photo-transistor at the tip of a finger. The small optical sensor with photodiodes and transistor operates similarly to the color measurements from Chapter 11. It is seen near the center of the breakout board and is connected to the NodeMCU through the two I2C lines SCL and SDA, attached to pins D1 and D2, respectively.

The third sensor, shown on the far left, is the bioimpedance network analyzer AD5933 that we already encountered in Chapter 12. We use it to measure the impedance between two electrodes attached to a circuit that models the impedance of a human body, called *Cole model*. The circuit with the eight-pin op-amp is the same we already used earlier; we only increase the value of the feedback resistor between pin 4 and 5 of the AD5933 to $10\,k\Omega$. At the same time, we add a $10\,k\Omega$ resistors to pin 1 of the op-amp in order to limit the current delivered to the "body" to around $100\,\mu A$. We daisy-chain the I2C lines coming from the oximeter to interface the AD5933 and the controller.

This circuit is brought to life with the following sketch that runs on the NodeMCU. It is a bit long, so we discuss it one part at a time. The first part closely follows the data acquisition system from Chapter 13 by providing WLAN credentials and loading the necessary libraries, only here we need to use the ones that are appropriate for the ESP8266-based NodeMCU. In order to control the MAX30102, we install the *DFRobot_MAX30102* library by issuing the command `git clone https://github.com/DFRobot/DFRobot_MAX30102.git` in the subdirectory `~/Arduino/libraries` and subsequently use it in our sketch by including its header file and creating an object `Sensor` to interact with the device. We then include the support files for I2C from Section 4.4.3 and for the AD5933 from Chapter 12. Then a number of variable declarations follow; many were already used in Chapter 13. We only point out that `mmode` is used to identify actions of the different sensors and `Rcal` specifies the value of the current limiting resistor in $k\Omega$.

```
// MediSense, V. Ziemann, 221106
const char* ssid     = "messnetz";
const char* password = "zxcvZXCV";
#include <ESP8266WiFi.h>
#include <ESP8266WebServer.h>
#include <WebSocketsServer.h>
#include <ArduinoJson.h>
#include <FS.h>
#include <SPI.h>
ESP8266WebServer server2(80);                        // port 80
WebSocketsServer webSocket = WebSocketsServer(81);  // port 81
#include <Ticker.h>
Ticker SampleSlow,ECGticker;
#include <DFRobot_MAX30102.h>
DFRobot_MAX30102 Sensor;
```

```
#include "I2Crw.h"
#include "AD5933.h"
int16_t Ival[512],Qval[512]; float mag50=1, phase50=0;
volatile uint8_t websock_num=0,info_available=0,output_ready=0;
int mmode=0,sample_period=100,samples[3],icounter=0,recording=0;
int ringbuf[50],ibuf=0,ped=0;   // pedestal subtraction for oximeter
char info_buffer[80];           // for websock messages
char out[300];                  // space for websocket messages
DynamicJsonDocument doc(300);
float Rcal=10.0;                // calibration resistor
uint16_t ecgtrace[200],ecg_counter=0;;
#define CS D8                   // chip select for SPI-DAC
```

In the following section we define a function to handle requests for unknown web pages. The function `cardio_action()` is the callback routine that reads the IO ports with signals related to the AD8232 and places them in the array `samples[]`, much the same way we did in Chapter 13. Likewise, in `oximeter_action()` we read the signals from the photo transistor of the MAX30102 illuminated by the two photodiodes. Here we also determine the pedestal of the raw signal, which varies widely. We then place the raw signal with pedestal subtracted and suitably scaled in order to fit it in the vertical range we can display. Note that every 200 samples we fill `sample[1]` with a large value that is zero otherwise. This causes a spike to appear every 200 samples on the display that allows us to easily convert the horizontal axis to time. In `impedance_action()` we program the AD5933 to produce a 50 kHz signal, read the I and Q signals, convert them to amplitude and phase and place the value, properly scaled into `samples[]`. You might want to adjust the scale factors to fill the available vertical range on the graph, which is 512 pixels. Moreover, the commented-out line stores the amplitude in logarithmic form and allows us to explore four orders of magnitude for the resitance on the same graph instead. Note that the three callback functions announce the availability of new data by setting `output_ready` to unity. This variable is checked and acted upon in the main program loop.

```
void handle_notfound() {  //..............http server error handling
    server2.send(404,"text/plain","not found, use http://ip-address/");
}
void cardio_action() {  //...................sampleslow_action
    samples[0]=analogRead(0)/2;   // adjust scale
    samples[1]=100*digitalRead(D3)+50*digitalRead(D4);
    samples[2]=0;
    output_ready=1;
}
void oximeter_action() { //........................oximeter_action
    float ir=0;
    if (mmode==3) {ir=(float)Sensor.getIR();
    } else if (mmode==4) {ir=(float)Sensor.getRed();}
    ringbuf[ibuf]=ir; ibuf++; if (ibuf>=50) {ibuf=0;}
    ped=ringbuf[0]; for (int k=1;k<50;k++) {ped=min(ped,ringbuf[k]);}
    samples[0]=100+(ir-ped)/4.0;   // adjust scale
    icounter=icounter+1;
    if (icounter==200){samples[1]=30000; icounter=0;} else {samples[1]=0;}
    samples[2]=0;
    output_ready=1;
}
```

```
void impedance_action() {  //....................impedance_action
  AD5933_sweep2(50e3,0.01,0,mode,pga,Ival,Qval);
  float mag=sqrt(Ival[0]*Ival[0]+Qval[0]*Qval[0]);
  float phase=atan2(Qval[0],Ival[0])*180/3.1415926;
  samples[0]=Rcal*(mag50/mag-1.0)*512.0;
  // samples[0]=128+128*log10(mag50/mag); // log scale
  samples[1]=256+phase-phase50;
  samples[2]=0;
  output_ready=1;
}
```

The function `set_dac()` receives the value to pass on to the MCP4921 DAC as input, adds the required configuration bits and transfers the value via SPI to the registers of the MCP4921. The function `ecg_action()` is the callback that continuously updates the DAC with values from the electrocardiogram trace stored in the array `ecgtrace[]`.

```
void set_dac(uint16_t val) {
  val|=(B0011 << 12);
  digitalWrite(CS,LOW);
  SPI.transfer(highByte(val));
  SPI.transfer(lowByte(val));
  digitalWrite(CS,HIGH);
}
void ecg_action() {   //.............................ecg_action
  set_dac(ecgtrace[ecg_counter]);
  ecg_counter++;
  if (ecg_counter==200) {ecg_counter=0;}
}
```

The function `send_samples()` receives the array `samples[]` as input, formats it as a JSON document `doc` and dispatches it via the websocket interface to be displayed on the web page. In a similar fashion, `sendMSG()` dispatches other websocket messages. Note that if the text starts with an exclamation point, it is sent immediately; otherwise, sending the message is handled later in the main program. This is necessary in order to send messages from within callback functions.

```
void send_samples(int samples[]) {  //...............send_samples
  for (int k=0;k<3;k++) {doc["ADC"][k]=samples[k];}
  serializeJson(doc,out); webSocket.sendTXT(websock_num,out,strlen(out));
}
void sendMSG(char *nam, const char *msg) { //..............sendMSG
  (void) sprintf(info_buffer,"{\"%s\":\"%s\"}",nam,msg);
  if (strstr(msg,"!")==msg) {
    webSocket.sendTXT(websock_num,info_buffer,strlen(info_buffer));
  } else {
    info_available=1;
  }
}
```

The function `webSocketEvent()` is called whenever a message arrives on the web socket. Most of the functionality equals that of the corresponding function from Chapter 13 and was discussed there. Here it differs, however, by making the actions dependent on the variables `mmode`. The command `START`, for example, behaves differently depending on mmode and starts the Ticker SampleSlow with either cardio_action, oximeter_action, or impedance_action. We point out that we limit the sample_rate for the oximeter, because the sensor cannot handle higher rates. If the command MMODE arrives, the sensors for the different modes are initialized suitably. The modes with numbers 200 and 201 turn recording on and off, respectively. Those with numbers 300 and 301 start and stop the ECG simulator, respectively.

```
void webSocketEvent(uint8_t num, WStype_t type, uint8_t * payload,
                    size_t length) {
  Serial.printf("webSocketEvent(%d, %d, ...)\r\n", num, type);
  websock_num=num;
  switch(type) {
    case WStype_DISCONNECTED:
      Serial.printf("[%u] Disconnected!\r\n", num);
      break;
    case WStype_CONNECTED:
      {
        IPAddress ip = webSocket.remoteIP(num);
        Serial.printf("[%u] Connected from %d.%d.%d.%d url: %s\r\n",
          num, ip[0], ip[1], ip[2], ip[3], payload);
      }
      sendMSG("INFO","ESP: Successfully connected");
      break;
    case WStype_TEXT:
    {
      Serial.printf("[%u] get Text: %s\r\n", num, payload);
      DynamicJsonDocument root(300);  //........parse JSON
      deserializeJson(root,payload);
      const char *cmd = root["cmd"];
      const int val = root["val"];
      if (strstr(cmd,"START")) {
        sendMSG("INFO","ESP: Received Start command");
        sample_period=val;
        Serial.print("sample_period = "); Serial.println(sample_period);
        if (mmode==0) {
          SampleSlow.attach_ms(sample_period,cardio_action);
        } else if (mmode==3 || mmode==4) {
          sample_period=max(50,sample_period);
          SampleSlow.attach_ms(sample_period,oximeter_action);
        } else if (mmode==8) {
          SampleSlow.attach_ms(sample_period,impedance_action);
        }
      } else if (strstr(cmd,"STOP")) {
        sendMSG("INFO","ESP: Received Stop command");
        SampleSlow.detach();
      } else if (strstr(cmd,"MMODE")) {
        mmode=val;
        if (mmode==3 & mmode==4) {
          if (!Sensor.begin()){sendMSG("INFO","ESP: No MAX30102");mmode=-1;
          } else {
            Sensor.sensorConfiguration(60,SAMPLEAVG_8,MODE_MULTILED,
```

```
                SAMPLERATE_400,PULSEWIDTH_411,ADCRANGE_16384);
        }
    } else if (mmode==5) {
        if (!Sensor.begin()){sendMSG("INFO","ESP: No MAX30102");mmode=-1;
        } else {
            sendMSG("INFO","ESP: MAX30102 is up and running");
            Sensor.sensorConfiguration(50,SAMPLEAVG_4,MODE_MULTILED,
                SAMPLERATE_100,PULSEWIDTH_411,ADCRANGE_16384);
        }
    } else if (mmode==200) {
        recording=1; sendMSG("INFO","ESP: Recording ON");
    } else if (mmode==201) {
        recording=0; sendMSG("INFO","ESP: Recording OFF");
    } else if (mmode==300) { // Start ECG output
        ECGticker.attach_ms(5,ecg_action);
        sendMSG("INFO","ESP: Starting ECG output");
    } else if (mmode==301) { // Stop ECG output
        ECGticker.detach();
        set_dac(0);
        sendMSG("INFO","ESP: Stopping ECG output");
    }
    } else {
    Serial.println("Unknown command");
    sendMSG("INFO","ESP: Unknown command received");
    }
  }
 }
}
```

The function `listdir()` lists the files on the internal flash disk on the NodeMCU and writes the result to the Serial line. We will use it to verify that the webpage, calibration data for the AD5933, and the recorded data are actually in place and to determine the size of the files.

```
void listdir() {   //....................Directory listing
  Serial.println("\nFile list:");
  sendMSG("INFO","ESP: File listing sent to serial line");
  Dir dir = SPIFFS.openDir("/");
  while (dir.next()) {
    Serial.print(dir.fileName()); Serial.print(" ");
    if(dir.fileSize()) {
      File f = dir.openFile("r");
      Serial.println(f.size());
    }
  }
}
```

In the function `setup()` we initialize the IO pins, connect to the WLAN, start the websocket server, and register the previously defined function `webSocketEvent()` as callback. Then we initialize access to the SPIFFS file system and, provided that it is available, start `server2` to publish the webpage `medisense.html`. Next we initialize the MAX30102 and define the frequency range for the AD5933, before loading calibration values for 50 kHz from the file `calib50kHz.dat` and the electrocardiogram trace from `ecg.dat`.

```
void setup() {    //...................................setup
  pinMode(D3,INPUT);
  pinMode(D4,INPUT);
  pinMode(D0,OUTPUT); digitalWrite(D0,HIGH); // LED on NodeMCU
  Serial.begin(115200);
  delay(1000);
  WiFi.begin(ssid, password);
  while (WiFi.status() != WL_CONNECTED) {delay(500); Serial.print(".");}
  Serial.print("\nConnected to ");  Serial.print(ssid);
  Serial.print(" with IP address: "); Serial.println(WiFi.localIP());
  webSocket.begin();
  webSocket.onEvent(webSocketEvent);
  if (!SPIFFS.begin()) {
    Serial.println("ERROR: no SPIFFS filesystem found");
  } else {
    server2.begin();
    server2.serveStatic("/", SPIFFS, "/medisense.html");
    server2.onNotFound(handle_notfound);
    Serial.print("SPIFFS file system OK and server started on port 80");
    listdir();
  }
  while (!Sensor.begin()) {Serial.println("No MAX30102"); delay(1000);}
  Sensor.sensorConfiguration(60,SAMPLEAVG_8,MODE_MULTILED,SAMPLERATE_200,
    PULSEWIDTH_411,ADCRANGE_16384);
  freq=1e4; finc=200; npts=511;                // AD5933 frequency range
  File f=SPIFFS.open("/calib50kHz.dat","r");   // 50kHz calibration data
  if (f) {
    float re,im; char line [80];
    f.readStringUntil('\n').toCharArray(line,80);
    sscanf(line," %g %g %g %g",&re,&im,&mag50,&phase50);
    f.close();
  }
  f=SPIFFS.open("/ecg.dat","r");
  if (f) {
    uint16_t v; char line[20];
    for (int k=0;k<200;k++) {
      f.readStringUntil('\n').toCharArray(line,80);
      sscanf(line,"%d",&v);// Serial.println(v);
      ecgtrace[k]=v;
    }
    f.close();
  }
  pinMode(CS,OUTPUT); digitalWrite(CS,HIGH);  // init SPI-CS
  SPI.begin(); SPI.setBitOrder(MSBFIRST);
}
```

Much of the function loop() resembles the one from Chapter 13; if info_available is unity the info_buffer is dispatched to the browser; if output_ready is unity and recording is zero, the array samples[] is dispatched, otherwise the file data.txt on the SPIFFS filesystem is opened and the measurement values are written to it before closing the file.

```
void loop() {
  server2.handleClient();    // http server
  webSocket.loop();          // websocket
  if (info_available==1) {
    info_available=0;
    webSocket.sendTXT(websock_num,info_buffer,strlen(info_buffer));
  }
  if (output_ready==1) {
    output_ready=0;
    if (recording==0) {
      send_samples(samples);
    } else if (recording==1) {  // record data on local file system
      File f=SPIFFS.open("/data.txt","a");    // open for append
      if (!f) {
        Serial.println("Unable To Open file");
      } else {
        f.print(samples[0]); f.print("\t"); f.println(samples[1]);
        f.close();
      }
      sendMSG("INFO","ESP: Recording...");
    }
  }
}
```

Then follows a section of code that handles the different actions that depend on the variable mmode. If it is five, a number of temporary variables are declared before reading the SPO2 value and an estimate for the heart rate HR from the pulse oximeter using a function from the DFRobot_MAX30102 library. We also receive values, ending with _OK that indicate whether the data appear to be valid. Then we build a message msg that we subsequently dispatch to the browser. Note, however, that despite reporting a successful measurement, the heart rate is very unreliable. It is much more accurate to directly measure it with the phototransistor and count the pulses visible on the left-hand side in Figure 15.2.

If mmode is six, we calibrate the AD5933 in the selected frequency range and store both the I and Q, as well as magnitude and phase in the file calib.dat on the SPIFFS file system. During the sweep we dispatch an INFO to the browser and illuminate the built-in LED on the NodeMCU to indicate that it is busy. In the next step we record the values at 50 kHz and store them in the file calib50kHz.dat. Since we only want to calibrate once, we set mmode to -1 once this task is complete.

For mmode==7 we perform a frequency sweep of the AD5933 before opening the file with the calibration data and looping (index k) over all frequencies. At each frequency, we calculate the magnitude mag and phase phase of the measurement, read the corresponding value from the calibration file, and fill sample[0] with the calibrated and scaled value that ranges between zero at the bottom of the display and $1 k\Omega$ at the top of the display. We then subtract the calibration phase ph from phase, scale the values to cover ± 10 degrees, and dispatch the values to the browser with send_samples(). Before leaving this section, we write the values to the serial line, which aids debugging.

```
if (mmode==5) {    // SPO2 rate displayed on status line
  char msg[50];
  int32_t SPO2,HR;        // values
  int8_t SPO2_OK,HR_OK;   // data OK?
  Sensor.heartrateAndOxygenSaturation(&SPO2,&SPO2_OK,&HR,&HR_OK);
  sprintf(msg,"!ESP: SPO2,HR = %d, %d  (%d,%d)",SPO2,HR,SPO2_OK,HR_OK);
```

```
    sendMSG("INFO",msg);
  } else if (mmode==6) {      // AD5933 calibrate
    digitalWrite(D0,LOW);  //turns LED on
    sendMSG("INFO","!ESP calibrating...");
    AD5933_sweep(freq,finc,npts,mode,pga,Ival,Qval);
    sendMSG("INFO","ESP: AD5933 Calibration complete");
    digitalWrite(D0,HIGH);  //turns LED off
    File f=SPIFFS.open("/calib.dat","w");
    for (int k=0;k<npts+1;k++) {
      f.print(Ival[k]); f.print("\t"); f.print(Qval[k]);
      f.print("\t"); f.print(sqrt(Ival[k]*Ival[k]+Qval[k]*Qval[k]));
      f.print("\t"); f.println(atan2(Qval[k],Ival[k])*180/3.1415926);
    }
    f.close();
    AD5933_sweep(50e3,0.01,0,mode,pga,Ival,Qval); // calibrate at 50 kHz
    f=SPIFFS.open("/calib50kHz.dat","w");
    f.print(Ival[0]); f.print("\t"); f.print(Qval[0]);
    f.print("\t"); f.print(sqrt(Ival[0]*Ival[0]+Qval[0]*Qval[0]));
    f.print("\t"); f.println(atan2(Qval[0],Ival[0])*180/3.1415926);
    f.close();
    mmode=-1;
  } else if (mmode==7) {      // AD5933 frequency sweep
    char line[80];
    float re,im,ab,ph;
    sendMSG("INFO","!ESP sweeping...");
    digitalWrite(D0,LOW);  //turns LED on
    AD5933_sweep(freq,finc,npts,mode,pga,Ival,Qval);
    digitalWrite(D0,HIGH);  //turns LED on
    sendMSG("INFO","!ESP: AD5933 Sweep complete");
    File f=SPIFFS.open("/calib.dat","r");
    for (int k=0;k<npts+1;k++) {
      float mag=sqrt(Ival[k]*Ival[k]+Qval[k]*Qval[k]);
      float phase=atan2(Qval[k],Ival[k])*180/3.1415926;
      f.readStringUntil('\n').toCharArray(line,80);
      sscanf(line," %g %g %g %g",&re,&im,&ab,&ph);
      samples[0]=Rcal*(ab/mag-1.0)*512.0;
      // samples[0]=128+128*log10(ab/mag);
      samples[1]=256+25.6*(phase-ph);
      samples[2]=0;
      send_samples(samples); send_samples(samples);
      Serial.print(ab/mag); Serial.print("\t"); Serial.print(phase-ph);
      Serial.print("\t"); Serial.println(samples[0]);
    }
    f.close();
    mmode=-1;
```

Finally, mmode==100 produces a SPIFFS file listing on the Serial line, and mmode==101 removes the file data file. With mmode=202 we cause the NodeMCU to read the data file, one line at a time, and dispatch the values to the browser. In this way all recorded samples are shown on the browser.

```
} else if (mmode==100) {    // Directory listing
  mmode=-1;
  listdir();
} else if (mmode==101) {    // Remove data file
  mmode=-1;
  SPIFFS.remove("/data.txt");
  sendMSG("INFO","ESP: Removing file /data.txt");
} else if (mmode==202) {                // replay data from local file
  Serial.println("Replaying data");
  File f=SPIFFS.open("/data.txt","r");    // open for read
  if (!f) {
    Serial.println("Cannot open file /data.txt");
    sendMSG("INFO","ESP: cannot open file /data.txt");
  } else {
    Serial.print("File size = "); Serial.println(f.size());
    char msg[80]; sprintf(msg,"ESP: File size of /data.txt = %d",f.size());
    sendMSG("INFO",msg);
    char line[80];
    samples[2]=0;
    while (f.position()<f.size()) {
      f.readStringUntil('\n').toCharArray(line,80);
      sscanf(line," %d %d",&samples[0],&samples[1]);
      send_samples(samples);
    }
  }
  f.close();
  mmode=-1;
}
yield();
}
```

In order to control the NodeMCU from the browser, we need to add features to the html file esp32-daq.html from Chapter 13. In the updated version medisense.html, we add the functionality to send JSON-formatted messages of type MMODE that set the variable mmode on the controller. This is accomplished by adding the following code snippet between the definition of the clear button and the display of the IP address. It defines a selection menu that sends different values of mmode to the NodeMCU with the JavaScript function sendSpecialCommand().

```
        :
<button id="clear" type="button" onclick="cleardisplay();">
    Clear</button>
<SELECT onchange="sendSpecialCommand(this.value);">
  <OPTION value="0">ECG live</OPTION>
  <OPTION value="3">Oximeter IR live</OPTION>
  <OPTION value="4">Oximeter Red live</OPTION>
  <OPTION value="5">Oximeter SP02 live</OPTION>
  <OPTION value="6">AD5933 Calibrate</OPTION>
  <OPTION value="7">AD5933 Frequency sweep</OPTION>
  <OPTION value="8">AD5933 at 50kHz</OPTION>
  <OPTION value="100">List directory</OPTION>
  <OPTION value="101">Remove data file</OPTION>
```

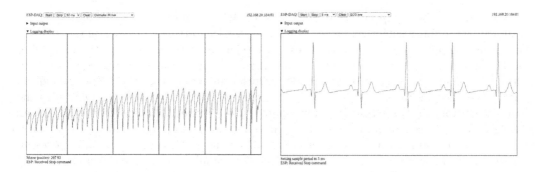

Figure 15.2 The raw signal of the IR diode from the pulse oximeter (left) and an electrocardiogram (right) with the pulses generated by the simulator.

```
    <OPTION value="200">Recording ON</OPTION>
    <OPTION value="201">Recording OFF</OPTION>
    <OPTION value="202">Replay recording</OPTION>
    <OPTION value="300">Start ECG output</OPTION>
    <OPTION value="301">Stop ECG output</OPTION>
  </SELECT>
  <A id='ip'>IP address</A>
                      :
```

At the end of <SCRIPT> in medisense.html we define the function sendSpecialCommand() with the following command that resembles the callback function for the menu to select the time.

```
function sendSpecialCommand(v) {
  toStatus("Special Command " + v + " ");
  websock.send(JSON.stringify({ "cmd" : "MMODE", "val" : v }));
}
```

Finally, as a convenience, we provide a JavaScript function that displays the pixel values where the mouse is clicked on the displayarea that shows the traces of measurement values. We use the method addEventListener() to register the callback function showCoordinates(), which displays the mouse coordinates in the status line below displayarea. Note that we use the lower-left corner as origin of the coordinate system.

```
document.getElementById("displayarea")
        .addEventListener('mousedown',showCoordinates, false);
function showCoordinates(event) {
  rect=document.getElementById("displayarea").getBoundingClientRect();
  toStatus("Mouse position: " + (event.clientX-rect.left) + " "
        + (rect.height-event.clientY+rect.top));
}
```

The rest of medisense.html is a straight copy of esp32-daq.html from Chapter 13. Please inspect the complete file listing that is available from this book's github site. Once the code is ready, we upload the sketch to the NodeMCU, place medisense.html into the data subdirectory of the sketch, and upload it to the NodeMCU with the command *ESP8266 Sketch Data Upload* that is available under the *Tools* menu of the Arduino IDE. Please refer to Chapter 8 about installing the uploader.

Once the system is up and running, the NodeMCU connects to the `messnetz` WLAN spanned by the Raspi such that we access its web page by starting a browser on the Raspi or any other computer connected to `messnetz`. We then direct the browser to the IP number of the NodeMCU and receive the `medisense.html` page in return. After placing a finger on the pulse oximeter, we select `Oximeter SPO2` which causes the value of SPO2 and the heart rate to be shown on the status line below the display area every few seconds. Observing the raw signals from the phototransistor illuminated by the infrared diode, we select `Oximeter IR live` from the selection menu on the right and then press the Start button. This starts the acquisition and after a few seconds during which the systems determines the pedestal, we see the signal appearing in the `displayarea`. It is similar to the one shown on the left-hand side in Figure 15.2. Recording the signals is accomplished by selecting `Recording ON` before selecting `Oximeter IR live` and pressing `Start`. To stop recording, we first `Stop` the acquisition before selecting `Recording OFF` and watch it with `Replay recording`. Note that we minimized the section labeled `Input output` on the webpage by clicking on the little triangle next to the text.

After connecting the three electrodes on the cable from the AD5933 to the corresponding wires (RL, RA, and LA) from the simulator, selecting `ECG live`, and pressing the `Start` button, the ECG shown on the right-hand side in Figure 15.2 appears on the display. We clearly see the narrow spike that signals the start of the heart's contraction and two smaller pulses, one on its left and one on its right. They indicate different stages of the complex process that a heart performs more than a billion times in a lifetime. Analyzing the traces further is beyond the scope of this book and requires special medical training.

In order to illustrate measurements of bioimpedances we simulate the impedance of the human body – typically between $500\,\Omega$ and $1\,\mathrm{k\Omega}$ – with an electronic analog, called the *Cole model*, shown on the bottom left in Figure 15.1 and, in more detail, in Figure 15.3. It consists of a $680\,\Omega$ resistor, which simulates the resistance of the extra-cellular fluids in the body. Connected in parallel, a $1\,\mathrm{k\Omega}$ resistor and a $2.2\,\mathrm{nF}$ capacitor simulates the resistance of the intra-cellular fluids and the capacitance of cell membranes, respectively. In order to limit the current though the "body", we place a $10\,\mathrm{k\Omega}$ resistor in series with the output from pin 1 of the MCP6002 op-amp. Before measuring the impedance, we calibrate the system by shorting the Cole circuit, such that only the $10\,\mathrm{k\Omega}$ resistor connects the output of the op-amp to the input on pin 5 of the AD5933. Then we reconnect the Cole circuit as shown in Figure 15.1 and sweep the AD5933, which results in the plot shown in Figure 15.4. The horizontal scale covers the frequency range between $10\,\mathrm{kHz}$ and $112.4\,\mathrm{kHz}$. The vertical scale covers zero to $1\,\mathrm{k\Omega}$. We see that the absolute value starts close to $680\,\Omega$ – click at the point and convert the pixel value back to Ohm (Ω) – at low frequencies and drops to $460\,\Omega$ at high frequencies, which is consistent with simple circuit analysis. The second trace

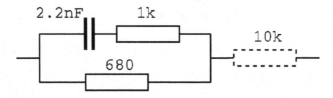

Figure 15.3 The schematics of the Cole model with the $680\,\Omega$ resistor modeling the resistance of the extra-cellur fluids, the $2.2\,\mathrm{nF}$ capacitor modeling the impedance of the cell membranes, and the $1\,\mathrm{k\Omega}$ resistor modeling the resistance of intra-cellular fluids. The dashed resistor limits the current through the "body."

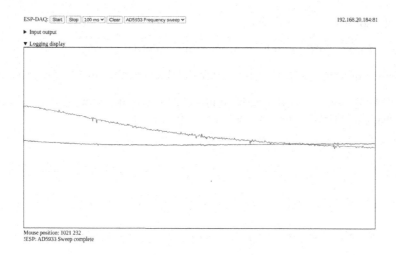

Figure 15.4 The measured impedance of the Cole model from Figure 15.3.

shows the phase, which hardly changes, because it is dominated by the large real part of the $10\,\mathrm{k\Omega}$ current-limiting resistor. The front end with an op-amp wired as a line buffer limits the resolution, which can be improved by using separate connections to feed the current to the unknown impedance and for measuring the voltage drop. A circuit with a so-called *Howland current source* and a separate instrumentation amplifier, as discussed for example in [32], will help.

After this excursion into medical matters, let us get back to physics and measure the width of a laser beam.

QUESTIONS AND PROJECT IDEAS

1. Find out how features of electrocardiograms are related to actions of the heart.

2. Place colored but mostly transparent films between your finger and the MAX30102 pulse oximeter. Observe how it affects the raw signals and the SPO2 value.

3. Use basic circuit theory to analyze the Cole model from Figure 15.3.

4. Find out the differences and benefits of a four-terminal impedance measurement over a two-terminal impedance measurement using only two wires like we do in this chapter.

5. Consult [32] and build a better analog front end for the AD5933.

6. In order to reduce noise in the raw calibration data, fit a quadratic polynomial to the calibration data (see Appendix B) and later use it like we did in the octave script in Chapter 12, instead of using the raw data points at every frequency.

7. Add an MLX90614 thermometer to the circuit in order to make contact-free measurements of the body's temperature.

8. Measure the impedance between the ends of a banana. Experiment with different types of electrodes.

9. Determine the impedance of a piece of wood. How much does it change when the wood gets wet?

10. Port the software to the ESP32.

11. Connect the AD5933 to an ESP-01 controller in order to make it the system portable.

12. Likewise, connect the MAX30102 and the MLX90614 to an ESP-01 controller.

13. The ESP-01 does not expose the built-in ADC to an external pin. Discuss how to interface the AD8232 anyway and then build the circuit.

## CHAPTER 16

Example: Profile of a Laser Beam

In this example we measure the transverse profile of the laser beam from a laser pointer by carefully moving an obstacle across the beam and observing the change of the signal from a photoresistor. Figure 16.1 illustrates the method. We use a laser module from a sensor kit that is mounted on a breadboard and exposes two wires, one for ground and one for the positive supply voltage. In order to adjust the intensity of the laser, we use pulse-width modulation of the positive supply voltage. The sensor is a voltage divider consisting of a light sensitive resistor (LDR) and a $10\,\mathrm{k\Omega}$ resistor. The obstacle is made of a piece of black plastic. As translation stage, we use the frame salvaged from an old CD-ROM drive, which moves a small wagon with a stepper motor back and forth over a distance of about $40\,\mathrm{mm}$, which is approximately the width of the readable area of a CD. The stepper motor drives a spindle, and that pulls and pushes the wagon with good precision.

Figure 16.2 shows the salvaged frame from the CD drive. The frame has plenty of holes, which are convenient to add screws that in turn are used to attach laser, sensor, and obstacle. The laser on its small breakout board is located at the top right with the stepper motor beneath and hidden from view, but the spindle is clearly visible, running right to left near the top of the frame. There it engages the white plastic part that is attached to the wagon and moves it, when the motor turns. On the wagon there is a black piece of plastic visible that intercepts the laser beam and prevents the laser from hitting the LDR on the lower right on the frame. The LDR is also mounted on a small breakout board that is part of a sensor kit. We use it because the mounting holes make assembly and attaching the LDR to the screws on the frame easy. As controller we use an Arduino UNO that drives the motor

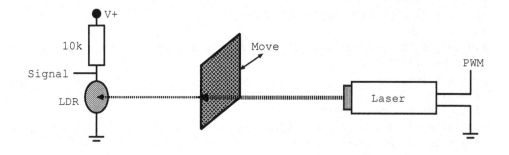

Figure 16.1 The schematic setup to measure the beam size of a laser pointer.

DOI: 10.1201/9781003341703-16

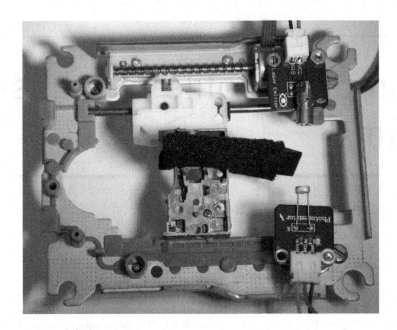

Figure 16.2 The chassis from the CD-ROM drive with the laser mounted on the top right and the photoresistor on the bottom right. The black obstacle is mounted on the carriage that can be moved via the spindle on the top by a small stepper motor that is located below the laser.

with an L293D H-bridge driver, uses pulse-width modulation on pin D9 to control the laser power, and reads the signal from the voltage divider with the LDR on analog input A0. Figure 16.3 shows the schematics. We use an external power supply to provide the voltage for the stepper motor that is connected to the terminals labeled PA,...,PD.

The next task is to program the UNO to control the power of the laser, move the stepper motor, and read the sensor, all in a well-orchestrated fashion. We base the sketch on the program to drive a stepper motor with an H bridge from Section 4.5.4 and augment the code with sections for the laser and the sensor, as shown below.

```
// Laser profile measurement, V. Ziemann, 170628
char line[30];
int settle_time=2, stepcounter=0;
int laser_power=10;
const int PA=2,PB=3,PC=4,PD=5,ENABLE=6;
const int LASER=9;
void set_coils_fullstep(int istep) { //.........set_coils
  bool patA[]={1,1,0,0};
  int pat_length=4;
  int ii;
  istep=istep % pat_length;
  if (istep < 0) istep+=pat_length;
  digitalWrite(PA,patA[istep]);
  ii=(istep+2) % pat_length;
  digitalWrite(PB,patA[ii]);
  ii=(istep+3) % pat_length;
```

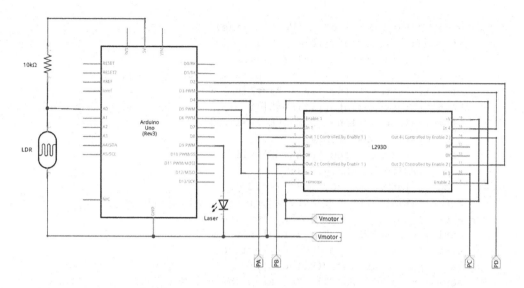

Figure 16.3 The schematics of the circuit. The stepper motor of the frame is connected to points labeled PA,...,PD[†].

```
    digitalWrite(PC,patA[ii]);
    ii=(istep+1) % pat_length;
    digitalWrite(PD,patA[ii]);
    delay(settle_time);
}
void setup() {  //..............................setup
    Serial.begin (9600);
    while (!Serial) {;}
    pinMode(PA,OUTPUT);
    pinMode(PB,OUTPUT);
    pinMode(PC,OUTPUT);
    pinMode(PD,OUTPUT);
    pinMode(ENABLE,OUTPUT);
    digitalWrite(ENABLE,HIGH);
    analogWrite(LASER,laser_power);
}
void loop() {  //...............................loop
    if (Serial.available()) {
      Serial.readStringUntil('\n').toCharArray(line,30);
      if (strstr(line,"FMOVE ")==line) {
        int steps=(int)atof(&line[6]);
        digitalWrite(ENABLE,HIGH);
        if (steps > 0) {
          for (int i=0;i<steps;i++) set_coils_fullstep(stepcounter++);
        } else {
          for (int i=0;i<abs(steps);i++) set_coils_fullstep(stepcounter--);
        }
      } else if (strstr(line,"STEPS?")==line) {
        Serial.print("STEPS "); Serial.println(stepcounter);
```

```
    } else if (strstr(line,"STEPS ")==line) {
      stepcounter=(int)atof(&line[6]);
    } else if (strstr(line,"WAIT?")==line) {
      Serial.print("WAIT "); Serial.println(settle_time);
    } else if (strstr(line,"WAIT ")==line) {
      settle_time=(int)atof(&line[5]);
    } else if (strstr(line,"DISABLE")==line) {
      digitalWrite(ENABLE,LOW);
    } else if (strstr(line,"ENABLE")==line) {
      digitalWrite(ENABLE,HIGH);
    } else if (strstr(line,"LDR?")==line) {
      Serial.print("LDR "); Serial.println(analogRead(A0));
    } else if (strstr(line,"POWER?")==line) {
      Serial.print("POWER "); Serial.println(laser_power);
    } else if (strstr(line,"POWER ")==line) {
      laser_power=(int)atof(&line[6]);
      Serial.print("POWER "); Serial.println(laser_power);
      analogWrite(LASER,laser_power);
    } else if (strstr(line,"FSCAN")==line) {
      int steps=(int)atof(&line[6]);
      digitalWrite(ENABLE,HIGH);
      delay(100);
      for (int i=0;i<abs(steps);i++) set_coils_fullstep(stepcounter++);
      delay(100);
      for (int i=0;i<abs(steps);i++) {
        set_coils_fullstep(stepcounter--);
        delay(50);
        unsigned long sum=0;
        for (int k=1; k<10; k++) {sum+=analogRead(A0); delay(10);}
        Serial.println(sum);
      }
      digitalWrite(ENABLE,LOW);
    } else if (strstr(line,"CALIBRATE")==line) {
      for (int power=0; power<256; power++) {
        analogWrite(LASER,power);
        delay(100);
        unsigned long sum=0;
        for (int k=1; k<10; k++) {sum+=analogRead(A0); delay(10);}
        Serial.println(sum);
      }
      analogWrite(LASER,10);
    } else {
      Serial.println("unknown");
    }
  }
}
```

At the top of the sketch, a number of variables are declared for the time to wait between stepper motor steps and the initial setting of the pulse-width modulation for the laser_power. Then the pins used for the stepper motor and the LASER are declared before defining the set_coils_fullstep() function, which is identical to the full-step version we used earlier in Section 4.5.4. We use full-step mode because the step resolution is sufficient, and in half-step mode there is a small visible disturbance on the sensor from the different currents drawn on alternate steps in half-step mode, where exciting one coil and two coils take turns. In the `setup()` function, we initialize the serial communication, declare the mode of the pins used for the motor, and initialize the laser power to be 10 (of up to 255). The `loop()` function is built in much the same way as before. It expects to receive single-line commands on the serial line. As in Section 4.5.4, the `FMOVE` command moves the motor by the specified number of steps. Then there are commands to set and return the `STEPS` and the time to `WAIT` or `ENABLE` the motor driver. The command `LDR?` reads the sensor value from analog pin A0, and the `POWER` command sets and reads the laser excitation. Finally we encounter the command `FSCAN nnn` where `nnn` is the scan range retrieved by reading the rest of `line` with the `atof()` function we used before. Here we assume that the scanner actually blocks the laser, and first needs to be retracted before collecting data. Inside the case block it first reports to the serial line what it is about to do, enables the motor driver, and moves the obstacle out of the way. Then comes a `for` loop, where the motor performs individual steps and within each step, takes ten measurements of the LDR on analog pin A0 and adds them up. At the end of each step the measurement value is written to the serial line. Once the number of `steps` is complete, the motor driver is disabled. The last command implemented is `CALIBRATE`, which measures the sensor response as a function of the laser power. The latter we may later use to convert the raw sensor readings to a linear intensity scale should the need arise.

As a first test of the system, we open the *Serial Monitor* in the Arduino IDE, make sure that the baud rate is set to the one specified in the `setup()` function, 9600 baud in our case, and query the LDR by sending `LDR?` to the UNO. The response has the format `LDR nnn`, where `nnn` is the raw value from the call to the `analogRead()` function. Then we query the laser power with `POWER?` and set it to a new value with `POWER nnn`, where `nnn` is a value between 0 and 255. The brightness of the laser spot should change accordingly. Next we move the stepper motor a few steps, for example, with `FMOVE 20`, and move it back to the starting position with `FMOVE -20`. The wagon with the obstacle should move back and forth. As a last point, we execute `FSCAN 60`, which will move the obstacle by 60 steps and then returns to the starting position step by step, while reading the sensor and reporting the value to the serial line. If all these initial tests complete satisfactorily, we progress to automatizing the process, and for that we chose octave.

Basically, we hook up the Arduino UNO to the Raspi, write an octave script to open a serial line, send commands to the UNO, and receive the response, very similar to what we did in previous examples. Before writing the octave script, we need to calibrate the motion of the obstacle in order to be able to show the width in mm instead of steps. In moving the wagon a large distance, I use 250 full steps, and, taking a photo of the setup with a ruler lying next to it before and after moving it, is easy to calculate the calibration constant `xscale` in mm/step. In my case, 250 steps moved the wagon by 39.2 mm, which explains the value for `xscale` near the top of the following script.

```
% scanplot2.m, V. Ziemann, 221103
close all; clear all
xscale=0.161;  % mm/fullstep
s=serialport('/dev/ttyACM0',9600);  % set device correctly
pause(3);
flush(s);        % flush the input queue
nsteps=60;
write(s,"FSCAN 60\n");
pause(5);
```

```
xx=zeros(1,nsteps); yy=xx;
for i=1:nsteps
   xx(i)=i*xscale;
   yy(i)=str2double(serialReadline(s));
end
clear s
yy=smooth3(yy);           % smooth data
subplot(2,1,1);
plot(xx,yy);              % raw sensor data
ylabel('arb. units');
subplot(2,1,2);
dy=yy(2:end)-yy(1:end-1);   % derivative
if (yy(1) > yy(end)) dy=-dy; end
plot(xx(1:end-1),dy);
xlabel('x [mm]'); ylabel('arb. units');
title(['FWHM = ', num2str(xscale*fwhm(dy),"%5.2f"), ' mm'])
print('laser_profile.png','-S1000,700')
```

The rest of the script should be familiar by now. After opening the serial line and flushing whatever characters are present, we define the number of steps and write FSCAN 60 to the UNO. Then we wait a short time and start retrieving the values from the serial line with the serialReadline() function we already encountered in Chapter 12. It reads characters from the serial line until the termination character is encountered, and returns the obtained characters as a character string. But back to the main script. While retrieving data and copying it to the array yy, we also fill the array xx with the properly scaled values for the horizontal axis. Once the loop completes, we close the serial device with clear s.

After all data are available in the script, we start processing it. Since we will calculate the derivative of the raw data later, which is a process that is very sensitive to noise, we weakly smooth the sensor values in array yy with the smooth() function. It averages three consecutive values and replaces the central value with the average. The octave script for smooth3() is

```
% smooth three consecutive points, V. Ziemann, 170628
function out=smooth3(y);
y=[y(1),y(1:end),y(end)];   % add extremities
f=ones(1,3)/3.0;            % filter function
out=conv(y,f);              % convolute
out=out(3:end-2);           % ensure same length
```

It first creates a new array with the extreme points doubled up to avoid ugly artefacts at the end points. In the next line we create the filter function f that consists of three values 1/3 and use it to convolute the data with the filter. Since the convolution creates an output array that is longer than the original, we remove the extreme points before returning the smoothed array as the value out. Once these initial preparations are complete, we show the smoothed raw values on the upper plot in Figure 16.4 and add a label to the vertical axis. Then we calculate the difference between consecutive values, which is motivated by the fact that in each step a little of the laser beam is obscured and the *change* in remaining intensity is proportional to the intensity in the small band that the obstacle transverses during that step. The result is a signal that is proportional to the transverse intensity profile of the laser beam shown in the bottom plot in Figure 16.4. Its title shows the full width at half maximum (FWHM) of the laser profile, which is about 0.8 mm. Finally, we produce an image file, which is precisely how we prepared Figure 16.4.

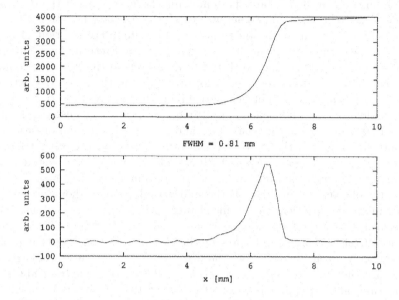

Figure 16.4 The raw sensor value as a function of the position of the obstacle and the derived laser beam profile, which shows a moderate asymmetry.

In this example, we use the FWHM rather than the standard deviation of the profile, because the FWHM is a rather robust measure and works more reliably with profiles that are non-Gaussian, asymmetric, and have tails that are moderately populated. The standard deviation is heavily biassed by the latter. The script for the FWHM is straightforward and is reproduced here.

```
% FWHM, V. Ziemann, 170628
function fwhm=fwhm(data)
N=length(data);
xmax = -1e30;
xmin=min(data);
imax=-1;
for i=1:N
  if (data(i) > xmax)
    xmax=data(i);
    imax=i;
  end
end
ileft=imax;
while (data(ileft) > (xmax+xmin)/2 && ileft>1)
  ileft=ileft-1;
end
iright=imax;
while (data(iright) > (xmax+xmin)/2 && iright<N-1)
  iright=iright+1;
end
fwhm=iright-ileft-1;
```

In the `fwhm()` function we first locate the minimum and maximum of the data points as well as the location of the maximum. Then we search toward the left until the value is less than halfway between maximum and minimum. This results in the location `ileft` of that point. Then we repeat the process on the right-hand side and return the difference of the right and left halfway points as the FWHM. The minus one in the last equation accounts for the fact that both searches overshoot before terminating the respective `while` loops. This is partially compensated by subtracting one from the final result.

In this example, we built a laser-beam profile monitor from scratch using a small stepper motor and an LDR to illustrate the basic steps and how to combine sensors and actuators to built an integrated system that is controlled from `octave` or any other language that communicates over serial devices, such as EPICS, LabView, or C. The system is far from perfect and many limitations and improvements come to mind. For example: Is diffraction on the edge of the obstacle a limitation? Is the step size adequate or should we use a better, microstepping motor-driver? Does the profile change with different laser intensities? Do we need to correct for intensity-related nonlinearities of the sensor system? This is actually the purpose of the `CALIBRATE` command to provide the base information. Does the profile show the same asymmetry when we scan from the other side? We may also want to know the vertical beam size and any cross-plane correlations to determine some basic aberrations. Is the profile the same after swapping position of LDR and $10\,\text{k}\Omega$ resistor? We might want to add a second motor moving the obstacle longitudinally in the direction of the laser beam. Adding a lens and measuring the beam profile at several places allows us to determine the M^2 of the laser beam, which is a measure of the laser quality and is unity for a diffraction-limited beam. But these exercises we leave for the interested reader to do at home, while we move on to build a robot that detects fire.

QUESTIONS AND PROJECT IDEAS

1. Use the frame from a salvaged CD-ROM drive to move a spoon back and forth to stir your cup of coffee or tea.

## CHAPTER 17

Example: Fire-Seeking Robot

In this final example, we discuss a small robot that detects heat radiation from a fire, moves towards the heat source, and starts beeping once it reaches the source of the radiation. This is, of course, a very simple prototype of an autonomous fire extinguisher. We require the robot to either move autonomously, or to be manually controlled by a remote controller. Despite the simple description, the project is rather ambitious and requires us to solve the following subtasks:

1. detect the heat source and its location;

2. control the speed and direction of motors;

3. detect collisions, preferably before they happen;

4. sense joysticks and switches on the remote control;

5. send messages between remote controller and robot.

We base the hardware of the robot on a ready-made chassis with two DC motors and an on-board battery pack with five AA cells providing up to 7.5 V. Onto that chassis we mount a full-size breadboard to house a NodeMCU controller, an H-bridge motor-driver, and some linear voltage converters to provide 5 and 3.3 V. To provide additional IO-pins we add a second Arduino that acts as a slave to the NodeMCU. On the robot chassis is a dedicated place for a model-servo onto which we mount a very small breadboard for the BPX38 IR phototransistors we use as flame sensors, and the HC-SR04 distance sensor. Figure 17.1 shows the chassis with the servo and the two breadboards already mounted. Mounting the sensors on the movable breadboard makes it possible to point them in different directions by turning the model-servo. We configure the NodeMCU as an access point that spans its own WLAN network. We also start a server to receive commands from the remote control. Apart from listening to the remote on the controller, we implement a simple state machine, to which we can hand the control of the robot, which subsequently runs autonomously.

The remote control is based on a second NodeMCU. It connects to the WLAN spanned by the first NodeMCU on the robot and to the server running on it. On the remote controller we connect a MCP3208 8-channel ADC via SPI to the NodeMCU, and set the input voltages of the ADC channels with two joysticks, two potentiometers, several voltage-dividers, and two switches. The ADC readings are then interpreted as motor speed and model-servo setting, and transmitted to the robot. The voltage on one ADC channel is set by a number of switches and voltage dividers, which allows us to use a larger number of switches than

DOI: 10.1201/9781003341703-17

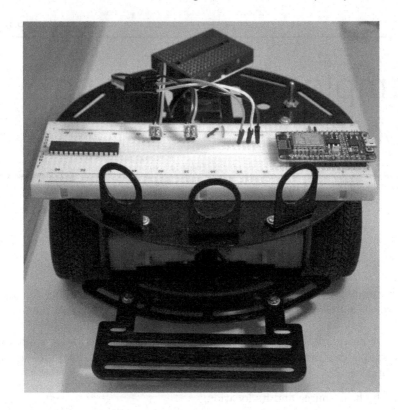

Figure 17.1 The chassis of the robot with two breadboards. The smaller one can be turned by operating a model-servo. Mounted on the larger breadboard are a NodeMCU on the right and a second chip on the left, a bare ATmega328 that was initially tested and later replaced by an Arduino NANO.

available digital IO pins, albeit at the expense of being able to sense only one button at a time.

We show a simplified version of the remote controller with less than all possible channels set up on a breadboard in Figure 17.2, where we see the NodeMCU on the right and the MCP3208 ADC on the left. The MCP3208 has the same pin assignment as the MCP3304 from Section 4.4.4, and is connected to the SPI port of the NodeMCU with CLK, MISO, MOSI, and CS lines, connected to D5, D6, D7, and D8, respectively. A joystick, visible immediately to the left of the breadboard, is connected to the power rails and the wiper to ADC channels A0 and A1. Likewise, the potentiometer, visible on the far left, is connected to ADC channel A5. Two buttons, visible near the bottom of Figure 17.2, connect voltage dividers consisting of a $10\,\mathrm{k\Omega}$ resistor to the positive supply voltage, as well as $68\,\mathrm{k\Omega}$ and $47\,\mathrm{k\Omega}$ resistors. By using additional resistors to ground with different values, it is possible to detect more buttons – one at a time. Pressing the buttons supplies a definite voltage level to ADC channel A7. Finally, there is a switch connected to IO pin D4. To clarify the wiring, we also show the schematic corresponding to the breadboard layout in Figure 17.3.

The sketch running on the remote controller that uses the hardware described in the previous paragraph is the following. Since the sketch is fairly long, we show and describe the preparatory sections first.

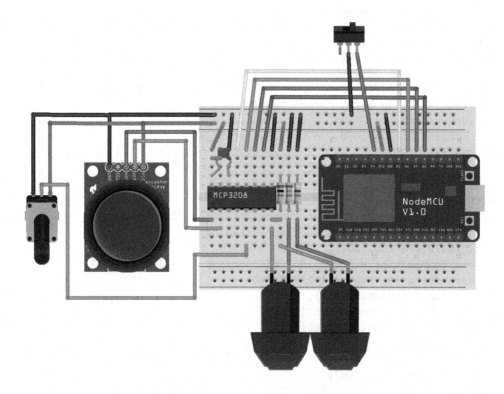

Figure 17.2 Simplified setup of the remote controller on a breadboard[†].

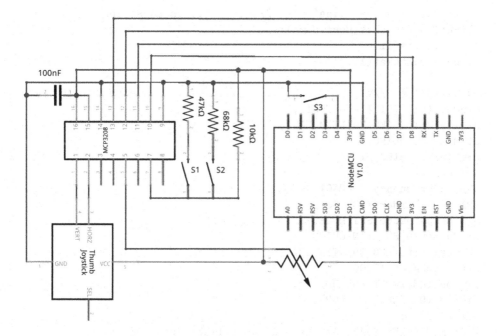

Figure 17.3 The schematic of the remote controller[†].

```
// RCsenderUDP, V. Ziemann, 170705
const char* ssid    = "FireBot";
const char* password = "..........";
const char* host = "192.168.4.1";
const int port=1137;
#include <ESP8266WiFi.h>
#include <WiFiUdp.h>
WiFiUDP client;
int adclast[8]={0,0,0,0,0,0,0,0},D4last=HIGH;
int adccalib[8];
//.................................................ADC
#include <SPI.h>
#define CS 15
int mcp3208_read_adc(uint8_t channel) { // 8 single ended
  int adcvalue=0, b1=0, hi=0, lo=0, reading;
  digitalWrite (CS, LOW);
  byte commandbits = B00001100; // Startbit+(single ended=1)
  commandbits |= ((channel>>1) & 0x03);
  SPI.transfer(commandbits);
  commandbits=(channel & 0x01) << 7;
  b1 = SPI.transfer(commandbits);
  hi = b1 & B00011111;
  lo = SPI.transfer(0x00);      // input is don't care
  digitalWrite(CS, HIGH);
  reading = (hi << 7) + (lo >> 1);
  return reading;
}
void send_string(char line[]) { //............send_string
  client.beginPacket(host,port);
  client.write(line);
  client.endPacket();
}
void setup() { //...............................setup
  pinMode(LED_BUILTIN,OUTPUT);
  digitalWrite(LED_BUILTIN,LOW);
  pinMode(CS,OUTPUT);
  digitalWrite(CS,HIGH);
  SPI.begin();
  SPI.setFrequency(100000);
  SPI.setBitOrder(MSBFIRST);
  SPI.setDataMode(SPI_MODE0);
  pinMode(LED_BUILTIN,OUTPUT);
  digitalWrite(LED_BUILTIN,LOW);
  Serial.begin(115200);
  pinMode(D3,INPUT_PULLUP);
  pinMode(D4,INPUT_PULLUP);
  delay(1000);
  WiFi.mode(WIFI_STA);  // needed for reliable communication
  WiFi.begin(ssid,password);
  while (WiFi.status() != WL_CONNECTED) {
    Serial.print("."); delay(500);
```

```
  }
  Serial.print("\nConnected to ");  Serial.print(ssid);
  Serial.print(" with IP address: ");
  Serial.println(WiFi.localIP());
  client.begin(port);
  for (int k=0;k<8;k++) {
    adccalib[k]=mcp3208_read_adc(k);
  }
  digitalWrite(LED_BUILTIN,HIGH);
}
```

At the top of the sketch, the WLAN name and passphrase are defined, as well as the robot's IP number, which defaults to 192.168.4.1 for the WLAN spanned by the robot. Then we include WiFi libraries and declare the **client**, which is the socket that connects to the server. Note that we use a UDP connection instead of TCP, because the latter uses extensive handshaking between server and client and this makes the response of the robot to remote commands very sluggish. UDP connections do not use handshaking, which makes the connection less reliable, but much more responsive. Since we continuously send commands to the robot, losing a single command is not very serious. Next we declare a number of variables that we discuss later, and define the **mcp3208_read_adc()** function to read a single channel from the ADC. It is very similar to the one we discussed in Section 4.4.4, except that we request single-ended ADC measurements, and we receive 12 data bits instead of 13. The function receives the requested channel as input parameter and returns the 12-bit ADC reading. The **send_string()** function encapsulates the construction and sending of UDP packages. In the **setup()** function, we configure a number of IO pins, and initialize the SPI communication and the serial line. Then we define the **WiFi.mode** to be **WIFI_STA** or a client, rather than an access point, the latter otherwise being the default, and connect to the WLAN. Once connected, we write the received IP number to the serial line and connect to the server on the robot with the call to the **client.begin()** function. Finally, all ADC channels are read and the values saved in the array **adccalib[]**, which is used to calibrate the center positions of the joysticks.

After the preparatory sections, we are ready to show and discuss the main part of the program, the **loop()** function.

```
void loop() { //................................loop
  char line[30];
  int adc[8];
  for (int k=0;k<8;k++) {adc[k]=mcp3208_read_adc(k);}
  if ((abs(adc[0]-adclast[0]) > 16) || (abs(adc[1]-adclast[1]) > 16)) {
    adclast[0]=adc[0]; adclast[1]=adc[1];
    int val0=(adc[0]-adccalib[0])*1023.0/2048.0;
    int val1=(adc[1]-adccalib[1])*1023.0/2048.0;
    sprintf(line,"RSPEED %d",(int)(val0+0.5*val1)); send_string(line);
    sprintf(line,"LSPEED %d",(int)(val0-0.5*val1)); send_string(line);
  }
  if (abs(adc[5]-adclast[5]) > 16) {
    adclast[5]=adc[5];
    int val=adc[5]*180.0/4095;
    sprintf(line,"SERVO %d",val); send_string(line);
  }
  if (abs(adc[7]-adclast[7]) > 5) {  //  check the buttons
    adclast[7]=adc[7];
```

```
      if (adc[7] < 1000) {
        Serial.println("Red button right pressed");
        sprintf(line,"FINDFIRE 1"); send_string(line);
      } else if (adc[7] < 1700) {
        Serial.println("Blue button right pressed");
        sprintf(line,"RANGE?"); send_string(line);
      } else if (adc[7] < 2250) {
        Serial.println("Joystick button right pressed");
        sprintf(line,"NEXTEVENT 0"); send_string(line);
      } else if (adc[7] < 3580) {
        Serial.println("Joystick button left pressed");
        sprintf(line,"NEXTEVENT 1"); send_string(line);
      } else if (adc[7] < 3660) {
        Serial.println("Blue button left pressed");
        sprintf(line,"BEEP 1000"); send_string(line);
        Serial.println(line);
      } else if (adc[7] < 3750) {
        Serial.println("Red button left pressed");
        sprintf(line,"NEXTEVENT 3"); send_string(line);
      }
    }
    if (digitalRead(D4) != D4last) {
      D4last=digitalRead(D4);
      sprintf(line,"DO %d",D4last); send_string(line);
    }
    yield();
    int packetsize=client.parsePacket();
    if (packetsize) {
      char line[30];
      int len=client.read(line,30); line[len]='\0';
      Serial.print("Message:"); Serial.println(line);
    }
    if (Serial.available()) {
      Serial.readStringUntil('\n').toCharArray(line,30);
      send_string(line);
    }
  }
```

In the loop() function, we first read all ADC channels and then test whether channels A0 and A1 have changed significantly with respect to the last time; the values are saved in the array adclast[]. This construction is useful, because it prevents continuous sending of commands, and only does that if a value has changed. If that is the case, the difference of the current ADC reading with respect to the calibration value is stored in variables val0 and val1. The first value is interpreted as the desired speed of the motors, and the second value as the direction. Thus we send commands to set val0+0.5*val1 as speed to one motor and val0-0.5*val1 to the other. Next the reading of ADC channel A5 is scaled to a value between 0 and 180 and transmitted to set the model-servo on the robot. ADC channel A7 is used to interpret the buttons pressed. In the current realization, six buttons are connected, and pressing them sends a number of different commands to the robot. Testing for ADC values progresses from smaller to larger values, such that the buttons are prioritized in a natural way. If a button connected to a smaller resistor is pressed, all buttons connected to higher-valued resistors are ignored. The details of what action the commands cause on the robot when a button is pressed we explain later, when we discuss the software running on the robot. The call to the `yield()` function ensures that all background processes for WLAN and serial communication can complete pending tasks. The call to the `client.parsePacket()` function checks whether a message from the robot has arrived. Here we only copy the received message to the serial line. But it is easy to envision other ways to handle this; for example, by showing them on an LCD display connected via a I2C interface to pins D1 and D2 on the NodeMCU controller. These pins are not used in the present circuit and are available for expansions. Finally, we check whether anything has arrived on the serial line and pass it on to the robot. The direct two-way communication with the robot is a very convenient way to debug the system, by sending commands from the serial console if the remote controller is connected via USB cable to the host computer.

And this brings us to the electronics on the robot. In Figure 17.4 we show the circuit diagram to implement the functionality discussed earlier in this chapter. Figure 17.5 shows the corresponding schematic. The central component is the NodeMCU microcontroller visible on the right of the solderless breadboard, with the L293D H-bridge motor driver and adjacent 7805 linear voltage regulator to its left. The voltage regulator receives power from the battery, whose negative pole is connected to system ground and the positive voltage to the input pin of the 7805. The positive battery voltage is also routed to the motor and logic power pins of the L293D motor driver. The 5 V output voltage of the 7805 voltage regulator is connected to the lower power rails from where it provides power to the Vin pin of the NodeMCU at its bottom right. The 5 V power is further routed to the model-servo and the HC-SR04 sonar sensor visible above the large breadboard. The 3.3 V voltage regulator on the NodeMCU provides power to the upper power rail, which carries 3.3 V and is routed to the respective circuits. The respective power rails have large electrolytic capacitors of $470\,\mu F$ connected, to buffer intermittent current requirements. Special care is needed to ensure that the three different voltages are routed correctly. Higher than permissible voltages can destroy some of the integrated circuits. The positive voltage from the battery only goes to the 7805 and L293D, and 5 V power the NodeMCU via its Vin pin, the servo, and the sonar. To facilitate the correct wiring, color images are made available on this book's github repository at `https://github.com/volkziem/HandsOnSensors2ed.git`.

Motor 1 is connected to the pins on the lower side of the L293D H-bridge motor driver, and the controlling input pins are connected to IO pins D3 and D4 on the NodeMCU. Likewise, motor 2 is connected to the upper half of the L293D, and the corresponding input pins are routed to pins D5 and D6 on the NodeMCU. The enable pins of the motor driver are permanently wired to 3.3 V. The model-servo is connected to the 5 V power rail, and its control wire connects to pin D7 on the NodeMCU, while the buzzer connects to pin D8. The HC-SR04 distance sonar is powered from the 5 V rail. It is triggered by a pulse from NodeMCU pin D1 and the echo is received on pin D2. Since the NodeMCU only operates on 3.3 V, we need to use a voltage divider made of a $10\,k\Omega$ and a $22\,k\Omega$ resistor to step the 5 V echo signal from the HC-SR04 to approximately 3.3 V. On the robot, the distance sensor is actually mounted on the small breadboard visible on the top left in Figure 17.4, which, in turn, is mounted on the movable axis of the model-servo that permits it to scan the surroundings for obstacles. Initially only the distance sensor and the middle BPX38 phototransistor with a pull-up resistor to the 3.3 V rail are mounted on the small breadboard to scan for both obstacles and sources of infrared radiation. The BPX38 is most sensitive around a photon wavelength of 880 nm, which is in the infrared part of the electromagnetic spectrum that is emitted as heat radiation by, for example, a fire or an old-fashioned light bulb. The emitter of the BPX38 is connected to ground, and the collector connects to the analog input pin A0 of the NodeMCU and via a $10\,k\Omega$ resistor to the 3.3 V power rail. In this configuration we need to continuously scan the IR and distance sensor with the servo

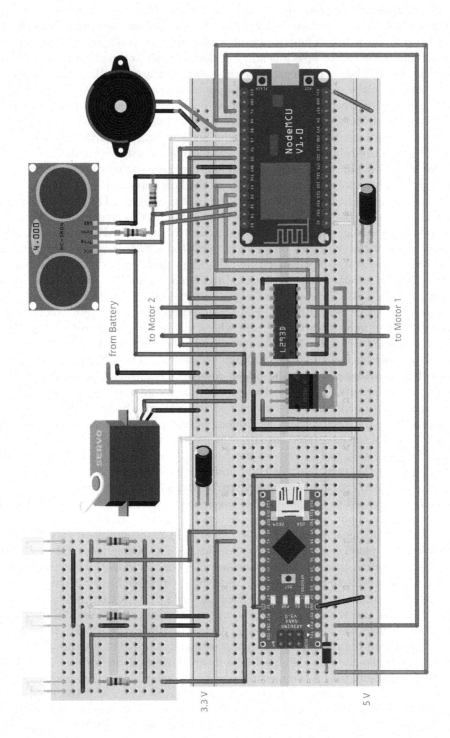

Figure 17.4 The electronics circuit of the robot (color version available online)[†].

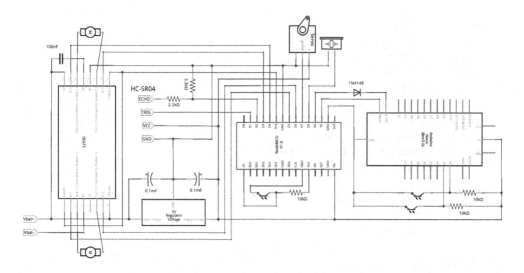

Figure 17.5 The schematic of the robot electronics[†].

in order to find the fire, which is a bit cumbersome. If we had two IR sensors to read out simultaneously, we could compare excitation and have instantaneous information about the location of the fire, without scanning with the servo. Therefore, we will add two extra IR phototransistors to the far left and far right on the small breadboard, wired in the same way as the middle one.

But we do not have enough analog input terminals on the NodeMCU, and have used almost all the IO pins. This significantly limits our ability add new functionality to the robot. We therefore add an Arduino-NANO to the large breadboard and program it to behave as a slave to the NodeMCU, and to communicate over the serial line. The wires that connect the respective TX and RX pins are crossed and behave similar to a null-modem cable. The NANO behaves almost like a UNO and is also programmed in the same way, by selecting *Arduino Nano* from the *Tools→Board* menu in the Arduino IDE. The only obviously visible difference is that the NANO has eight instead of six analog input pins, while there are also 13 digital IO pins. Since we power the NANO with 5 V supplied to the pin labeled "5 V", all IO pins are operating on 5 V logic levels. In order not to damage the NodeMCU that operates at 3.3 V, we use a reverse-biased diode in the connection from the TX pin of the NANO to the RX pin of the NodeMCU. The diode blocks the 5 V from reaching the NodeMCU, but if the TX is pulled low, the signal on the input pin of the NodeMCU is also pulled low. The prototype circuit works with a normal switching 1N4148 diode, but ideally one should use a Schottky diode, which has a smaller voltage drop. Once the essential communication between NodeMCU and the slave-NANO works, we connect the two additional phototransistors from the small breadboard to analog input pins A0 and A1 on the slave NANO. This completes the description of the hardware on the robot chassis, and we can turn to programming the NodeMCU.

The code that runs on the robot is the following, again split in two blocks because it is rather long. First we show and discuss the preparatory sections.

```
// RCreceiver, V. Ziemann, 170701
#define Max(a,b) ((a)>(b)?(a):(b))
#define Min(a,b) ((a)<(b)?(a):(b))
const char *ap_ssid = "FireBot";
```

```
const char *ap_password = "..........";
const int port=1137;
#include <ESP8266WiFi.h>
#include <WiFiUdp.h>
WiFiUDP server;
#include <Servo.h>
Servo myServo;
int servo_pos=90,servo_inc=5;
int right_sensor=-1,left_sensor=-1;
long lasttime=0,sleeptime=1000,nextevent=1;
void send_string(char line[]) {  //.......send_string
  server.beginPacket(server.remoteIP(),port);
  server.write(line);
  server.endPacket();
}
int range() { //...........................range
  digitalWrite(D1,LOW);
  delayMicroseconds(2);
  digitalWrite(D1,HIGH);
  delayMicroseconds(10);
  digitalWrite(D1,LOW);
  int val=(int)(0.017*pulseIn(D2,HIGH));
  if (val<10) tone(D8,1000,200);
  return val;
}
void motor_speed(int left, int right) {  //..........motor_speed
  left=Max(-1023,Min(1023,left));
  analogWrite(D3,0);
  analogWrite(D4,0);
  if (left<0) {analogWrite(D3,abs(left));}
    else {analogWrite(D4,abs(left));}
  right=Max(-1023,Min(1023,right));
  analogWrite(D5,0);
  analogWrite(D6,0);
  if (right<0) {analogWrite(D5,abs(right));}
    else {analogWrite(D6,abs(right));}
}
void motor_stop() { //..................motor_stop
  analogWrite(D3,0);
  analogWrite(D4,0);
  analogWrite(D5,0);
  analogWrite(D6,0);
}
void setup() { //............................setup
  pinMode(LED_BUILTIN,OUTPUT);
  digitalWrite(LED_BUILTIN,LOW);
  pinMode(D1,OUTPUT);  // HCSR04-TRIG
  digitalWrite(D1,LOW);
  pinMode(D2,INPUT);   // HCSR04-ECHO
  pinMode(D3,OUTPUT);
  pinMode(D4,OUTPUT);
```

```
        analogWrite(D3,0);
        analogWrite(D4,0);
        pinMode(D5,OUTPUT);
        pinMode(D6,OUTPUT);
        analogWrite(D5,0);
        analogWrite(D6,0);
        Serial.begin(38400);
        WiFi.softAP(ap_ssid,ap_password);
        IPAddress myIP = WiFi.softAPIP();
        server.begin(port);
        Serial.print("\nAccess point and server started at address: ");
        Serial.print(myIP); Serial.print(" and port: "); Serial.println(port);
        Serial.print("with SSID: "); Serial.println(ap_ssid);
        pinMode(D7,OUTPUT);   // D7=Servo, D8=Tone
        myServo.attach(D7);
        digitalWrite(LED_BUILTIN,HIGH);
        tone(D8,880,500);
        lasttime=millis();
    }
```

This code runs on the NodeMCU on the robot, and first defines the Max() and Min() functions to determine the larger and smaller of two input values. Then it defines the name and passphrase of the WLAN as well as the used port number before including WiFi libraries for UDP and defining the server. We also include support for model-servos and declare one instance myServo, as well as variables needed for scanning the servo, reading the IR phototransistors, and the state machine. We then declare the convenience function send_string() to send a UDP package back to the remote controller. The range() function controls the HC-SR04 distance sensor in the way we discussed in Section 4.4.5. It ensures that the trigger pin D1 is LOW before pulling it HIGH for 10 μs, and then waits for the echo to arrive on pin D2 with the pulseIn() function. The factor 0.017 converts the duration of the echo in microseconds to distance in cm. If an obstacle is closer than 10 cm, the buzzer briefly beeps before returning the distance as the function value. The motor_speed() function receives the speed, with sign indicating the direction, for the motors as input values. First it clamps the values to be within ±1023, turns both motors briefly off, and then, depending on the sign, sets the pulse-width modulation period of one or the other control pin to the desired value with a call to analogWrite(). It first handles one motor controlled by pins D3 and D4, and then the other motor, controlled by D5 and D6. The motor_stop() function turns all relevant IO pins off.

In the setup() function we mostly configure the IO pins as input or output according to their purpose. The LED_BUILTIN pin is actually D0 and controls an LED on the NodeMCU circuit board, which is convenient for debugging. Pulling D0 LOW lights the LED, and pulling it HIGH turns it off again. D1 and D2 are connected to the trigger and and echo pin of the distance sensor, and D3 to D6 to control the motors. Then we initialize the serial line to communicate at 38400 baud to accommodate the capabilities of the slave NANO before starting to span the WLAN with the call to the WiFi.softAP() function, which configures the NodeMCU as an access point with the supplied name and passphrase. The call to WiFi.softAPIP() returns the IP number of the access point, normally 192.168.4.1. Next we start the server to listen on the selected port for packets to arrive from the remote controller. Finally, we attach a servo controller to pin D7, turn the LED off, sound a short tone, and remember the elapsed time in the variable lasttime, which is needed for scheduling the state machine; but more on that topic later.

Having declared all variables and defined all preparatory functions, we can continue to discuss the `loop()` function of the sketch.

```
void loop() { //................................loop
  char line[30];
  int packetsize=server.parsePacket();
  if (packetsize) {
    int len=server.read(line,30);
    line[len]='\0';
    if (strstr(line,"LSPEED ")==line) {
      int val=(int)atof(&line[7]);
      val=Max(-1023,Min(1023,val));
      analogWrite(D3,0);
      analogWrite(D4,0);
      if (val<0) {analogWrite(D3,abs(val));}
        else {analogWrite(D4,abs(val));}
    } else if (strstr(line,"RSPEED ")==line) {
      int val=(int)atof(&line[7]);
      val=Max(-1023,Min(1023,val));
      analogWrite(D5,0);
      analogWrite(D6,0);
      if (val<0) {analogWrite(D5,abs(val));}
        else {analogWrite(D6,abs(val));}
    } else if (strstr(line,":")==line) {
      line[len]='\n'; line[len+1]='\0';
      Serial.println(line);
    } else if (strstr(line,"D0 ")==line) {
      int val=(int)atof(&line[3]);
      if (val==0) {
        digitalWrite(LED_BUILTIN,HIGH);
      } else {
        digitalWrite(LED_BUILTIN,LOW);
      }
    } else if (strstr(line,"SERVO ")==line) {
      int val=(int)atof(&line[6]);
      myServo.write(val);
    } else if (strstr(line,"BEEP ")) {
      int val=(int)atof(&line[5]);
      Serial.print("BEEP val= "); Serial.println(val);
      tone(D8,440,val);
    } else if (strstr(line,"A0?")) {
      int val=analogRead(A0);
      Serial.print("A0 "); Serial.println(val);
      sprintf(line,"A0 %d",val); send_string(line);
    } else if (strstr(line,"RANGE?")) {
      int val=range();
      sprintf(line,"RANGE %d",val); send_string(line);
    } else if (strstr(line,"SCANRANGE ")==line) {
      int val=(int)atof(&line[10]);
      int minval=2000,minpos=-1;
      if (val>0) {
```

```
      myServo.write(10);
      delay(1000);
      for (int k=10;k<170;k+=5) {
        myServo.write(k); delay(200);
        val=range();
        if (val<minval) { minval=val; minpos=k;}
        sprintf(line,"SCANRANGE %d %d",k,val); send_string(line);
      }
      myServo.write(minpos);
      sprintf(line,"MINIMUM at %d",minpos); send_string(line);
    } else {
      myServo.write(90);
    }
  } else if (strstr(line,"FINDFIRE ")==line) {
    int val=(int)atof(&line[9]);
    int minval=2000,minpos=-1;
    if (val>0) {
      myServo.write(10);
      delay(1000);
      for (int k=10;k<170;k+=5) {
        myServo.write(k); delay(200);
        val=analogRead(A0);
        if (val<minval) { minval=val; minpos=k;}
        sprintf(line,"FINDFIRE %d %d",k,val); send_string(line);
      }
      myServo.write(minpos);
      sprintf(line,"MINIMUM at %d",minpos); send_string(line);
    } else {
      myServo.write(90);
    }
  } else if (strstr(line,"NEXTEVENT ")==line) {
    nextevent=(int)atof(&line[10]);
  } else if (strstr(line,"SLEEPTIME ")==line) {
    sleeptime=(int)atof(&line[10]);
  } else {
    Serial.println("unknown");
  }
}
yield();
if (millis()>lasttime+sleeptime) {  // next scheduled event
  switch (nextevent) {
    case 1:  // determine range
      sprintf(line,"RANGE %d",range()); send_string(line);
      sprintf(line,"A0 %d",analogRead(A0)); send_string(line);
      nextevent=1;
      break;
    case 2:  // scan with servo
      servo_pos+=servo_inc;
      if (servo_pos>170) {servo_inc=-servo_inc;}
      if (servo_pos<10) {servo_inc=-servo_inc;}
      myServo.write(servo_pos);
```

```
              nextevent=2;
          case 3:  // request direction sensors
            if (range()<10) {
              motor_stop();
              nextevent=1;
            } else {
              Serial.println(":A0?\n:A1?");
              sleeptime=50; nextevent=4;
            }
            break;
          case 4:  // read direction sensors and take action
            if ((right_sensor>0) && (left_sensor>0)) { // new data
              if ((left_sensor<250) || (right_sensor<250)) {
                sprintf(line,"SENSORS %d %d",left_sensor,right_sensor);
                send_string(line);
                int val0=(int)(600+0.2*(right_sensor-left_sensor));
                val0=Max(-1023,Min(1023,val0));
                int val1=(int)(600-0.2*(right_sensor-left_sensor));
                val1=Max(-1023,Min(1023,val1));
                motor_speed(val0,val1);
              } else {
                motor_stop();
              }
              right_sensor=-1; left_sensor=-1;
              sleeptime=1000; nextevent=3;
            }
          default:
            break;
        }
        lasttime=millis();
      }
      yield();
      if (Serial.available()) {
        Serial.readStringUntil('\n').toCharArray(line,30);
        send_string(line);
        if (strstr(line,".A0 ")==line) {
          right_sensor=(int)atof(&line[3]);
        } else if (strstr(line,".A1 ")==line) {
          left_sensor=(int)atof(&line[3]);
        }
      }
    }
```

Here we first check whether a UDP packet has arrived from the remote controller with the call to the `server.parsePacket()` function. If a packet of size `packetsize` is available, it is read with `server.read()` and its length is determined. To avoid ugly output, we ensure that the last character is a NULL character, to indicate the end of a string and start testing what type of command has arrived. If it is LSPEED we interpret the rest of the line as the speed value with the call to `atof()`, clamp the values to the acceptable range, and set the speed of the motor, in the same way we do in the `motor_speed()` function. If RSPEED is received, the speed of the other motor is adjusted. If the line starts with a colon : it is passed on to the serial line. In this way we can send commands to the slave NANO. If the command starts with D0 we turn the built-in LED on and off. The SERVO command sets the angle of the model-servo, and BEEP nnn makes a sound of 440 Hz for the duration of nnn ms. The A0? command returns the analog value read from the analog pin on the NodeMCU to the serial line and to the remote controller, and the RANGE? command likewise returns the distance to an obstacle as determined by the HC-SR04 sensor. The following command, SCANRANGE 1, starts the servo to move, which causes the distance sensor to point towards different directions, while the HC-SR04 scans the distance to an obstacle, and simultaneously records the direction and distance to the closest object. Finally, the servo is moved to point towards the minimum distance, and returns the direction at which the closest object is found. Calling SCANRANGE with argument 0 causes the servo to move to its middle position. The FINDFIRE command performs the same action, but scans the middle IR photodiode connected to analog pin A0 on the NodeMCU to determine the direction of a heat source. Note that the phototransistor pulls the signal line towards ground, which causes the voltage on pin A0 to approach zero, if a heat source is detected. The last two commands, NEXTEVENT and SLEEPTIME, allow us to set variables related to the state machine that governs the autonomous running of the robot. They are attached to buttons on the remote controller and can be changed by pressing these buttons.

The next section of the code, following the yield() command, implements this very simple state machine to operate asynchronously with all other actions that the NodeMCU performs. First we test whether the current time exceeds the value for the next scheduled event, lasttime+sleeptime, has elapsed. If that is the case, we branch according to the value of the nextevent variable. If it is 1 we only determine the distance from the HC-SR04 sensor with the range() function and the reading of the middle phototransistor. Then we set nextevent=1, which will repeat the same action after sleeptime has elapsed. If nextevent is 2, we sweep the servo position back and forth across its range, one step at a time. The next two event codes, 3 and 4, implement the autonomous motion of the robot towards a heat source. If nextevent is 3, we first test whether an object is closer than 10 cm, stop the motors, and branch to event code 1 discussed above. If there is no close obstacle, the commands :A0? and :A1? are sent on the serial line. Here we employ the convention that a command prepended with a colon is interpreted by the slave NANO, which in this case is requested to read its analog pins A0 and A1, to which the two outer phototransistors on the small breadboard are connected. Since we have two transistors, the difference of the reading will provide information about the direction of the heat source. The command is only dispatched in event code 3, but by setting the sleeptime to 50 ms and nextevent=4, we will execute event code 4 about 50 ms later. The response from the slave NANO is recorded asynchronously, and we discuss the sketch running on the NANO a little later. After 50 ms, event code 4 is executed, and there we test whether the phototransistors detected a valid signal, and whether one of the signals is sufficiently small to be interpreted as a heat source. In that case, the sensor readings are sent to the remote controller, and the motor speed of one motor is set to a constant, not-too-large value, here 600, plus a small contribution proportional to the difference between the sensor readings. This small contribution is added to the speed of one motor and subtracted from the other. In this way we implement a very simple proportional control loop that feeds the error signal to the motors, to make the robot turn towards the heat source. If no heat source is detected, we stop the motors. Before leaving this part of the program, we set the variables right_sensor and left_sensor to an invalid value to indicate that no new value is present. Finally, we make the state machine sleep for 1000 ms and restart with event code 3 in order to repeat the process of reading the sensors and feeding the difference of their reading to the motors. Before leaving this part of the program, we update the lasttime variable in order to know when to handle the next event code. After the call to yield(), we test whether any characters are available on the serial line, and if the response starts with .A0 or .A1, we interpret the numerical value as the right_sensor or the left_sensor reading. We point out that handling the communication via UDP packets and on the serial line is handled asynchronously, and interleaved with the state machine executing the events enumerated by the variable nextevent.

The serial line serves a dual purpose: First, it displays debugging information on the host computer if that is connected, and second, it communicates with the slave NANO. In order to distinguish the latter, we adapted the query-response protocol used earlier by the convention that all communication *to* the slave NANO starts with a colon ":," and all communication *from* the slave starts with a period "." such that those commands can be easily filtered out from the serial line. The sketch that runs on the slave NANO is the following.

```
// Slaveduino, V. Ziemann, 170723
void setup() { //.......................setup
  Serial.begin(38400);
  while (!Serial) {;}
  pinMode(13,OUTPUT);
  digitalWrite(13,LOW);
  pinMode(8,INPUT_PULLUP);
}
void loop() { //.........................loop
  char line[30];
  if (Serial.available()) {
    Serial.readStringUntil('\n').toCharArray(line,30);
    if (strstr(line,":A0?")==line) {
      Serial.print(".A0 "); Serial.println(analogRead(A0));
    } else if (strstr(line,":A1?")==line) {
      Serial.print(".A1 "); Serial.println(analogRead(A1));
    } else if (strstr(line,":D13 ")==line) {
      int val=(int)atof(&line[4]);
      if (val==0) {digitalWrite(13,LOW);} else {digitalWrite(13,HIGH);}
    } else if (strstr(line,":D8?")==line) {
      Serial.print(".D8 "); Serial.println(digitalRead(8));
    }
    delay(3);
  }
}
```

This sketch follows earlier examples, and in the setup() function we configure the serial line and the used pins to be OUTPUT or INPUT, in this case even with internal pull-up resistor enabled. In the loop() function we test whether data is available on the serial line and then test the different requests the slave is able to handle, namely responding to :A0?, :A1?, :D13, and :D8?, which all start with a colon, while any other query is silently ignored. Note also that any reply back to the NodeMCU via the serial line is prepended by a period. In the code running on the NodeMCU we only react to :A0? and :A1? and their respective response, but extending the code running on the slave NANO is easy to implement.

In Figure 17.6 we show an early prototype of the operational robot from the back and from the front. On the left-hand image we see the larger breadboard with the NodeMCU on the right. In the center of the breadboard, hidden behind the wires, is the L293D, and next to it the voltage regulator. In this prototype, instead of NANO, we use an ATmega328 that is programmed in an Arduino UNO, removed from its socket, and inserted in the breadboard on the robot. After adding a 16 MHz crystal and two 22 pF ballast capacitors,

Figure 17.6 The operational robot from the back (left) and from the front (right).

it works equally well on the breadboard. On the right we see the front of the robot with the small breadboard slightly turned to the side. On it the distance sensor and below it the three phototransistors are visible. Owing to the breadboards, the wiring has a distinct "spaghetti-flavor," which makes the robot rather fragile, but also very convenient for developing and testing new functionality.

At this point we can control our simple robot with the remote controller and also make it follow a simple algorithm to find a heat source autonomously. We have to admit that the performance of the prototype system is less than impressive; the heat source has to be rather close to be detectable and the motion towards it is awkward. But our main purpose is to show how to implement all the functionality in a simple prototype system. Building a marketable system requires significantly more effort.

The hardware with one large breadboard directly mounted on the robot chassis and a smaller one movable on the model-servo makes the system very versatile and extendable; for example, connecting other sensors or making the robot follow other algorithms. We use a slave NANO as an IO extender, but also other extenders are available, such as the MCP23017 that provides up to 16 digital IO pins, or the PCA9685 that controls up to 16 pulse-width modulated outputs or model-servos. Both extenders are controlled via the I2C-port on the NodeMCU and require two IO-pins only. The system is open for further experiments, and to expand the system. Other improvements comprise fine-tuning the algorithms to make the robot more robust, and whatever else come to mind.

Now that we have discussed the hardware and software for our projects, we are ready to use the equipment in experiments, gather measurements, and interpret them. Once we reach firm conclusions, we want to present the results in a seminar and eventually write a report about the experiment. These two points, related to presenting our work, are the topic of the final chapter.

QUESTIONS AND PROJECT IDEAS

1. Discuss the pros and cons of using TCP versus UDP.

2. Add a second state-machine (thread) to the robot that periodically reads out an MQ-x gas-sensor and sounds a special alarm, once it reports a significant presence of a gas.

3. Add an LCD display with I2C interface to the remote controller, to show status messages from the robot.

4. *Build a remote-controlled boat* instead of a robot. The boat can be powered by a fan salvaged from a computer. We can steer it by directing the air stream behind the propeller with fins controlled by a model-servo.

5. *Build a line-follower* that uses LDR or phototransistors pointed towards the floor. Program it to follow a white (or black) line made of masking tape.

6. *Build a vending machine* that detects the size and weight of inserted coins and that pushes a chewing gum or other desirable items from a safe place to the bin where we can pick it up.

7. Construct the *model crane* from question 13 in Chapter 3. Equip the containers with a marker, such as a periodically flashing LED, and invent a scheme to run the crane autonomously.

8. Contemplate how to remotely control a *sailboat*. What actuators do you need? In case you want to add autonomous control, which sensors do you need?

## CHAPTER 18

Presenting and Writing

After having gone through the basic electronics and programming examples, both on Arduinos and on the Raspi, and after completing several projects, we need to communicate our activities to our colleagues and condense our activities into a well-motivated and concise sequence of descriptions. This can be in the form of a presentation with slides, or as a report for a thesis or a journal. As a template for the contents of either presentation or report, we use the weather station example, just to illustrate the concepts with a specific example. We start with the discussion of a presentation with slides.

18.1   PREPARING A PRESENTATION

The key issues when preparing a presentation are *motivation* and a good *story line*, as well as substantial subject matter. Why is that so? The audience of a seminar usually is rather heterogenous and we have to provide a funnel to guide them from their different backgrounds to the subject matter of the seminar. Often a good start is stating a problem that is easy to understand and then pointing out how our project solves or at least alleviates the problem. The weather station came up in my home lab, because we use pressure vessels filled with liquid helium as well as high-voltage equipment. The potential effect of barometric pressure on the pressure vessels and the relevance of humidity for high voltage is intuitively understandable for most people. So that is a viable motivation as to why we care about the weather station as a device that helps us to correlate weather-based data with other measurement data.

Once we have captured the attention of the audience with a catchy motivation, we need to keep it alive by following a well-conceived *story line* that places the relevant topics – the subject matter of the seminar – in a logically coherent sequence where one detail follows the previous in a natural way. My suggestion is to start with easily understandable facts and then progressively increase the complexity of the discussed material. We must avoid situations where the audience starts wondering *why* we talk about a particular topic and loses track of our story line. Since the story line addresses the overall organization of the presentation it needs to be sorted out beforehand.

A method to organize a presentation that I found useful is the following: I start by estimating the number of slides that fit into the allotted time. An average slide typically requires 2–3 minutes for the audience to absorb. Thus, for a 20-minute presentation I target about 8–10 slides as a rule of thumb. Then I prepare a title page with a catchy title that encapsulates the essential point of the seminar, followed by one slide that motivates the problem I intend to address. This I follow up with a brief and qualitative description of the idea of how to solve it that I will elaborate in the remainder of the seminar. At this point the audience should have a good idea about the scope of my seminar and should

DOI: 10.1201/9781003341703-18

be convinced that the idea I pursue has a decent chance of addressing the problem in a meaningful way. On the subsequent slides it is useful to concentrate on the *flow* of some quantity or some information from one stage to the next. In the example with the weather station, the information flows from the sensor via the microcontroller to the host computer where it is presented, either on a web page or accessible via control system. This flow of information we can mimic in the organization of the slides.

For the bulk of the presentation, I recommend preparing the allotted number of empty slides and giving each slide a title. The title should address one key issue per slide and the sequence of slides should follow the logical flow we have identified beforehand. In the weather-station example we have the issues: hardware with sensors, microcontroller, host computer, then software running on the respective computing devices and protocols used to interface the devices. Once we have the slides with their titles, we can shuffle them around until we are satisfied with our story line.

In the next step I place one or several pictures or graphs on each slide to illustrate the topic on the slide. Once all, or at least most, slides are equipped with a picture, I add a few bullet-points with keywords to each slide. They remind me of what I intend to say when presenting. They help me to understand my slides after 6 months, and those in the audience to remember what I said after 2 months. The slides are not a substitute for a self-contained report, but serve as illustrations for my presentation during a seminar.

Once all slides contain a picture and a few keywords, it is time to review the story line again and see whether linking of the slides works. This addresses the flow from one slide to the next and whether it comes in a natural and well-motivated way. If many forward references are necessary, I consider reordering the slides to achieve a more natural flow where information required at some stage is already discussed on a previous slide. Sometimes forward references are difficult to avoid, but I try to minimize them.

When almost all slides are completed, I suggest preparing a final slide with a clear synopsis of the main results and possibly some comments about how to extend the work. After a final pass through the slides, with a check for coherence to ensure that the title and final slides act as parentheses to enclose the subject matter, the presentation is ready.

Just for convenience I summarize the above guidelines in a presentation cookbook:

- Motivation: the problem and intended way to address it.

- Think of a story line and the logical flow of a concept, preferably with increasing complexity.

- Write a title on each slide and sort to follow the story line.

- Add pictures to slides.

- Add keywords to slides.

- Add slide with conclusions and outlook.

- Final check of coherence and linking. Done!

Naturally, my presentation cookbook is subjective, and should not keep you from using a working way of preparing slides, but if you get stuck very early on in the process of preparing a seminar, my guidelines may help you to get started.

Presenting one's work in a seminar is the first step, and writing it up in a report or a thesis is the next.

18.2   PREPARING A REPORT

When preparing a report or a thesis in general, I follow the same guidelines used for the preparation of a presentation. I need to *motivate* what I am about to describe and then adhere to a *story line* that follows the *flow* of a concept. As a first step I normally try to formulate a catchy title and an abstract of 100 to 200 words to explain to myself what I intend to write in the report. Often this is a reasonable starting point even for the final version of the abstract.

For the main part of the report, instead of starting with empty slides and filling the title line, I start by identifying a logical sequence of chapters, sections, or subsections, and give each one a title before ordering them in a sequence that follows the flow of the argument. In the next step I add pictures or other illustrative material to the respective sections. This normally results in a valid skeleton for the report that I need to flesh out in the next step. I start this process by adding bullet lists to the sections with topics that are relevant in the respective section. I do this for all originally identified chapters or sections, and often in this stage I note that I need to split sections into two or more or that I can combine sections with similar contents into a single one. As a guideline I assume that each topic in the bullet list will require one paragraph with about 200 words or so to discuss adequately. I also try to roughly balance the length of sections, but this is of minor concern. Once I know what contents should show up in the respective sections, I start writing the text and "fill in the blanks." Since key components – images, graphs, or tables – are already in place, I start each section with a short description of how it fits into the flow and then describe the material. I also try to write each section from start to end, because that helps to maintain a logical flow. At other times, if I do not come up with a decent start, I just start at a place where I have a good idea about how to present it and retrofit the missing parts later. In that case, however, I need to pay special attention later to ensure the flow.

A particularly important section is the *Introduction*, where I need to convince the reader that the report treats a relevant and interesting subject by stating the chosen problem and how to solve it. But ideally the problem should also be put into the context of previous work. What have other researchers done before on related problems? How does their work differ from mine? This section, typically one or two paragraphs long, requires a number of key references from the published literature. At the end of the introduction, I sometimes give a brief outline of the report in the fashion of a commented table of contents. By this time I have hopefully convinced the reader to continue with the remainder of my report. And this is the prime task of the introduction: a brief statement of the problem, the context, and a brief outline of things to come.

The *main part* of the report should follow the story line and the flow I initially decided upon. I write one or two paragraphs for each topic in the bullet list and make sure that the paragraphs are well connected in the sense that the reader knows how the following paragraph links to the present one. I try to avoid having the reader wonder what the lines he is presently reading have to do with the overall story line. Placing crosslinks with references to topics addressed earlier and how they connect to the present paragraph are helpful to make the text denser, and creates additional associations for the reader. Phrases such as, "where we use the result from page..." or "as discussed in Section..." illustrate the idea.

In the *Conclusions*, I give a synoptic review of the main part and the results of the report. I try to organize the introduction and summary jointly as an executive summary. Often, readers first read those two sections before deciding to spend time on the report in its entirety. The most important part of the conclusions is a concise summary of the key results. I sometimes follow this up with a number of questions that came up during the research and that may lead to further work, either by me or some other researcher.

I like to point out that the organizational triplet of introduction, main body, and conclusion vaguely resembles the form of a sonata, where first the themes are introduced. In the main body of the sonata the themes are elaborated and in the coda they often reappear in the original form once again. In this section we discussed the structure of the report and my recommendations are again subjective, but may help in case you are stuck.

After the discussion of the overall organization of the report, we now look at some of the ingredients and start with the presentation of data, often using graphics and plots.

18.3 PRESENTING DATA

Many aspects of scientific work are presented with the help of figures describing the experimental setup, or graphs showing the data, sometimes with comparison to a model. In [33] E. Tufte names three guiding principles to achieve graphical excellence.

Clarity is the first principle, and it requires carefully explaining the experiment from which the data originate, describing what quantities are plotted on a graph, and properly labeling the axes. Moreover, using a legible font of adequate font-size and avoiding low-contrast colors such as yellow or light green on white background, is mandatory.

Precision, the second principle, dictates presenting data honestly and truthfully. If data points are excluded, the reason and method to select the discarded points need to be explained. The graphics data of a manuscript must be proofread in the same way as the main text. Make sure to catch all errors, such as omitted exponents or lost or incorrect axis labels. Choose axes adequately. A bad example is to display the average temperature on Earth in Kelvin with zero displayed. Global warming is invisible on such a plot.

Efficiency, the third principle, aims at presenting information in a parsimonious way: to present the largest possible insight with the least effort. In particular, display only data that advance the main argument, but then explain *all* displayed features in the plot, either in the caption or in the text. If the plot is part of a seminar, the speaker may explain it as well. On the other hand, any unnecessary data, which Tufte calls "chart-junk," should be omitted from a graph. A bad example is to present all available measurement data, even though a well-chosen subset suffices to make a point.

In *The Elements of Graphing Data,* W. Cleveland discusses a large number of guidelines to produce graphical data, with many examples. He summarizes the guidelines in a concise list of "rules" such as

- to make the data the most prominent feature and avoid cluttering the data area;

- to keep the number of tick marks limited and preferably outside the data area;

- to have compatible scales when comparing two data sets;

and many more. Consulting the book is highly recommended in case of questions about graphing data.

Scientific journals have a strong interest that their authors produce high-quality articles, and those often include graphical data. They often expand the general terms of Tufte and Cleveland and provide comprehensive style guides, both for written and for graphical material. One good example is [34], which inspired the following points.

- Ensure your audience understands four things about the data points: what *quantity* they represent, their physical *unit,* their *magnitude,* and the *uncertainty.*

- Explain the error bars: Are they standard errors or confidence limits?

- Always label axes with quantity plotted and with units. I normally display units in square brackets. In journals, the use of title text is discouraged in favor of text in the caption.

- If there are several plots in the same graph, preferably label them, use a legend, or explain them in the figure caption.

- Choose the vertical scale such that at least 80 % of the range is used.

- If an axis covers more than 2 orders of magnitude, consider using logarithmic scale, if appropriate.

- Avoid "eye-candy" such as 3-D bar-graphs or pie-charts.

- When only a few data points need to be presented, say 10 or less, a table is often preferable to a plot.

These basic guidelines should provide a starting point to produce presentable plots.

The last point in the above list refers to data presented in *numerical form* in tables or in the main text. Under no circumstances should you state more than a sensible number of significant figures! If a measurement generates data that is accurate to 1 %, two or maybe three significant figures are adequate. Just because a computer displays results in double precision with 10 or more figures, this does not mean all figures are significant. And finally, error bars in numeric form should never be displayed with more than one or sometimes two significant figures.

And this brings us to writing the main text. Therefore, a few words about writing good English are in order.

18.4 GOOD ENGLISH

There are a number of style guides for the English language available, from the classic *Elements of Style* [35], the MLA handbook [36], S. Pinker's *Sense of Style,* [37] to S. King's *On Writing* [38]. Apart from these more general guides, several scientific journals such as *Nature* [39] or *Reviews of Modern Physics* [40] make style guides available for their authors. In particular, the latter style guide is very readable, and its Appendix A on "Writing a better scientific article," is highly recommended. In this section I highlight some of the topics in more detail.

- In one of the style guides [39] we find the sentence "Nature journals prefer authors to write in the active voice. . . ." The *active voice* usually makes the presentation clearer, more vigorous, and often more concise. Experiments are not done by themselves, but we, the experimenters, perform them. Moreover, often the subject matter is already difficult to understand, so we should make the presentation as clear as possible without obfuscating who does what. Science does not become more objective just because the person who did the experiment hides behind the passive voice.

- Authors should strive for *economy* in their presentation and avoid unnecessarily complicated constructions such as "owing to the fact that," which we can usually replace with "because." Often replacing passive constructions with active ones helps to disentangle complex sentences.

- Avoid overloading the reader with sentences containing a large number of subclauses to explain every conceivable exception in one sentence. At the end of the sentence, the reader may have forgotten the beginning and what the subject of the sentence is.

– Be friendly to your reader and invite her into your intellectual world by more or less addressing her directly, using phrases such as "let us" or "we now return." This helps to make the report more accessible by mimicking normal colloquial speaking patterns. The reader is not alienated by a stiff and abstract presentation but feels welcome to mentally participate in your exposé.

– Commit yourself to what you write and avoid unnecessary hedging by conditionals. This is easily done by trying to replace any occurrence of "might be" or "could be" with "is." Other hedging phrases to look out for are "may" and "could," or anything that hints at you being scared of writing what you really mean.

– If in doubt, use English instead of Latin phrases. In that sense, "first" is a better choice than "initial," use "place" instead of "location."

– Avoid acronyms unless a long term with a commonly used acronym appears in several places throughout the report. In that case, introduce the acronym at the first occurrence by writing out the full term first and adding the acronym in brackets. Henceforth use the acronym consistently!

– Avoid jargon! Someone uninitiated in the jargon specific to your particular project might want to read – and understand – your report.

– Pay special attention to grammatical correctness and especially to agreement errors. Ask an English-speaking colleague to proofread your report.

– Before releasing your report, make sure to run it through a spell-checker, and read it a final time to verify that all is correct!

These guidelines are certainly incomplete, but hopefully will help you to write an interesting and readable report. I suggest you also read one of the more comprehensive style guides such as [40]. A further source of advice regarding writing and presenting scientific work in general is *Nature* publication's web page, *English communication for scientists* [41]. A wonderful source of synonyms and antonyms is the *WordNet browser* [42]. You may consider using either the online version or a version installed on your computer in order to make your writing more lively.

18.5   POSTSCRIPTUM

And at this point we have reached the end of the book. In the process we talked about various sensors, both analog and digital, connected them to microcontrollers, massaged the data into a common format, and passed them on to a host computer. There we postprocessed the measurements, stored them in databases, and prepared them for a presentation or report. I hope that you found useful, interesting, and inspiring topics between Preface and Postscriptum that will help you in your projects that have a data-acquisition aspect.

I tried out all circuitry and programming presented in the book, but the odd bug may have crept in. If you find one, please do not keep the bug, but share it with me so I can improve future versions of the book. Of course, the general disclaimer applies, namely that the code in the book is provided *as is* and users are advised to use caution, but are in any case responsible themselves for using the program code.

Basic Circuit Theory

As a reminder, we will briefly review the basic theory of electronic circuits. We start by considering *Ohm's law,* which states that the current I in a segment of a circuit is proportional to the voltage U with the resistance R as the proportionality constant, or $U = IR$. This is a consequence of the balance of the electric field accelerating electrons in the *resistor* and the friction force experienced by electrons due to scattering with vibration modes, phonons, of the ions that make up the material. The force pulling on the electrons is proportional to U and the friction force is proportional to the electron's drift velocity v, or by $-\alpha v$ where α is a material-dependent friction coefficient. Under stationary conditions, the two forces need to balance, and we have $v = U/\alpha$. But the current I is proportional to the drift velocity v and we find $I \propto v = U/\alpha$, which is the essence of Ohm's law, and the friction coefficient α is proportional to the resistance R. In case the voltage is varying moderately slowly compared to the relaxation time of the electrons, the current directly follows the applied voltage; the current I and the voltage U are in phase.

For *capacitors* the situation is different, because the two plates of a capacitor are electrically separated and do not allow constant transport of charges under stationary conditions. They can, however, store a charge Q on the plates, and the capacitance C is the proportionality constant between Q and the applied voltage U, given by $Q = CU$. Since the charge Q changes as a consequence of current I flowing onto the plates, we have $U = (1/C)\int^t I dt'$, or by differentiation, $dU/dt = (1/C)I$. A sinusoidally oscillating voltage $U \propto e^{i\omega t}$ with frequency ω will therefore be related to the current by $U = (1/i\omega C)I$, and we can identify $Z_C = 1/i\omega C$ as the generalized resistance, the *impedance,* of the capacitor. Here i is the imaginary unit.

Inductors are made of coils that store energy in their magnetic field, and if we turn off the current, a voltage develops. Expressing this behavior in a formal way, we have $U \propto dI/dt$, with the inductance L of the coil as the proportionality constant. If we again assume that the coil is excited by a sinusoidal current $I \propto e^{i\omega t}$ with frequency ω, we find that we have $U = (i\omega L)I$ and identify $Z_L = i\omega L$ as the impedance of the inductor.

Now that we have resistances and impedances of common elements found in simple circuits, we may ask how to combine them to networks. This question is answered by *Kirchhoff's laws,* of which the first states that under stationary conditions, all currents flowing into a network node must add up to zero. This is a statement about the preservation of charges, and what comes in must also come out, because under stationary conditions piling up charges is not allowed. The second law states that the voltage differences around a loop in the network have to add up to zero, which is the requirement that the voltages at each node of the network with respect to a reference node are unique.

We immediately use these laws to investigate the resistance or impedances connected in series, as shown on the left in Figure A.1. The currents in node B need to balance, and this

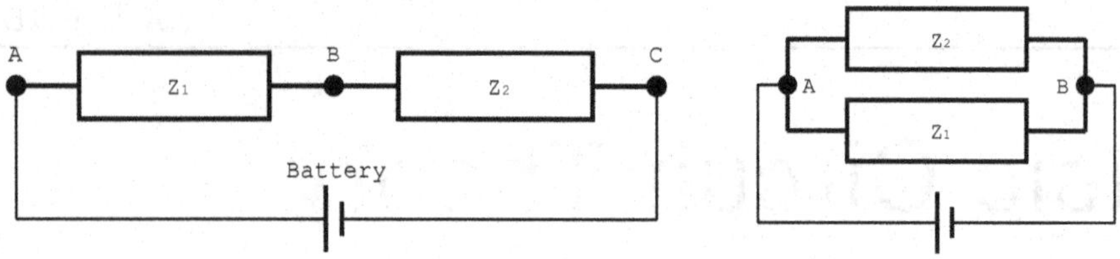

Figure A.1 Two impedances connected in series (left) and in parallel (right).

implies that the current I_1 passing through Z_1 is the same as the current $I_2 = I_1$ passing through Z_2. At the same time, the voltage U provided by the battery is the same that is dropped across the two impedances. We thus find $U = I_1 Z_1 + I_2 Z_2 = I_1(Z_1 + Z_2)$, and we find that the total impedance of the circuit equals the sum of the impedances coupled in series.

Considering the circuit on the right in Figure A.1, we know that the currents through the two branches have to add up: $I_t = U/Z_t = I_1 + I_2 = U/Z_1 + U/Z_2$. We can simplify this to $1/Z_t = 1/Z_1 + 1/Z_2$, which implies that the individual impedances add as reciprocals if the impedances are coupled in parallel.

Note that the argument in the previous two paragraphs is valid for normal resistors and constant (DC) voltages and currents, but works as well for AC voltages if we use the complex frequency-dependent impedances for inductors and capacitances. We also note that all relations among currents and voltages are *linear,* such that we can use the super-position principle and analyze circuits for each voltage or current source independently and then add the contributions in the end.

Apart from the linear circuit elements, the impedances, there are also nonlinear elements, and *diodes* are among them. Their voltage-current behavior follows an exponential dependence given by $I \propto \left(e^{(eU - E_g)/kT} - 1\right)$, where $E_g \approx 1.2\,V$ is the bandgap energy of the semiconductor material, here silicon. We see that for negative voltages the current is very small, while for positive currents it grows exponentially once the threshold of the bandgap voltage is passed. Practically, diodes conduct current in one direction and block current in the opposite direction. This makes them perfect to rectify voltages in the way we discussed in Section 2.2.5. A diode is conducting if it is forward biased and the cathode is at a more negative voltage than the anode.

Transistors are other nonlinear elements based on two diodes coupled antiparallel; they come in two types, dependent on whether the central tap, called the base terminal, is an anode or a cathode. Injecting an additional current into the base terminal enables a larger current flow through the two outer terminals, called collector and emitter. In the main body of the text we use transistors for switching applications, but with suitable ancillary circuitry they can be used as amplifiers as well. That goes beyond the scope of this appendix and is covered in books on electrical engineering, such as [17].

Least-Squares Fit

In Chapter 12 we used the `linfit()` function on the Arduino to fit a straight line through a number n of data points sampled at equidistant times. Here we briefly discuss the inner workings of that function that determines the slope a and intercept b by fitting data to a straight line. These parameters are determined by the requirement to minimize the sum of squared residuals $r_i = y_i - at_i - b$ for a number of data points (t_i, y_i). We can restate the problem in the form of a linear equation that needs to be inverted in the least-squares sense

$$\begin{pmatrix} \vdots \\ y_i \\ \vdots \end{pmatrix} = \begin{pmatrix} \vdots & \vdots \\ t_i & 1 \\ \vdots & \vdots \end{pmatrix} \begin{pmatrix} a \\ b \end{pmatrix} \tag{B.1}$$

which we write in abbreviated form

$$y = Ax \tag{B.2}$$

where A is the $n \times 2$ matrix in the previous equation and $x = (a, b)^T$. Here the superscripted T denotes the transpose, and for the squared sum of the residuals $\sum_{i=1}^{n} r_i^2$ we write χ^2. The latter we express as

$$\begin{aligned} \chi^2 &= \sum_{i=1}^{n} r_i^2 = (y^T - x^T A^T)(y - Ax) \\ &= y^T y - x^T A^T y - y^T Ax + x^T A^T Ax \; . \end{aligned} \tag{B.3}$$

Minimizing this expression with respect to x results in a condition for the sought solution vector x. The condition for a minimum is the requirement that the gradient with respect to the fit parameters x^T is zero. We somewhat sloppily write it as

$$0 = \frac{\partial \chi^2}{\partial x^T} = -2A^T y + 2A^T Ax \tag{B.4}$$

where we used $x^T Ay = y^T A^T x$ because the expressions are scalars and any matrix of the form $A^T A$ is symmetric. This allows us to combine terms and we arrive at the previous equation. For a more detailed derivation see [43]. Left-multiplying with the inverse of $(A^T A)$, provided the matrix is nonsingular, isolates the sought-after fit-parameter x and we obtain

$$\begin{pmatrix} a \\ b \end{pmatrix} = x = \left(A^T A\right)^{-1} A^T y \; . \tag{B.5}$$

The right-hand side is often called pseudo-inverse of the matrix A, and we need to evaluate this expression in order to find a and b. If the t values are equidistant, we can absorb the

DOI: 10.1201/9781003341703-B

step size in a redefinition of a, and the matrix A obtains the form

$$A = \begin{pmatrix} 1 & 1 \\ 2 & 1 \\ 3 & 1 \\ \vdots & \vdots \end{pmatrix} \tag{B.6}$$

such that $A^T A$ becomes

$$A^T A = \begin{pmatrix} \sum_{k=1}^{n} k^2 & \sum_{k=1}^{n} k \\ \sum_{k=1}^{n} k & \sum_{k=1}^{n} 1 \end{pmatrix} = \begin{pmatrix} n(n+1)(2n+1)/6 & n(n+1)/2 \\ n(n+1)/2 & n \end{pmatrix} \tag{B.7}$$

which accounts for the definition of S0,S1,S2 in the linfit routine. Inverting this 2×2 matrix is trivial, and calculating $A^T y$ is done in the loop where ay0 and ay1 are calculated. The linfit function then returns the parameter a, the slope, and in the main program we need to rescale the result by multiplying the slope with the step size. It is easy to cross-check the results with octave, and the fitting is done with the polyfit() function, but if we want to do the calculation on the Arduino, we need methods like the one shown in this appendix.

Note also that generalizing the method to higher-order polynomials is simple. We only need to add columns in Equation B.1 with t_i^n and can rescale the fit parameters $a, b, \ldots$ by the appropriate power of the step size, and we are left with adding powers of positive integers for which closed expressions exist. This means that the matrix $A^T A$ can always be calculated in closed form for any number of data points n. Only the inversion of a matrix with rank of the number of fit parameters remains. Being able to precompute most of the matrices and possibly inverting them on the host computer for fixed n makes this method rather suitable for microcontrollers.

QUESTIONS AND PROJECT IDEAS

1. Fit a parabola given by $y = at^2 + bt + c$ to equally-spaced data points $(i\Delta t, y_i)$.

Where to Go from Here?

After our tour-de-force through the world of sensors, electronics, microcontrollers, and computers I am sure you want to dig deeper. Here's how.

A great overview over sensors that are available on the market are websites of distributors for electronic components, such as `www.digikey.com`, `www.rs-online.com`, `www.distrelec.com`, or `www.farnell.com`. Just search for the word *sensor* on their sites. Typically a large number of subcategories appear with an even larger number of individual sensors beneath. On the pages for the specific sensors datasheets with detailed information invite further exploration. Just browsing around can be very inspiring.

The classic reference for all things electronic is Horowitz's and Hill's *The Art of Electronics*. It goes into great depth on practically all aspects of electronics, including discussions of circuits that do *not* work. Make sure to get the most recent (third, at the time of writing) edition. It addresses a mature audience, whereas the companion book *Learning the Art of Electronics, a Hands-On Lab Course* by Hayes and Horowitz addresses students. After presenting components in quite some detail it goes step-by-step through many worked examples. I found the *ARRL Handbook for radio-communications* from the American Radio Relay League a great source of inspiration. It is a source book of solutions for the electronics that is relevant to ham-radio amateurs. An updated edition is released every other year.

A first step towards exploring additional microcontrollers and their specific features is perusing `www.arduino.cc` and, in particular `store.arduino.cc/collections/boards`. All those boards can be programmed with the Arduino IDE. Other microcontroller families, such as the ATtinys and PICs from Microchip or the MSP430s from Texas Instruments are often optimized for specific purposes, such as low-power consumption. Moreover, they come in a large variety of sizes and capabilities, such as number of IO pins, DAC and ADC channels, built-in op-amps, on-board WLAN or Bluetooth. Occasionally they must be programmed in assembler, but nowadays they often come with a software development kit (SDK) including controller-specific libraries for writing code and uploading programs to the controllers. Even the Raspis have gotten a smaller brethren in the form of Raspberry Pi Picos, available with and without built-in WiFi. They are based on the RP2040, a dual-core microcontroller that is also used on some of the Arduino boards mentioned on page 58. The Picos can be programmed using their native SDK or with MicroPython, a Python interpreter that runs on the Picos. Initial tests are usually described in "Getting started" guides, but mastering all features requires careful study of the datasheets, often many hundreds of pages long. Considering the recent shortage of components, the time to adapt a project to a different controller, however, might be well spent.

Likewise the recent shortage of Raspberry Pis can be alleviated by moving projects to a different computer. Even desktop or laptop computers with common operating systems can be used as a substitute, as long as they have a working Python3 or Octave installation. Usually also databases and webservers, such as apache2 are open source and available for all common operating systems. Porting the software from this book should then work out-of-the-box or require only moderate effort to adapt it. But sometimes the small size of a Raspi is important and luckily there are other small boards on the market, such as the *ODROID-C4*, which has similar specifications as Raspis or the *Beagleboard*, which is geared more towards data acquisition and robotics than Raspis. Even more geared towards data acquisition is the *Red Pitaya*. It features two DACs and two ADCs that operate at speeds of up to 125 MHz and that are directly connected to a Field-Programmable Gate Array which itself is accessible from two general-purpose processor cores running Linux. The *NVIDIA Jetson Nano* is geared towards machine-learning applications. It features a dual-core processor running Linux that is connected to a 128-core Graphical Processing Unit, which supports the *CUDA* programming language that speeds up the machine-learning algorithms.

Let me end on a lighter note and suggest sources for additional fun projects with Arduinos and Raspis. Apart from the many hobbyist magazines that feature Arduinos and Raspberry Pis, many cool projects can be found on `create.arduino.cc/projecthub` and in the more than 100 issues of the MagPi magazine `magpi.raspberrypi.com` that are available online free of charge (though they appreciate a donation). Two final sources for inspiration of what else to do with Arduinos and Raspis are `www.instructables.com/circuits` and `hackaday.com`. Recreating some of these projects and realizing your own should put and end to boredom for the foreseeable future.

Bibliography

[1] Fritzing. Project website: http://fritzing.org/.

[2] ATLAS collaboration. Project website: https://cern.ch/atlas.

[3] CMS collaboration. Project website: https://cern.ch/cms.

[4] Large Hadron Collider. Project website: https://cern.ch/lhc.

[5] CERN the European Center for Nuclear Research. Project website: https://cern.ch.

[6] MQTT. Project website: http://mqtt.org/.

[7] EPICS. Project website: http://www.aps.anl.gov/epics.

[8] Arduino. Project website: https://www.arduino.cc.

[9] ESP8266 microcontroller. Project website: https://www.espressif.com/en/products/socs/esp8266.

[10] ESP32 microcontroller. Project website: https://www.espressif.com/en/products/socs/esp32.

[11] Raspberry Pi. Project website: https://www.raspberrypi.org.

[12] J. Fraden. *Handbook of Modern Sensors.* Springer Verlag, Berlin, third edition, 2004.

[13] J. Wilson, editor. *Sensor Technology Handbook.* Elsevier, Amsterdam, 2005.

[14] R. B. Northrop. *Introduction to Instrumentation and Measurements.* CRC Press, Boca Raton, third edition, 2014.

[15] M. Coplan, J. Moore, and C. Davies. *Building Scientific Apparatus.* Cambridge University Press, Cambridge, UK, fourth edition, 2009.

[16] C. Kittel. *Introduction to Solid State Physics.* Wiley, Hoboken, NJ, eighth edition, 2005.

[17] T. Giuma and P. Peebles. *Principles of Electrical Engineering.* McGraw-Hill, New York, 1991.

[18] P. Horowitz and W. Hill. *The Art of Electronics.* Cambridge University Press, Cambridge, UK, 1990.

[19] A. Peyton and V. Walsh. *Analog Electronics with OP Amps.* Cambridge University Press, Cambridge, UK, 1996.

[20] D. Lancaster. *Active Filter Cookbook.* Newnes, Oxford, second edition, 1996.

[21] G. Franklin, J. Powell, and A. Emami-Naeni. *Feedback Control of Dynamic Systems.* Pearson, Boston, seventh edition, 2015.

[22] M. Margolis. *Arduino Cookbook*. O'Reilly, Sebastopol, CA, 2011.

[23] Standard Commands for Programmable Instruments SCPI. Project website: `https://www.ivifoundation.org/scpi/default.aspx`.

[24] S. Monk. *Raspberry Pi Cookbook*. O'Reilly, Sebastopol, CA, 2014.

[25] Scientific Programming Language GNU Octave. Project website: `https://www.gnu.org/software/octave/`.

[26] Python Software Foundation. Project website: `https://www.python.org/`.

[27] MariaDB Database. Project website: `https://mariadb.org/`.

[28] RRDtool. Project website: `https://oss.oetiker.ch/rrdtool/`.

[29] Apache web server. Project website: `https://httpd.apache.org/`.

[30] Installing EPICS on the Raspberry Pi. `https://prjemian.github.io/epicspi/`.

[31] Sparkfun AD8232 heart rate monitor. Project website: `https://learn.sparkfun.com/tutorials/ad8232-heart-rate-monitor-hookup-guide`.

[32] J. Munoz, V. Mosquare, and C. Rengifo. A low-cost, portable two-dimensional bioimpedance distribution estimation system based on the ad5933 impedance converter. *Hardware X*, 11:e00274, 2022.

[33] E. Tufte. *The Visual Display of Quantitative Information*. Graphic Press, Cheshire, 1983.

[34] C. Mack. How to write a good scientific paper: Figures, part 1. *J. Micr/Nanolith*, 12:040101–1, 2013.

[35] W. Strunk and E. White. *The Elements of Style*. Harcourt, 1920. Online available from Project Gutenberg at `https://www.gutenberg.org/ebooks/37134`.

[36] J. Gibaldi. *MLA Handbook for Writers of Research Papers*. Modern Language Association of America, New York, NY, sixth edition, 2003.

[37] S. Pinker. *The Sense of Style*. Penguin Books, London, 2015.

[38] S. King. *On Writing*. Hodder, London, 2012.

[39] Nature style guide. `http://www.nature.com/authors/author_resources/how_write.html`.

[40] K. Friedman. Reviews of modern physics style guide. `http://journals.aps.org/files/rmpguide.pdf`.

[41] Nature. English communication for scientists. `http://www.nature.com/scitable/ebooks/english-communication-for-scientists-14053993/contents`.

[42] C. Fellbaum, editor. *WordNet: An Electronic Lexical Database*. MIT Press, Cambridge, MA. Online version: `http://wordnet.princeton.edu`.

[43] W. Press et al. *Numerical Recipes*. Cambridge University Press, Cambridge, second edition, 1992.

Index

Printed in the United States
by Baker & Taylor Publisher Services